my revision notes

Edexcel A Level
MATHEMATICS (PURE)

Sophie Goldie

Consultant editor
Elaine Lambert

Every effort has been made to trace all copyright holders, but if any have been inadvertently overlooked, the Publishers will be pleased to make the necessary arrangements at the first opportunity.

Although every effort has been made to ensure that website addresses are correct at time of going to press, Hodder Education cannot be held responsible for the content of any website mentioned in this book. It is sometimes possible to find a relocated web page by typing in the address of the home page for a website in the URL window of your browser.

Hachette UK's policy is to use papers that are natural, renewable and recyclable products and made from wood grown in well-managed forests and other controlled sources. The logging and manufacturing processes are expected to conform to the environmental regulations of the country of origin.

Orders: please contact Hachette UK Distribution, Hely Hutchinson Centre, Milton Road, Didcot, Oxfordshire, OX11 7HH. Telephone: +44 (0)1235 827827. Email education@hachette.co.uk
Lines are open from 9 a.m. to 5 p.m., Monday to Friday.
You can also order through our website:www.hoddereducation.com

ISBN: 978 1 5104 1751 9

© Sophie Goldie, Elaine Lambert and Diana Boynova

First published in 2018 by

Hodder Education,
An Hachette UK Company
Carmelite House
50 Victoria Embankment
London EC4Y 0DZ

www.hoddereducation.co.uk

The authorised representative in the EEA is Hachette Ireland, 8 Castlecourt Centre, Dublin 15, D15 XTP3, Ireland (email: info@hbgi.ie)

Impression number   10 9 8 7 6 5 4 3 2

Year   2023

All rights reserved. Apart from any use permitted under UK copyright law, no part of this publication may be reproduced or transmitted in any form or by any means, electronic or mechanical, including photocopying and recording, or held within any information storage and retrieval system, without permission in writing from the publisher or under licence from the Copyright Licensing Agency Limited. Further details of such licences (for reprographic reproduction) may be obtained from the Copyright Licensing Agency Limited, www.cla.co.uk

Cover photo © ASAF_ELIASON/123RF.COM

Typeset in Bembo Std Regular 11/13 by Integra Software Services Pvt. Ltd., Pondicherry, India

Printed and bound by CPI Group (UK) Ltd, Croydon, CR0 4YY

A catalogue record for this title is available from the British Library.

Get the most from this book

Welcome to your Revision Guide for the **Pure Mathematics content of the Edexcel A Level Mathematics** course. This book will provide you with reminders of the knowledge and skills you will be expected to demonstrate in the exam with opportunities to check and practice those skills on exam-style questions. Additional hints and notes throughout help you to avoid common errors and provide a better understanding of what's needed in the exam.

In order to revise the Applied Mathematics (Mechanics and Statistics) content of the course, you will need to refer to My Revision Notes: **Edexcel A Level Mathematics (Applied)**.

The material in this book also covers the Edexcel AS Level Mathematics exam, however you may prefer to use the Revision Guide for the Pure Mathematics content of the Edexcel Year 1/AS Level Mathematics course which just covers all the pure mathematics needed for that exam.

Included with the purchase of this book is valuable online material that provides full worked solutions to all the 'Target your revision', 'Exam-style questions' and 'Review questions', as well as full explanations and feedback to each answer option in the 'Test yourself' multiple choice questions. The **online material** is available at www.hoddereducation.co.uk/MRNEdexcelALPure.

Features to help you succeed

Target your revision

Use these questions at the start of each of the three sections to focus your revision on the topics you find tricky. **Short answers** are at the back of the book, but use the **worked solutions online** to check each step in your solution.

About this topic

At the start of each chapter, this provides a concise overview of its content.

Before you start, remember

A summary of the key things you need to know **before** you start the chapter.

Key facts

Check you understand all the key facts in each subsection. These provide a useful checklist if you get stuck on a question.

Worked examples

Full worked examples show you what the examiner expects to see in order to ensure full marks in the exam.

Hint

Expert tips are given throughout the book to help you do well in the exam.

Common mistakes

Your attention is drawn to typical mistakes students make, so you can avoid them.

Test yourself

Succinct sets of multiple-choice questions test your understanding of each topic. Check your **answers online**. The **online feedback** will explain any mistakes you made as well as common misconceptions, allowing you to try again.

Exam-style questions

For each topic, these provide typical questions you should expect to meet in the exam. **Short answers** are at the back of the book, and you can check your working using the **online worked solutions**.

Review questions

After you have completed each of the three sections in the book, answer these questions for more practice. **Short answers** are at the back of the book, but the **online worked solutions** allow you to check every line in your solution.

Edexcel A Level Mathematics (Pure)   iii

Additional help

At the end of the book, you will find some useful information:

Exam preparation

Includes hints and tips on revising for the A Level Mathematics exam, and details the exact structure of the exam papers.

Make sure you know these formulae for your exam

Provides a succinct list of all the formulae you need to remember and the formulae that will be given to you in the exam.

Please note that the formula sheet as provided by the exam board for the exam may be subject to change.

Integration at a glance

Provides an overview of all the different integration methods so you can see when to use each one.

During your exam

Includes key words to watch out for, common mistakes to avoid and tips if you get stuck on a question.

My revision planner

SECTION 1

Target your revision (Chapters 1–5)

Chapter 1 Proof
    4 Proof

Chapter 2 Indices, surds and logs
    9 Surds and indices
    13 Exponential functions and logarithms
    18 Modelling curves

Chapter 3 Algebra
    24 Quadratic equations
    31 Simultaneous equations
    35 Inequalities
    38 Working with polynomials and algebraic fractions
    43 The factor theorem and curve sketching

Chapter 4 Coordinate geometry
    47 Straight lines
    52 Circles

Chapter 5 Functions
    57 Functions
    62 Graphs and transformations
    66 Inverse functions
    70 The modulus function

Review questions (Chapters 1–5)

SECTION 2

Target your revision (Chapters 6–8)

Chapter 6 Trigonometry
    78 Working with trigonometric functions
    83 Triangles without right angles
    87 Radians and circular measure
    91 Reciprocal trig functions and small angle approximations
    95 Compound angle formulae
    99 The forms $r\cos(\theta \pm \alpha)$ and $r\sin(\theta \pm \alpha)$

Edexcel A Level Mathematics (Pure)

Chapter 7 Differentiation
    103 Tangents and normals
    107 Curve sketching and stationary points
    114 First principles and differentiating $\sin x$, $\cos x$ and $\tan x$
    117 Differentiating $\ln x$ and e^x
    119 The chain rule
    124 The product and quotient rules
    127 Implicit differentiation

Chapter 8 Integration
    131 Integration as the reverse of differentiation
    134 Finding areas
    140 Integration by substitution
    144 Integrating trigonometric functions
    148 Integration by parts

Review questions (Chapters 6–8)

SECTION 3

Target your revision (Chapters 9–14)

Chapter 9 Sequences and series
    156 Definitions and notation
    160 Sequences and series

Chapter 10 Further algebra
    163 Partial fractions
    167 The binomial theorem

Chapter 11 Parametric equations
    172 Parametric equations
    176 Calculus with parametric equations

Chapter 12 Differential equations
    181 Solving differential equations
    185 Differential equations and problem solving

Chapter 13 Vectors
    188 Vectors

Chapter 14 Numerical methods
    192 Solving equations numerically
    196 Numerical integration

Review questions (Chapters 9–14)

201 Exam preparation

202 Make sure you know these formulae for your exam

206 Integration at a glance

208 Formulae that will be given

210 During your exam

213 Answers to Target your revision, Exam-style questions and Review questions

Go online for:
- full worked solutions and answers to the Test yourself questions
- full worked solutions to all Exam-style questions
- full worked solutions to all Review questions
- full worked solutions to the Target your revision questions

www.hoddereducation.co.uk/MRNEdexcelALPure

SECTION 1

Target your revision (Chapters 1–5)

1. **Use the symbols $\Rightarrow$, $\Leftarrow$ and $\Leftrightarrow$**
   Complete each of the following statements by selecting the best connecting symbol ($\Leftrightarrow$, $\Leftarrow$ or $\Rightarrow$) for each box.
   Explain your choice, giving full reasons.
   i   For integer n: n is odd $\square$ $n^2 + 1$ is even
   ii ABCD has 4 equal sides $\square$ ABCD is a square
   iii $x > 3$ $\square$ $(x-1)(x-3) > 0$
   (see page 4)

2. **Use proof by exhaustion**
   Prove that $n^2 + n + 41$ is prime for all positive integers less than 8.
   (see page 4)

3. **Use proof by deduction**
   Prove that the product of any two odd numbers is odd.
   (see page 4)

4. **Use proof by contradiction**
   Prove that $\sin\theta + \cos\theta \geqslant 1$ when $0° \leqslant \theta \leqslant 90°$.
   (see page 4)

5. **Disprove by counter example**
   Prove or disprove the following conjecture:
   Given a and b are irrational and $a \neq b$, then ab is also irrational.
   (see page 4)

6. **Use and manipulate surds**
   Show that $\sqrt{80} - \sqrt{20}$ can be written in the form $a\sqrt{b}$ where b is the smallest integer possible.
   (see page 9)

7. **Rationalise the denominator of a surd**
   Show that $\dfrac{1 - 4\sqrt{2}}{1 + \sqrt{2}}$ can be expressed in the form $a + b\sqrt{2}$, where a and b are integers.
   (see page 9)

8. **Use the laws of indices**
   Simplify $\dfrac{18a^3 b^4 c^3}{(3ab^2 c^3)^2}$.
   (see page 9)

9. **Understand negative and fractional indices**
   Write $3^{\frac{1}{2}} + 3^{\frac{3}{2}} - \dfrac{1}{3^{\frac{1}{2}}}$ in the form $k\sqrt{3}$.
   (see page 9)

10. **Sketch the graphs of logarithms and exponentials**
    Sketch the graphs of
    i $y = 2 + \ln x$ ii $y = 1 + e^{-x}$.
    Give the equations of any asymptotes and the coordinates of any intersections with the axes.
    (see page 13)

11. **Simplify expression involving logs**
    Express $2\log x + \log 3 - \log \sqrt{x}$ as a single logarithm.
    (see page 13)

12. **Solve equations involving logs and exponentials**
    i Given $8^x = 27$ find the value of 4^x.
    ii Given that $2\log_{10} x - \log_{10} 10 = \log_{10} 20$, find the exact value of x.
    (see page 13)

13. **Use logs in modelling**
    The relationship between y and t is modelled by $y = A \times 10^{kt}$, where A and k are constants.
    i Show that the graph of $\log y$ against t is a straight line.

    The straight-line graph obtained when $\log y$ is plotted against t passes through the points (1, 2.8) and (4, 4.3).
    Find
    ii the value of A and of k
    iii the value of y when $t = 10$
    iv the value of t when $y = 100$, giving your answer correct to one decimal place.
    (see page 18)

14. **Work with quadratic equations**
    Use factorising to solve $6x^2 - 11x + 3 = 0$.
    Hence sketch the curve $y = 6x^2 - 11x + 3$.
    (see page 24)

15. **Complete the square**
    Write $y = x^2 - 6x + 5$ in the form $y = (x + a)^2 + b$.
    (see page 24)

16. **Use the discriminant**
    Find the values of k so that the equation $3x^2 + kx + 3 = 0$ has one repeated root.
    (see page 24)

17 Find the coordinates of the point where two lines intersect
The lines $3x - 2y = 5$ and $y = 4 - 3x$ intersect at the point X.
Find the coordinates of X.
(see page 31)

18 Solve simultaneous equations where one equation is quadratic
Find the coordinates of the points where the line $y = 3x - 2$ intersects the curve $y^2 = 4x + 8$.
(see page 31)

19 Find the coordinates of the point where two curves intersect
Find the coordinates of the point(s) where $(x+1)^2 + y = 6$ and $y = 3x^2 - 1$ intersect.
(see page 31)

20 Solve linear inequalities
Solve $-8 < 2(3 - 2x) \leq 10$.
(see page 35)

21 Solve quadratic inequalities
Solve $x^2 - 3x - 10 \geq 0$.
(see page 35)

22 Represent inequalities graphically
Show graphically the region represented by $x^2 + x \leq 3 - x$.
(see page 35)

23 Add, subtract, multiply and divide polynomials
You are given that $f(x) = 2x^3 - 7x^2 + 7x - 2$, $g(x) = 3x^3 - 2x + 5$ and $h(x) = x - 2$.
Find
  i $f(x) + g(x)$ ii $g(x) - f(x)$
  iii $h(x) \times g(x)$ iv $f(x) \div h(x)$.
(see page 38)

24 Simplify rational expressions
Simplify
  i $\dfrac{x^2 - 1}{2x^2 + x - 1}$ ii $\dfrac{2}{x^2 - 1} - \dfrac{x}{x + 1}$.
(see page 38)

25 Use the factor theorem and sketch the graph of a polynomial
You are given $f(x) = -3x^3 + 4x^2 + 5x - 2$.
  i   Show that $x = -1$ is a root of $f(x) = 0$.
  ii Show that $(3x - 1)$ is a factor of $f(x)$.
  iii Factorise $f(x)$ fully and hence solve $f(x) = 0$.
  iv Sketch the graph of $y = f(x)$.
(see page 43)

26 Calculate the length, midpoint and gradient of a line segment
Two points A and B have coordinates (−2, 3) and (4, −5) respectively.
Find
  i   the midpoint of AB
  ii the gradient of AB
  iii the distance AB.
(see page 47)

27 Use the relationship between the gradients of parallel lines and find the equation of a line
The line l passes through the point (2, −5) and is parallel to the line $x + 2y = 6$.
  i   Find the equation of the line l, giving your answer in the form $y = mx + c$.
  ii Find the coordinates of the points where l crosses the x and y-axes.
(see page 47)

28 Use the relationship between the gradients of perpendicular lines
The points A and B have coordinates (3, 7) and (−6, 1).
Find the gradient of a line perpendicular to AB.
(see page 47)

29 Use the equation of a circle
Find the equation of the following circles:
  i   centre (−2, 1), radius $\sqrt{5}$
  ii passing through the points A(−4, 3) and B(2, 7) with diameter AB.
(see page 52)

30 Solve problems involving circles
The point P(6, −1) lies on the circle $(x - 3)^2 + (y + 5)^2 = 25$.
Find the equation of the tangent to the circle at P.
(see page 52)

31 Find a composite function
The functions $f(x) = 3x - 2$ and $g(x) = x^2$ are defined for all real numbers.
  i   Find
     A   ff(2)   **B**   gf(x)
  ii Solve gf(x) = g(x).
(see page 57)

32 Use transformations to sketch graphs

The diagram shows a sketch of the graph of the function $y = f(x)$.

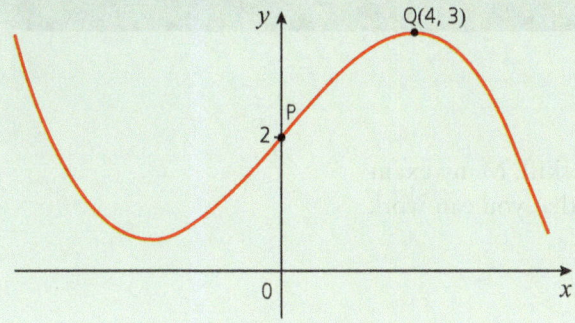

The graph intersects the y-axis at the point P(0, 2) and the point Q(4, 3) is the maximum turning point.

For each of the following transformations of $y = f(x)$, find the coordinates of the points to which the points P and Q are transformed.

i $\quad y = -f(2x)$
ii $\quad y = 3f(x) - 5$
iii $\quad y = f(2x + 1)$

(see page 62)

33 Find an inverse function and sketch its graph

You are given that $f(x) = 1 + e^{2x}$.

i Find the domain and range of f(x).
ii Find $f^{-1}(x)$ and the domain and range of $f^{-1}(x)$.
iii Draw the graph of $y = f(x)$ and $y = f^{-1}(x)$ on the same pair of axes.

(see page 66)

34 Draw graphs of modulus functions

Draw the graphs of

i $\quad y = |2x + 3|$
ii $\quad y = 2|x| + 3$.

(see page 70)

35 Solve equations involving modulus functions

Solve

i $\quad |4 - 2x| = 6$
ii $\quad |4 - 2x| = 3x$
iii $\quad |4 - 2x| = |3x + 6|$.

(see page 70)

36 Solve inequalities involving modulus functions

i Express $3 < x < 7$ in the form $|x - a| < b$, where a and b are to be determined.
ii Solve the inequality $|2x - 1| \geq 3$.

(see page 70)

Short answers on page 213–214

Full worked solutions online

CHECKED ANSWERS

Chapter 1 Proof

About this topic

Constructing a mathematical argument is an important skill. Many exam questions require you to prove results and demonstrate that you can work logically and systematically.

Before you start, remember

- different types of number: integer, multiple, factor, square, cube, prime, rational and irrational
- an irrational number is a number that can't be written as a fraction $\frac{a}{b}$, where a and b are integers (whole numbers). For example, $\sqrt{2} = 1.414213\ldots$ is an irrational number as it can't be expressed as a fraction — the decimal part continues for ever and never repeats
- how to use the laws of indices
- geometry from GCSE maths.

Proof

REVISED

Key facts

1. $P \Rightarrow Q$ means P **implies** Q or P **leads to** Q.

   P is a **sufficient** condition for Q.

   For example, ABCD is a square $\Rightarrow$ ABCD is a quadrilateral.

2. $P \Leftarrow Q$ means P **is implied by** Q or P **follows from** Q.

   P is a **necessary** condition for Q.

3. $P \Leftrightarrow Q$ means P **implies and is implied by** Q or P **is equivalent to** Q.

   P is a **necessary and sufficient** condition for Q.

   For example: the last digit of an integer is 0 or 5 $\Leftrightarrow$ it is divisible by 5.

4. The **converse** of $P \Rightarrow Q$ is $P \Leftarrow Q$.

5. A **conjecture** is a mathematical statement which appears likely to be true, but has not been formally proved to be true.

   You can prove a conjecture by:
   - proof by exhaustion
   - proof by deduction or direct argument
   - proof by contradiction.

6. Sometimes it is easier to disprove a conjecture by finding a **counter example**.

> ABCD is a quadrilateral does not imply ABCD is a square – it could be a rectangle! So you can't say ABCD is a square $\Leftarrow$ ABCD is a quadrilateral.

> You test every possible case – to exhaust all possibilities.

> Start from a known result and then construct a logical argument as to why the conjecture must be true. This type of proof often uses algebra.

> Start by assuming the conjecture you are trying to prove is false and then show that this leads to something that is clearly not true (a contradiction), meaning that the conjecture must actually be true.

> You only need to find a single counter example to disprove a statement.

Worked examples

1 Using the symbols ⇒, ⇐ and ⇔

In each case choose one of the statements

$P \Rightarrow Q$ $P \Leftarrow Q$ $P \Leftrightarrow Q$

to describe the complete relationship between P and Q.

  i P: $x = 3$ and Q: $x^2 = 9$
 ii Given that n is an integer
    P: $n^3 - 1$ is an odd integer and Q: n is an even integer
iii ABCD is a quadrilateral.
    P: ABCD has two pairs of parallel sides.
    Q: ABCD is a rectangle.

Solution

  i When $x = 3$ then $x^2 = 9$ so $P \Rightarrow Q$ is true.
     When $x^2 = 9$ then $x = 3$ or -3 so the converse $Q \Rightarrow P$ or $P \Leftarrow Q$ is not true.
     So the relationship is $P \Rightarrow Q$.

 ii When $n^3 - 1$ is an odd integer then n^3 is even ⇒ n is even.
     So $P \Rightarrow Q$ is true.
     When n is an even integer then n^3 is even ⇒ $n^3 - 1$ is odd.
     So the converse $Q \Rightarrow P$ or $P \Leftarrow Q$ is also true.
     So the relationship is $P \Leftrightarrow Q$.

iii ABCD has two pairs of parallel sides ⇒ ABCD is a rectangle or a parallelogram.
     So $P \Rightarrow Q$ is not true.
     ABCD is a rectangle ⇒ ABCD has two pairs of parallel sides.
     So the converse $Q \Rightarrow P$ or $P \Leftarrow Q$ is true.
     So the relationship is $P \Leftarrow Q$.

Hint: Make sure you set out your reasoning clearly. Don't forget to check the converse!

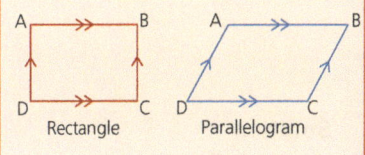
Rectangle   Parallelogram

2 Proof by exhaustion

Prove that no square number has a final digit that is an 8.

Hint: Proof by exhaustion is effective when there are only a limited number of cases to check.

Solution

The final digit in any square number comes from squaring the last digit.

Test all single digit numbers.
$0^2 = 0$, $1^2 = 1$, $2^2 = 4$, $3^2 = 9$, $4^2 = 16$, $5^2 = 25$
$6^2 = 36$, $7^2 = 49$, $8^2 = 64$, $9^2 = 81$, $10^2 = 100$

No single digit square number has a final digit of 8, and hence no square number can end in an 8.

Think about working out, say, 123 × 123 using long multiplication. The resulting number will have 9 as a final digit.

Edexcel A Level Mathematics (Pure)

3 Proof by direct argument using algebra

Prove that the difference between consecutive square numbers is always an odd number.

Solution

If n is a general integer then n^2 and $(n+1)^2$ are consecutive square numbers.

This means that the difference between consecutive square numbers is of the form

$$(n+1)^2 - n^2 = n^2 + 2n + 1 - n^2$$
$$= 2n + 1.$$

$2n + 1$ is not divisible by 2, so the difference between consecutive square numbers must always be odd.

Hint: To construct this proof, you need a general expression for the difference between consecutive square numbers. To test whether a number is odd, you need to use the definition of an odd number, i.e. a number that is not divisible by 2.

4 Proof by direct argument using geometry

Prove that the angle subtended by an arc of a circle at the centre is twice the angle subtended at the circumference by the same arc, i.e. that $y = 2x$ in the diagram shown.

You may assume the result that the angles in a triangle add up to 180°.

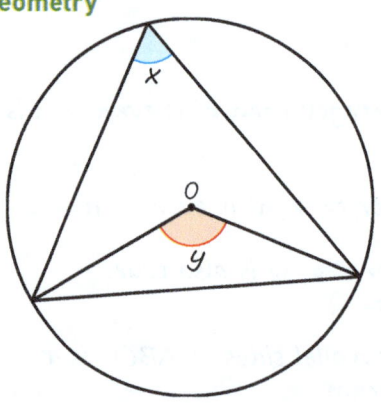

Solution

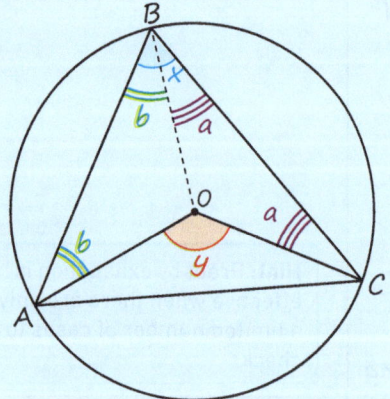

The angle subtended by arc AC at the circumference is $\angle ABC = x$.

The angle subtended by arc AC at the centre is $\angle AOC = y$.

Triangles OBC and OBA are isosceles. ◀

Hint: For proofs of geometric properties, it is essential to draw a clear diagram. It often helps to add construction lines, like line OB in this example, so that you can use known properties of shapes (such as isosceles triangles having two equal angles) to help construct your proof.

Hint: The angles around the centre point add to 360°.

The angles in a triangle add to 180°.

They each have two sides that are radii of the circle.

So ∠OBC = ∠OCB = a and ∠OAB = ∠OBA = b

So ∠BOC = 180° − 2a and ∠AOB = 180° − 2b

⇒ 180° − 2a + 180° − 2b = 360° − y

⇒ 2a + 2b = y

⇒ y = 2(a + b)

but x = a + b so y = 2x.

Hence, the angle subtended at the centre is twice the angle subtended at the circumference.

5 Proof by contradiction

Prove that $\sqrt{3}$ is irrational.

Solution

Assume that $\sqrt{3}$ is rational, so it can be written in the form $\sqrt{3} = \frac{a}{b}$, where a and b are integers with no common factors.

$$\sqrt{3} = \frac{a}{b}$$

⇒ $3 = \frac{a^2}{b^2}$

⇒ $3b^2 = a^2$

⇒ a^2 is a multiple of 3, so a is a multiple of 3, i.e. $a = 3k$, where k is an integer.

⇒ $3b^2 = (3k)^2$

⇒ $3b^2 = 9k^2$

⇒ $b^2 = 3k^2$

⇒ b^2 is a multiple of 3, so b is a multiple of 3.

The proof began by assuming that a and b have no common factors and then showed that, if $\sqrt{3}$ is rational, 3 must be a factor of both a and b. This is a contradiction and therefore $\sqrt{3}$ must be irrational.

6 Disproof by counter example

Is it true that $x^2 \geqslant x$ for all real numbers?

Solution

$1^2 = 1$

$50^2 = 2500$

$(-4)^2 = 16$

However,

$\left(\frac{1}{2}\right)^2 = \frac{1}{4}$ and $\frac{1}{4} < \frac{1}{2}$, showing that it is not true that

$x^2 \geqslant x$ for all real numbers.

Hint: You can think of proof by contradiction as being a similar argument to an alibi in a criminal case. For example, if you want to prove that Mrs X is innocent, you could begin by assuming that she is guilty but then show that this is a contradiction, because she was in Glasgow at the time when the murder was committed in London, so Mrs X must be innocent.

If a^2 is divisible by 3 then 3 is a prime factor of a^2. However, all square numbers are the product of square prime factors, for example $30 = 2 \times 3 \times 5$ and $30^2 = 2^2 \times 3^2 \times 5^2$. So 3^2 is a factor of a^2 and 3 is a factor of a.

If b is a multiple of 3 then 3 is a factor of b.

Hint: Definitions are often important in mathematical proofs. Here the argument is based upon the definition of a rational number.

Hint:
- To show that a conjecture is false, only one case where it does not work, a counter example, is required. Counter examples can be hard to find, but the technique is very powerful because only one is needed to prove a conjecture is false.
- For conjectures involving numbers, it's worth testing to see whether they work for negative numbers and fractions. In worked example 6, the conjecture that $x^2 \geqslant x$ only fails when $0 < x < 1$.

Edexcel A Level Mathematics (Pure)

Test yourself

1. 'For all values of n greater than or equal to 1, $n^2 + 3n + 1$ is a prime number.'
   Which value of n gives a counter example which disproves this conjecture?
   A $n = 7$
   B $n = 2$
   C $n = 8$
   D $n = 6$

2. Below is a proof that appears to show that $2 = 0$.
   The proof must contain an error. At which line does the error occur?

   Let $\quad a = b = 1$
   $\Rightarrow \quad a^2 = b^2$ [Line 1]
   $\Rightarrow \quad a^2 - b^2 = 0$ [Line 2]
   $\Rightarrow \quad (a + b)(a - b) = 0$ [Line 3]
   $\Rightarrow \quad a + b = 0$ [Line 4]
   $\Rightarrow \quad 2 = 0$ [Line 5]

   A Line 1   B Line 2   C Line 4   D Line 3   E Line 5

3. Below is an attempt to prove that if an integer, p, is even, p^2 is also even.
   Is the proof correct, or does it contain an error? If it contains an error, at which line does the error occur?

   If p is even then $p = 2k$, where k is an integer. [Line 1]
   $\Rightarrow p^2 = (2k)^2 = 4k^2$ [Line 2]
   $\Rightarrow p^2 = 2(2k^2)$ [Line 3]
   $\Rightarrow p^2$ is even [Line 4]

   A There are no errors in the proof.   B Line 2   C Line 3
   D Line 4   E Line 1

4. Below is an attempt to prove that there are an infinite number of primes.

   Step 1: If there are a finite number of prime numbers, there must be a largest prime number, p_n.

   Step 2: If $p_1, p_2, p_3, \ldots, p_{n-1}$ are all the primes less than p_n then
   $p_1 \times p_2 \times p_3 \times \ldots \times p_{n-1} \times p_n + 1$ will leave a remainder of 1 when it is divided by any of $p_1, p_2, p_3, \ldots, p_{n-1}, p_n$

   Step 3: Therefore $p_1 \times p_2 \times p_3 \times \ldots \times p_{n-1} \times p_n + 1$ is prime.

   Step 4: Since $p_1 \times p_2 \times p_3 \times \ldots \times p_{n-1} \times p_n + 1$ must be bigger than p_n, we have a contradiction because we began by stating that p_n was the largest prime.
   Therefore there must be an infinite number of prime numbers.

   The following 5 statements refer to this attempted proof. Four of them are false and one of them is true. Which one is true?
   A $n = 5$ gives a counter example because $p_1 \times p_2 \times p_3 \times p_4 \times p_5 + 1 = 121$.
   B $n = 3$ gives a counter example because $p_1 \times p_2 \times p_3 + 1 = 106$ and 106 is not prime.
   C $n = 6$ gives a counter example because $30\,031 = 59 \times 509$.
   D $n = 4$ gives a counter example because $2 \times 3 \times 5 \times 7 + 1 = 211$ and 211 is not prime.
   E There are no errors in the proof.

Full worked solutions online

Exam-style question

Prove by contradiction that $\sqrt{2}$ is irrational.

Short answers on page 214

Full worked solutions online

Chapter 2 Indices, surds and logs

About this topic

This topic deals with indices (powers), surds and logarithms. You will need to use the laws of indices to help you manipulate and simplify expressions.

A surd is a type of irrational number which can be written exactly in square root form such as $\sqrt{5}$ or $4 - 2\sqrt{3}$.

Exponential functions are functions where the index (or power) is a variable. The inverse of an exponential function is a logarithm.

Before you start, remember
- how to expand brackets
- how to simplify expressions
- how to use the laws of indices from GCSE maths.

Surds and indices

REVISED

Key facts

1. In the expression a^m, a is the **base** and m is the **index** or **power** to which the base is raised.

2. The **laws of indices** are:
   - $a^m \times a^n = a^{m+n}$ — Multiplication
   - $\dfrac{a^m}{a^n} = a^{m-n}$ — Division
   - $(a^m)^n = a^{mn}$. — Power of a power

3. Remember that any non-zero number to the **power zero** is equal to 1.
   $3^0 = 1 \qquad (-2)^0 = 1 \qquad 2.6^0 = 1$ — 0^0 is undefined.

4. For negative and fractional powers
   - $a^{-m} = \dfrac{1}{a^m}$ — A negative index indicates a reciprocal.
   - $a^{\frac{1}{m}} = \sqrt[m]{a}$
   - $a^{\frac{m}{n}} = \sqrt[n]{a^m}$. — A fractional index is a root.

5. A **surd** is an expression containing an irrational root, such as $5 + \sqrt{3}$ or $2 - \sqrt[3]{7}$. — Keep the rational numbers and the roots separate.

6. A surd is in its **simplest form** when the number under the square root has no square factors.
   - $\sqrt{12}$ is not in simplest form
   - $2\sqrt{3}$ is in simplest form.

   Remember that $\sqrt{}$ means the **positive square root** only.

7. You can **add and subtract surds** to **simplify them** in the same way as other algebraic expressions.

8. When you multiply surds, remember:
   $\sqrt{x} \times \sqrt{x} = x \qquad \sqrt{xy} = \sqrt{x}\sqrt{y}$.

Edexcel A Level Mathematics (Pure)

9 When a fraction has a surd in the denominator it is not in its simplest form.
   You simplify it by **rationalising the denominator**.
   For fractions in the form:
   - $\dfrac{1}{\sqrt{a}}$ multiply the top and bottom lines by $\sqrt{a}$
   - $\dfrac{1}{a+\sqrt{b}}$ multiply the top and bottom lines by $a-\sqrt{b}$.
   Remember $(a+b)(a-b) = a^2 - b^2$ and $(\sqrt{a}+\sqrt{b})(\sqrt{a}-\sqrt{b}) = a-b$.

Worked examples

1 Simplifying expressions involving indices

Simplify

i $\dfrac{3a^4 b \times (2ab^2)^3}{4a^2 b^9}$

ii $\sqrt[6]{c}\sqrt{c}$

Solution

i
$$\dfrac{3a^4 b \times (2ab^2)^3}{4a^2 b^9} = \dfrac{3a^4 b \times 2^3 a^3 b^6}{4a^2 b^9}$$

Use the laws of indices to remove the brackets: $(a^m)^n = a^{mn}$.

$$= \dfrac{3 \times 2^3 \times a^4 \times a^3 \times b \times b^6}{4a^2 b^9}$$

$$= \dfrac{3 \times 8 \times a^7 \times b^7}{4a^2 b^9}$$

Using $a^m \times a^n = a^{m+n}$.

$$= 6 \times a^5 \times b^{-2}$$

Using $\dfrac{a^m}{a^n} = a^{m-n}$.

$$= \dfrac{6a^5}{b^2}$$

Using $a^{-m} = \dfrac{1}{a^m}$.

ii $\sqrt[6]{c}\sqrt{c} = c^{\frac{1}{6}} c^{\frac{1}{2}}$

Using $a^{\frac{1}{m}} = \sqrt[m]{a}$.

$c^{\frac{1}{6}+\frac{1}{2}} = c^{\frac{2}{3}}$

$= \sqrt[3]{c^2}$

Using $a^{\frac{m}{n}} = \sqrt[n]{a^m}$.

2 Using index notation

Find the value of x in each case.

i $32\sqrt{2} = 2^x$

ii $\dfrac{2^x}{\sqrt{2}} = \dfrac{1}{4}$

Solution

i $32\sqrt{2} = 2^5 \times 2^{\frac{1}{2}}$

Using $a^{\frac{1}{m}} = \sqrt[m]{a}$.

$= 2^{5+\frac{1}{2}}$

$= 2^{\frac{11}{2}}$

$2^x = 2^{\frac{11}{2}} \Rightarrow x = \dfrac{11}{2}$

Common mistake: Don't forget to answer the question!

ii $\dfrac{2^x}{\sqrt{2}} = \dfrac{2^x}{2^{\frac{1}{2}}}$

$= 2^{x-\frac{1}{2}}$

Write the left-hand side as a single power of 2.

$$\frac{1}{4} = 2^{-2}$$

Write the right-hand side as a single power of 2.

$$\Rightarrow 2^{x-\frac{1}{2}} = 2^{-2}$$

Equate the powers.

$$\Rightarrow x - \frac{1}{2} = -2$$

$$\Rightarrow x = -\frac{3}{2}$$

3 Simplifying surds

Simplify

i $\frac{4}{\sqrt{5}}$

This is not in the simplest form because the bottom line is a surd.

ii $(\sqrt{7} - \sqrt{5})(\sqrt{7} + \sqrt{5})$

iii $\sqrt{54} - \sqrt{24}$.

Solution

i $\frac{4}{\sqrt{5}} = \frac{4}{\sqrt{5}} \times \frac{\sqrt{5}}{\sqrt{5}} = \frac{4\sqrt{5}}{5}$

You have rationalised the denominator.

ii $(\sqrt{7} - \sqrt{5})(\sqrt{7} + \sqrt{5}) = \sqrt{7}\sqrt{7} + \sqrt{7}\sqrt{5} - \sqrt{5}\sqrt{7} - \sqrt{5}\sqrt{5}$

$$= 7 - 5$$

Using $(\sqrt{a} + \sqrt{b})(\sqrt{a} - \sqrt{b}) = a - b$.

$$= 2$$

iii $\sqrt{54} - \sqrt{24} = \sqrt{9 \times 6} - \sqrt{4 \times 6}$

$$= \sqrt{9}\sqrt{6} - \sqrt{4}\sqrt{6}$$

$$= 3\sqrt{6} - 2\sqrt{6}$$

$$= \sqrt{6}$$

Hint: To simplify a surd look for factors that are square numbers.

4 Rationalising the denominator

Rationalise the denominator $\frac{14 + 7\sqrt{2}}{5 - 3\sqrt{2}}$.

Solution

$$\frac{14 + 7\sqrt{2}}{5 - 3\sqrt{2}} = \frac{14 + 7\sqrt{2}}{5 - 3\sqrt{2}} \times \frac{5 + 3\sqrt{2}}{5 + 3\sqrt{2}}$$

Multiply by $\frac{5+3\sqrt{2}}{5+3\sqrt{2}}$ to make the bottom line a whole number.

$$= \frac{14 \times 5 + 35\sqrt{2} + 42\sqrt{2} + 21 \times 2}{5^2 - (3\sqrt{2})^2}$$

$$= \frac{112 + 77\sqrt{2}}{25 - 18}$$

$$= \frac{112 + 77\sqrt{2}}{7}$$

$$= 16 + 11\sqrt{2}$$

Common mistake: You must multiply both the top and bottom by the same thing (so you are really multiplying by 1), otherwise you will change the value of the fraction.

Test yourself

TESTED

Make sure you can work these out without a calculator!

1. Find the value of $\left(\frac{1}{3}\right)^{-2}$

   A -9 B $\frac{1}{9}$ C 9 D $-\frac{1}{9}$ E $-\frac{2}{3}$

2. Find the value of $\dfrac{36^{\frac{1}{2}}}{16^{\frac{3}{4}}}$, giving the answer in its simplest form.

   A $\frac{3}{4}$ B $\frac{3}{2}$ C $\left(\frac{36}{16}\right)^{-\frac{1}{4}}$ D 3 E $\frac{6}{8}$

3. Simplify $(2 - 2\sqrt{3})^2$, giving your answer in factorised form.

   A 16 B $8(2 - \sqrt{3})$ C $-8(1 + \sqrt{3})$ D $16 - 8\sqrt{3}$ E $4(4 - \sqrt{3})$

4. Simplify $\dfrac{(2x^4 y^2)^3}{10(x^3 \sqrt{y^5})^2}$

   A $\frac{1}{5}x^6 y$ B $\frac{2}{25}x^6 y$ C $\frac{4x^6}{5y^4}$ D $\frac{4}{5}x^6 y$ E $\dfrac{8x^{12} y^6}{10x^6 y^5}$

5. Find the exact answer to $\sqrt{54 \times 48}$ simplifying your answer as much as possible.

   A 50.9 B $36\sqrt{2}$ C $12\sqrt{18}$ D 36 E $\sqrt{2592}$

Full worked solutions online   CHECKED ANSWERS

Exam-style question

In this question you must show detailed reasoning.

Find the value of a and b in each case.

i $\dfrac{4 \times 2^a \times 3^{2b}}{3} = \dfrac{2}{\sqrt{3}}$

ii $\dfrac{3\sqrt{5} - \sqrt{3}}{\sqrt{5} + \sqrt{3}} = a - 2\sqrt{b}$

Short answers on page 214

Full worked solutions online   CHECKED ANSWERS

Exponential functions and logarithms

REVISED ☐

Key facts

1. An **exponential function** is a function which has the variable as the power, such as $f(x) = 2^x$.

   An alternative name for power is **exponent**.

   > Exponential functions are in the form $f(x) = a^x$ where a is a constant and $a \neq 1$. They obey all the usual rules of indices, see page 9.

2. $f(x) = e^x$ is **the exponential function** where e is the number 2.718281...

3. Exponential functions model real-life situations and all graphs follow a similar pattern to one of the curves below.

   > **Hint:** Note **all** curves in the form $y = a^{kx}$:
   > - pass through (0, 1)
   > - lie above the x-axis (so y is positive for all values of x)
   > - have a gradient that is proportional to the y coordinate, so $\frac{dy}{dx} \propto y$.

   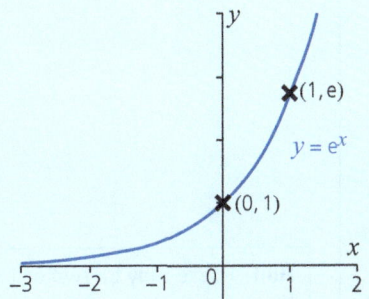
   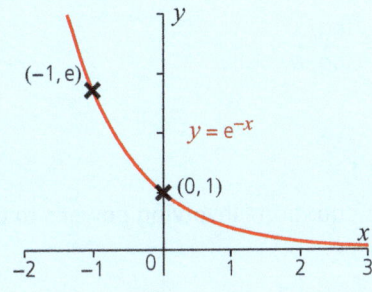

   > In general, the graph of $y = a^x$, where a is constant and $a > 1$, has a gradient that is increasing. This is described as **exponential growth**.

   > In general, the graph of $y = a^{-x}$, $a > 1$ has a gradient that is decreasing. This is described as **exponential decay**.

   For the blue curve, $y = e^x$, the gradient at any point equals the y coordinate, so $\frac{dy}{dx} = e^x$

   $$y = e^{kx} \Rightarrow \frac{dy}{dx} = ke^{kx}$$

4. Logarithm is another word for **index** or **power**.

   A logarithm is the **inverse** of the exponential function.

   So $y = \log_2 x$ is the inverse function of $y = 2^x$

   > The graph of $y = \log_a x$ is a reflection of $y = a^x$ in the line $y = x$.

   This is true for any base a, not just 2:
   $y = \log_a x \Leftrightarrow a^y = x$

   So $\log_a(a^x) = x$ and $a^{\log_a x} = x$.

5. The **natural logarithm** of x is written as $\ln x$ or $\log_e x$.

   e^x and $\ln x$ are inverse functions.

   > $y = \ln x \Leftrightarrow x = e^y$.

   The graph of $y = \ln x$ is a reflection of $y = e^x$ in the line $y = x$.

   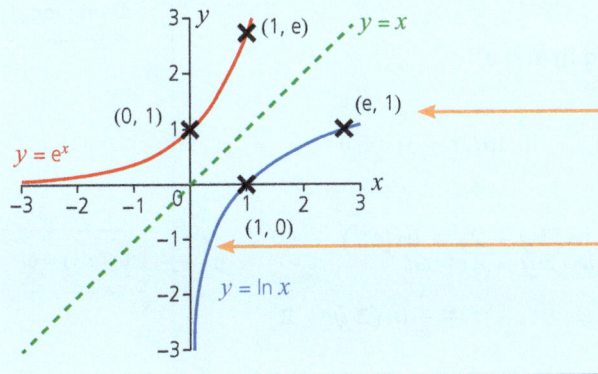

   > The negative x-axis is an asymptote for $y = e^x$. The negative y-axis is an asymptote for $y = \ln x$.

   > Notice that for values of x between 0 and 1, $\ln x$ is negative.

6 Rules for logarithms to any base (including natural logarithms):
- multiplication: $\log xy = \log x + \log y$
- division: $\log \frac{x}{y} = \log x - \log y$
- powers: $\log x^n = n \log x$
- roots: $\log \sqrt[n]{x} = \frac{1}{n} \log x$
- logarithm of 1: $\log 1 = 0$
- reciprocals: $\log \frac{1}{x} = -\log x$
- logarithm to its own base: $\log_a a = 1$.

So $\ln e = 1$.

- If a logarithmic expression is true for any base, then the base will often be omitted.

7 Change of base formula: $\log_a x = \dfrac{\log_b x}{\log_b a}$

For example: $\log_3 5 = \dfrac{\log_{10} 5}{\log_{10} 3}$

$= 1.46$ (to 3 s.f.)

8 Logarithms can be used to solve equations involving powers to any level of accuracy:

$5^x = 50$
$\Rightarrow \log 5^x = \log 50$
$\Rightarrow x \log 5 = \log 50$
$\Rightarrow x = \dfrac{\log 50}{\log 5} = 2.43$ (to 3 s.f.)

Hint: There may be two or three logarithm buttons on your calculator, depending on how sophisticated it is. You need to use the basic log button which may be labelled just 'log' or 'log$_{10}$' and the natural log button 'ln'.

Worked examples

1 Evaluating logarithms

Find the values of the following.

i $\log_5 125$ ii $\log_2 \left(\dfrac{1}{16}\right)$ iii $\log_9 3$

Solution

i $\log_5 125 = \log_5 5^3 = 3 \log_5 5 = 3$

Write 125 as a power of 5 and then use $\log x^n = n \log x$.

ii $\log_2 \dfrac{1}{16} = -\log_2 16 = -\log_2 2^4 = -4 \log_2 2 = -4$

Using $\log \dfrac{1}{x} = -\log x$.

iii $\log_9 3 = \log_9 \sqrt{9} = \log_9 9^{\frac{1}{2}} = \dfrac{1}{2} \log_9 9 = \dfrac{1}{2}$

Using $\log_a a = 1$.

2 Rearranging equations involving ln and e^x

Make t the subject.

i $3y - 2 = e^{6t}$ ii $\ln(2t + 3) = 5y$

Solution

i Take the ln of both sides: $\ln(3y - 2) = \ln(e^{6t})$
$\Rightarrow \ln(3y - 2) = 6t$
$\Rightarrow t = \dfrac{1}{6} \ln(3y - 2)$

$\ln(e^{6t}) = 6t$

ii Apply the exponential function to both sides:
$$e^{\ln(2t+3)} = e^{5y}$$
$$\Rightarrow 2t + 3 = e^{5y}$$
$$\Rightarrow 2t = e^{5y} - 3$$
$$\Rightarrow t = \frac{e^{5y} - 3}{2}$$

3 Solving equations containing logarithms

Given that $2\log_a x = \frac{1}{2}\log_a 64 + \log_a 32 - 2\log_a 4$, find the value of x.

Solution

$$2\log_a x = \frac{1}{2}\log_a 64 + \log_a 32 - 2\log_a 4$$
$$\Rightarrow \log_a x^2 = \log_a 64^{\frac{1}{2}} + \log_a 32 - \log_a 4^2 \quad \text{Using } n\log x = \log x^n.$$
$$\Rightarrow \log_a x^2 = \log_a \sqrt{64} + \log_a 32 - \log_a 16$$
$$\Rightarrow \log_a x^2 = \log_a 8 + \log_a \frac{32}{16} \quad \text{Using } \log x - \log y = \log \frac{x}{y}.$$
$$\Rightarrow \log_a x^2 = \log_a 8 + \log_a 2$$
$$\Rightarrow \log_a x^2 = \log_a (2 \times 8) \quad \text{Using } \log x + \log y = \log xy.$$
$$\Rightarrow x^2 = 16$$
$$\Rightarrow x = 4.$$

Common mistake: The option $x = -4$ has been rejected since the original equation contained $\log x$ and this is only defined for positive values of x.

4 Using logarithms to evaluate a power

Solve

i $3^{2x} = 5^{x+1}$

ii $e^{2x} - 10e^x + 9 = 0$.

Solution

i $3^{2x} = 5^{x+1} \Rightarrow \log 3^{2x} = \log 5^{x+1}$ — Take logs of both sides.
$$\Rightarrow 2x \log 3 = (x+1)\log 5$$
$$\Rightarrow 2x \log 3 = x \log 5 + \log 5$$
$$\Rightarrow 2x \log 3 - x \log 5 = \log 5 \quad \text{Write all the terms involving } x \text{ on one side and then factorise.}$$
$$\Rightarrow x(2\log 3 - \log 5) = \log 5$$
$$\Rightarrow x = \frac{\log 5}{(2\log 3 - \log 5)}$$
$$\Rightarrow x = 2.74 \text{ (to 3 s.f.)}$$

It doesn't matter which base you use for the logarithm. Check for yourself that using ln would give the same answer.

ii $e^{2x} - 10e^x + 9 = 0 \Rightarrow (e^x)^2 - 10e^x + 9 = 0$ — This is a quadratic equation in disguise.

Let $z = e^x \Rightarrow z^2 - 10z + 9 = 0$
$$\Rightarrow (z-9)(z-1) = 0$$
$$\Rightarrow z = 9 \text{ or } z = 1$$
$$e^x = 9 \Rightarrow x = \ln 9 \quad \text{Taking ln of both sides.}$$
$$\Rightarrow x = 2.20 \text{ (to 3 s.f.)}$$
$$e^x = 1 \Rightarrow x = 0$$

5 Modelling exponential growth

Following the decision by a major international company to set up its base in the UK a new town is being developed and it is estimated that the growth of population over the first five years will be modelled by the equation $P = 10\,000 \times 10^{0.05t}$, where t is the time in years.

i Calculate the increase in the population during the first year, giving your answer to 3 s.f.

ii When will the population exceed 20 000?

iii Why is this not a suitable model in the long term?

Solution

i The initial population is when $t = 0$:
$P = 10\,000 \times 10^{0.05 \times 0} = 10\,000$ *(Any number to the power of 0 is 1.)*

After 1 year: $P = 10\,000 \times 10^{0.05 \times 1} = 11\,220$. *(Substitute $t = 1$ into $P = 10\,000 \times 10^{0.05t}$.)*

So the population increase is $11\,220 - 10\,000 = 1220$ to 3 s.f.

ii Solve the inequality $10\,000 \times 10^{0.05t} > 20\,000$.

$$\Rightarrow 10^{0.05t} > 2$$
$$\Rightarrow 0.05t > \log 2$$ *(Taking logs of both sides.)*
$$\Rightarrow t > \frac{\log 2}{0.05}$$
$$\Rightarrow t > 6.02\ldots$$
$$\Rightarrow t = 7$$

Hint: It is a good idea to check your answer.

Check: After 6 years, $P = 10\,000 \times 10^{0.05 \times 6} = 19\,952$.

After 7 years, $P = 10\,000 \times 10^{0.05 \times 7} = 22\,387$.

So the population will exceed 20 000 during the 7th year.

iii After 50 years the model predicts the population of the town would be $P = 10\,000 \times 10^{0.05 \times 50} = 3\,160\,000$ which is unrealistically high.

It is likely that the population will increase exponentially for a few years and then gradually stabilise.

6 Sketching graphs of exponential functions

Sketch the graph of $y = 1 + 2e^{-x}$.

Solution

When $x = 0$, $y = 1 + 2e^0 = 1 + 2 = 3$

When $x \to \infty$, $y \to 1 + 2 \times 0 = 1$

So the line $y = 1$ is a horizontal asymptote.

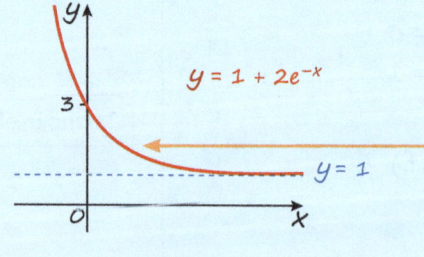

The curve slopes downwards because the power of e is negative.

Test yourself

TESTED

1. Which of the following is the correct answer when writing $\log 12 - 3\log 2 + 2\log 3$ as a single logarithm?

   A $\log 12$ B $\log 13.5$ C $\log \frac{1}{6}$ D $\log 10\frac{2}{3}$

2. Use logarithms to base 10 to solve the equation $2.5^x = 1000$ to 2 d.p.

   A 2.90   B 3.00   C 7.54   D 7.53882

3. Simplify $\frac{1}{2}\log 64 - 2\log 2$ writing your answer in the form $\log x$

   A $\log 8$ B $\log 2$ C $\log 32$ D 0.301

4. Express $\log\sqrt{x} + \log x^{\frac{7}{2}} - 2\log x$ as a single logarithm.

   A $2\log x$ B $\log x^2$ C $\log\left(x^{-\frac{3}{2}} + x^{\frac{3}{2}}\right)$ D $\log\left(\dfrac{x^{\frac{1}{2}} \times x^{\frac{7}{2}}}{x^2}\right)$

5. The value of an investment varies according to the formula $V = Ae^{0.1t}$.
   The investment is predicted to be worth £10 000 after 5 years.
   Find the value of A to the nearest £.

   A £16 487   B £6065   C £7358   D £6065.31

6. The graph shows the curve $y = \ln x$.

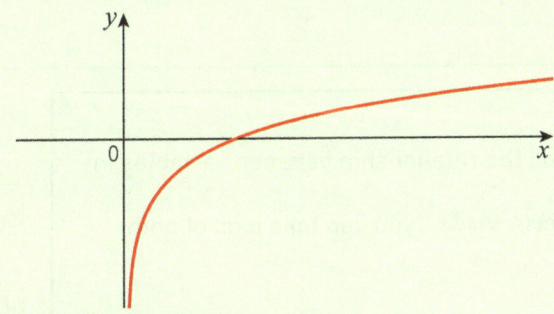

   Four of the following statements are false and one is true. Which one is true?
   A The graph crosses the y-axis at $(0, -1)$.
   B The graph crosses the x-axis at $(e, 0)$.
   C The graph passes through the point $(e, 1)$.
   D The graph flattens out for large values of x and approaches a horizontal asymptote.
   E If you draw the graph for negative values of x, it is the same curve reflected in the y-axis.

7. You are given that $M = 100 + 300e^{-0.1t}$. Find the value of t when $M = 250$.

   A 4.7   B 50.1   C 0.0693   D 6.93   E 16.5

8. Which of the following is the equation of this curve?

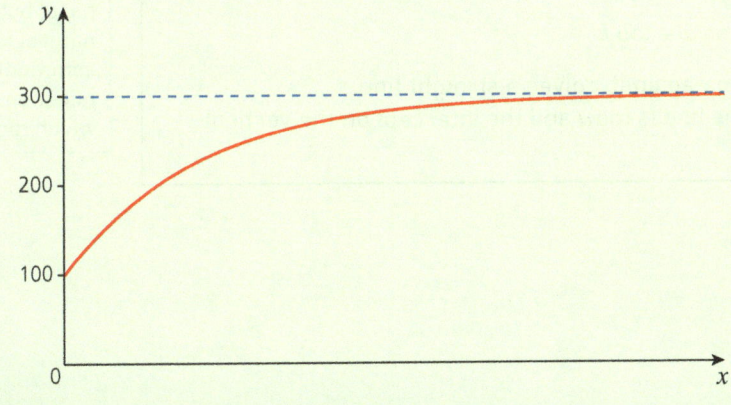

   A $y = 300 - 200e^{-x}$ B $y = 100 + 200e^{-x}$ C $y = 300 - 200e^x$

   D $y = 100 + 200e^x$ E $y = 200 - 100e^{-x}$

Full worked solutions online   CHECKED ANSWERS

Exam-style question

A certain type of parrot is found only in Australia apart from a population which live on a remote island in the south Pacific. It is believed that two of the parrots escaped from a passing ship long ago and established the island's population. The number of parrots on the island, P, has been studied for many years and has been found to be well modelled by the equation

$$P = 12000 - 8000e^{-0.005T}$$

where T is the number of years that have passed since 1850.

   i   Find the number of parrots on the island in **a** 2000 **b** 1850.
   ii   In what year should there be 10 000 parrots?
   iii   Sketch the graph of the number of parrots against T for $T \geq 0$.
   iv   Use the equation for P to estimate the year when the two original parrots arrived on the island and give one reason why this might not be very accurate.

Short answers on page 214

Full worked solutions online

CHECKED ANSWERS

Modelling curves

REVISED

Key facts

Logarithms can be used to find the relationship between variables in two situations:

1. For relationships of the form $y = kx^n$, you can take logs of both sides and write

$$\log y = \log kx^n$$

$\Rightarrow \quad \log y = \log x^n + \log k$ *(Using law: $\log ab = \log a + \log b$.)*

$\Rightarrow \quad \log y = n \log x + \log k$ *(Using power law: $\log x^n = n \log x$.)*

and so plotting $\log y$ against $\log x$ gives a straight line. The **gradient** of the line is n and the **intercept** on the vertical axis is $\log k$.

(Notice that this is in the form $y = mx + c$ with gradient n and y-intercept $\log k$.)

2. For relationships of the form $y = ka^x$, you can take logs of both sides and write

$$\log y = \log ka^x$$

$\Rightarrow \quad \log y = \log a^x + \log k$

$\Rightarrow \quad \log y = x \log a + \log k$ *(This is also in the form $y = mx + c$.)*

and so plotting $\log y$ against x gives a straight line. The **gradient** of the line is $\log a$ and the **intercept** on the vertical axis is $\log k$.

Hint: You can use log to any base, but base 10 and the natural logarithm, ln, are most commonly used. The exam question will make it clear which log to use.

Worked examples

1 Plotting log y against log x

In an experiment the temperature $\theta°C$ of a cooling liquid is measured every 2 minutes. The table shows the results.

Time in minutes (t)	2	4	6	8	10
Temperature (θ)	185	114	86	70	60

i Plot the graph of $\ln\theta$ against $\ln t$ and draw a line of best fit.
ii Use the graph to find the relationship between θ and t
iii Use the equation in part ii to predict the temperature of the liquid after 20 minutes, giving your answer to the nearest degree.
iv At what time, to the nearest minute, will the temperature of the liquid be 30°?

Solution

i

ln t	0.69	1.39	1.79	2.08	2.30
ln θ	5.22	4.74	4.45	4.25	4.09

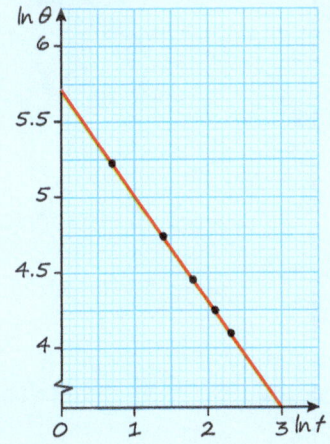

ii The graph is a straight line with an equation of the form

$\ln\theta = n\ln t + \ln k$
$\Rightarrow \ln\theta = \ln t^n + \ln k$
$\Rightarrow \ln\theta = \ln kt^n$
$\Rightarrow \theta = kt^n$

The line cuts the vertical axis at 5.7 so:
$\ln k = 5.7$
$\Rightarrow k = e^{5.7}$
$\Rightarrow k = 298.8... = 300$ (to 1 s.f.)

The line passes through the points (0, 5.7) and (2, 4.3). The gradient of the line is $\frac{5.7 - 4.3}{0 - 2} \Rightarrow n = -0.7$.

So the relationship between θ and t is $\theta = 300t^{-0.7}$.

Hint: The graph is a straight line, so you can use it to find a relationship between θ and t since you know that the equation of a straight line is $y = mx + c$.

Common mistake: Real data will not fit a straight line exactly so you need to draw a line of best fit. Try to draw the line of best fit as accurately as you can as this will affect your answer to the other parts of the question.

Use any two points from your graph.

iii To find the temperature after 20 minutes, substitute $t = 20$ into the equation of the line:

$$\theta = 300 \times 20^{-0.7} = 36.8°$$

The temperature will be 37° to the nearest degree after 20 minutes.

iv To find the time at which the temperature is 30°, substitute $\theta = 30$ into the equation of the line.

$$30 = 300t^{-0.7}$$

$$\Rightarrow \frac{30}{300} = t^{-0.7}$$

$$\Rightarrow \ln 0.1 = \ln t^{-0.7} \quad \text{← Take logarithms to base e of both sides.}$$

$$= -0.7 \ln t$$

$$\Rightarrow \ln t = -\frac{\ln 0.1}{0.7}$$

$$= 3.289...$$

$$\Rightarrow t = e^{3.289...} \quad \text{← Since the inverse of } y = \ln x \text{ is } y = e^x.$$

$$= 26.8$$

So the temperature of the liquid will be 30° after 27 minutes (to the nearest minute).

2 Plotting log y against x

The population, P, of bats in a large cavern is modelled by $P = ka^t$, where t is the time in years. The table shows the population over a five year period.

Year (t)	1	2	3	4	5
Population	1800	2700	4050	6075	9110

i Plot the graph of $\log_{10} P$ against t and use the graph to find the equation for P in terms of t.
ii Use the equation to find the population after 8 years
iii After how long will the population be greater than one hundred thousand according to this model?
iv Is this model suitable in the long term? Give a reason for your answer.

Solution

i

t	1	2	3	4	5
$\log_{10} P$	3.26	3.43	3.61	3.78	3.96

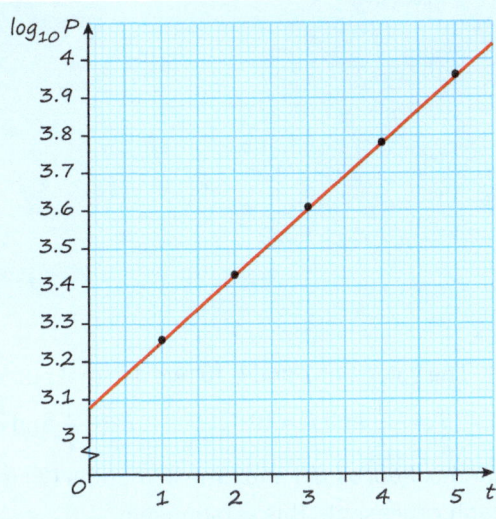

The line cuts the vertical axis at $\log_{10} k = 3.08$

so $k = 10^{3.08}$

$\Rightarrow k = 1200$ (to 3 s.f.)

The line goes through (5, 3.96) and (0, 3.08) so its gradient, $\log a$, is given by

$\log a = \dfrac{3.96 - 3.08}{5 - 0} = 0.176$

$\Rightarrow a = 10^{0.176} = 1.50$ (to 3 s.f.)

So the relationship between P and t is
$P = 1200 \times 1.50^t$.

Use any two points on your line.

ii To find the population after 8 years, substitute $t = 8$ into the equation of the curve.

$P = 1200 \times 1.50^8$

$\quad = 30\,755$ (nearest whole number)

iii To find when the population will be greater than 100 000, substitute $P = 100\,000$ into the inequality.

$$1200 \times 1.50^t > 100\,000$$

$$\Rightarrow \quad 1.50^t > \dfrac{100\,000}{1200}$$

$$\Rightarrow \quad 1.50^t > \dfrac{250}{3}$$

$$\Rightarrow \quad t \log 1.50 > \log\left(\dfrac{250}{3}\right)$$

Take logs to base 10 of both sides.

$$t > \dfrac{\log\left(\dfrac{250}{3}\right)}{\log 1.50}$$

$$t > 10.9$$

So the population will be greater than one hundred thousand after 11 years to the nearest year.

iv The population will not continue increasing exponentially as the size of the cavern will restrict the total number of bats. Also, the population will be restricted by the amount of food (moths etc.) nearby.

Test yourself

TESTED

1. The graph shows the result of plotting $\log_{10} y$ against x.

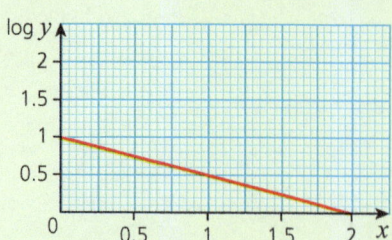

The relationship between x and y is of the form $y = k \times T^x$. The values of T and k, to two decimal places, are

A $T = 0.32$ and $k = 10$ B $T = 0.32$ and $k = 1$ C $T = -\frac{1}{2}$ and $k = 10$ D $T = -\frac{1}{2}$ and $k = 1$.

2. The relationship between a company's profits in thousands of euros (P) and time in months (T) is found to be $P = 20T^{-0.65}$ (all numbers to 2 s.f.). Which graph represents this relationship?

A
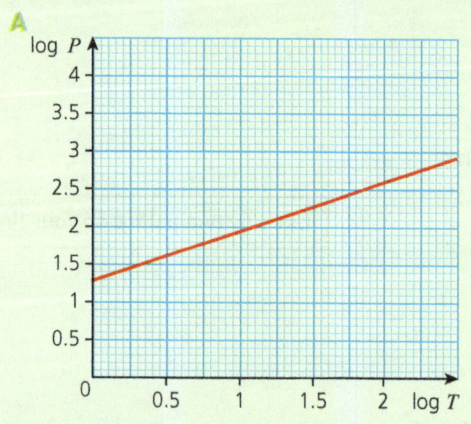

B
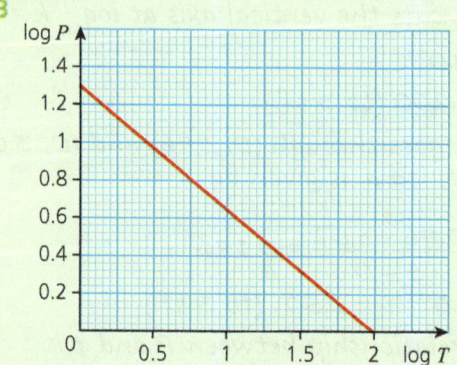

C
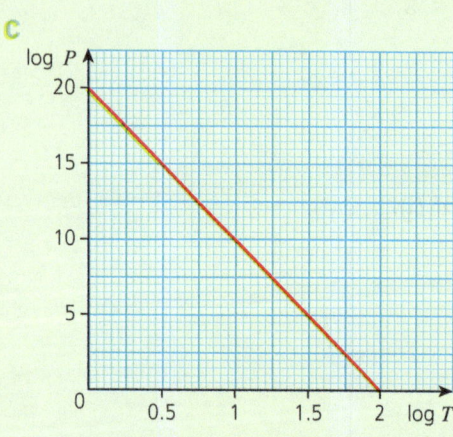

D
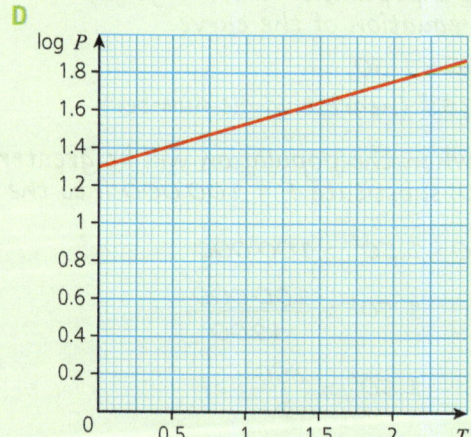

3. In an experiment, a variable, y, is measured at different times, t. The graph shows $\log_{10} y$ against $\log_{10} t$.

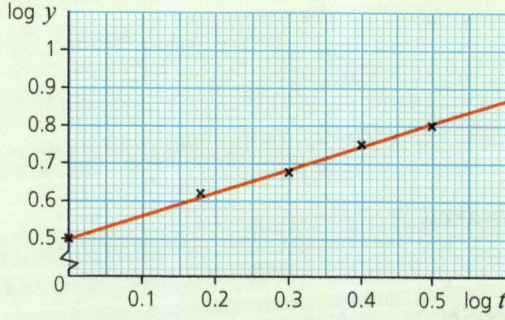

The relationship between y and t is

A $y = 3.16t^{1.67}$ B $y = 3.16t^{0.6}$ C $y = 0.5t^{0.6}$ D $y = 3.16t^{-0.22}$

4 The area of a patch of mould grows over time. Scientists measure the area, A cm^2, at 3 hourly intervals. The table shows the results of these measurements.

Time (hours)	3	6	9	12	15
Area (cm^2)	13	19	24	28	31

What is the relationship between area and time?

A $A = 8.5 + 1.5t$ B $A = 10.5 \times 1.075^t$ C $A = \frac{1}{2} \times t^8$ D $A = 8\sqrt{t}$

Full worked solutions online CHECKED ANSWERS

Exam-style question

The table shows a small firm's annual profits for the first six years, to the nearest £100.

Year (x)	1	2	3	4	5	6
Profit (P)	7800	9400	11 200	13 500	16 200	19 400

The firm's profits are modelled by $P = ka^x$, where a and k are constants.

i   Copy and complete the table below and plot $\log_{10} P$ against x. Draw a line of best fit for the data.

Year (x)	1	2	3	4	5	6
Profit (P)	7800	9400	11 200	13 500	16 200	19 400
$\log_{10} P$						

ii   Use your graph to find an equation for P in terms of x.
iii Using this model, predict the profit for year 10 to the nearest £100.
iv   After how many years does the model predict that the firm will make a profit of over a million pounds? Comment on the validity of your answer.

Short answers on page 215

Full worked solutions online CHECKED ANSWERS

Chapter 3 Algebra

About this topic

Being able to confidently manipulate and solve equations and inequalities is an essential part of A Level maths, and indeed all mathematics. This chapter revises these vital skills.

Before you start, remember

- how to expand brackets
- how to factorise expressions.

Quadratic equations

REVISED

Key facts

1. A **quadratic equation** can be written in the form $ax^2 + bx + c = 0$ where $a \neq 0$ and b and c can be any number. ◄── b and c can be positive, negative or zero.

2. Some quadratic equations can be solved by **factorising**. ◄── See Worked examples 1 and 2 for a reminder of this.

   Remember:
   - a **perfect square** is an expression in the form $x^2 + 2ax + a^2 = (x+a)^2$
   - the **difference of two squares** is an expression in the form $x^2 - a^2 = (x+a)(x-a)$.

3. The graph of a quadratic equation is a curve called a **parabola**.

   To sketch the quadratic equation $y = ax^2 + bx + c$
   - look at the sign of the x^2 term as this tells you which way up the curve is.

   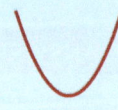 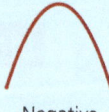

   Positive   Negative

   - find where the curve cuts the x-axis by solving $ax^2 + bx + c = 0$
   - remember the curve cuts the y-axis at c. ◄── When $x = 0$ then $y = c$.

4. **Completing the square** means write a quadratic expression in the form $a(x-p)^2 + q$.

5. **Completing the square method** ◄── You can only use this method when the coefficient of x^2 is 1. See Worked example 2.

   Example:
   $x^2 - 12x + 25$

   - Halve the coefficient of x — Half of -12 is -6
   - Square it — $(-6)^2$ is 36
   - Add this to the first two terms and subtract it from the constant term — $x^2 - 12x + 36 + 25 - 36$
   - Factorise the first three terms, to make a perfect square — $(x-6)^2 - 11$

6. You can use completing the square to solve quadratic equations.
   Example: $x^2 - 12x + 25 = 0 \Rightarrow (x-6)^2 - 11 = 0$
   $$\Rightarrow (x-6)^2 = 11$$
   $$\Rightarrow x - 6 = \pm\sqrt{11}$$
   $$\Rightarrow x = 6 \pm \sqrt{11}$$

   Don't forget the negative square root.

7. Completing the square tells you a lot about the position of the graph.

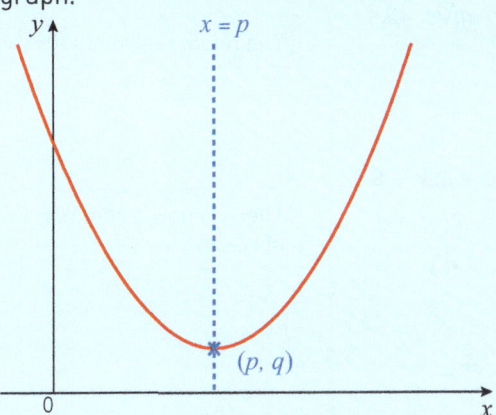

   The graph of $y = (x-p)^2 + q$ has a **stationary point** (or **vertex**) at (p, q).
   The curve is symmetrical about the stationary point so the line $x = p$ is a **line of symmetry**.

8. The **quadratic formula** is $x = \dfrac{-b \pm \sqrt{b^2 - 4ac}}{2a}$.
   You can use the quadratic formula to solve quadratic equations written in the form $ax^2 + bx + c = 0$.

9. The **discriminant** is $b^2 - 4ac$.
   The sign of the discriminant tells you how many real roots to expect.

   Hint: You should use the quadratic formula to solve equations which can't be factorised.

Discriminant	Positive $b^2 - 4ac > 0$	Zero $b^2 - 4ac = 0$	Negative $b^2 - 4ac < 0$
Number of real roots	2	1 (repeated)	0

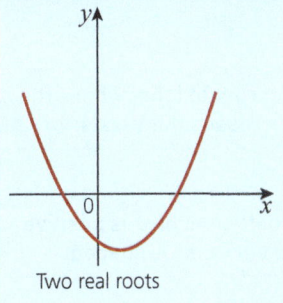
Two real roots

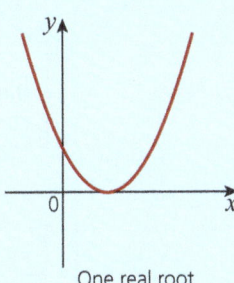
One real root

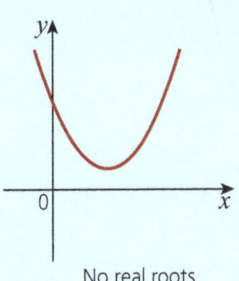
No real roots

Edexcel A Level Mathematics (Pure)

Worked examples

1 Solving a quadratic equation by factorising

Solve $3x^2 + 10x - 8 = 0$.

Solution

Step 1: Find the product of the two outside numbers: -24 ◀ $3 \times (-8) = -24$

Step 2: Look for two numbers which multiply to give -24 and add to give $+10$.   ◀ The middle term is $+10x$.

These numbers are $+12$ and -2.

Step 3: Split the middle term:
$$3x^2 + 10x - 8 = 3x^2 + 12x - 2x - 8$$

Step 4: Factorise in pairs: $= 3x(x+4) - 2(x+4)$ ◀ There is now a common factor of $(x+4)$

$= (3x - 2)(x + 4)$

Step 5: Now solve: $(3x - 2)(x + 4) = 0$

$$\Rightarrow x = \frac{2}{3} \text{ or } x = -4$$

2 Sketching the graph of a quadratic

Sketch the graph of

i $y = x^2 - 12x + 36$

ii $y = 81 - 9x^2$

Solution

i When $x = 0$ then $y = 36$

When $y = 0$ then $x^2 - 12x + 36 = 0$ ◀ This is a perfect square as it is in the form $x^2 - 2ax + a^2 = (x-a)^2$.

$\Rightarrow (x - 6)^2 = 0$

$\Rightarrow x = 6$ ◀ There is one repeated root, so the curve just touches the x-axis.

[Graph showing $y = x^2 - 12x + 36$, a U-shaped parabola touching the x-axis at $x=6$ and crossing the y-axis at 36.]

Don't forget to label where the curve crosses the y-axis.

The coefficient of x^2 is positive so the curve is $\cup$-shaped.

ii When $x = 0$ then $y = 81$

When $y = 0$ then $81 - 9x^2 = 0$ ◀ This is the difference of two squares as 81 is square and so is $9x^2$.

$\Rightarrow (9 + 3x)(9 - 3x) = 0$

$\Rightarrow x = -3 \text{ or } x = 3$

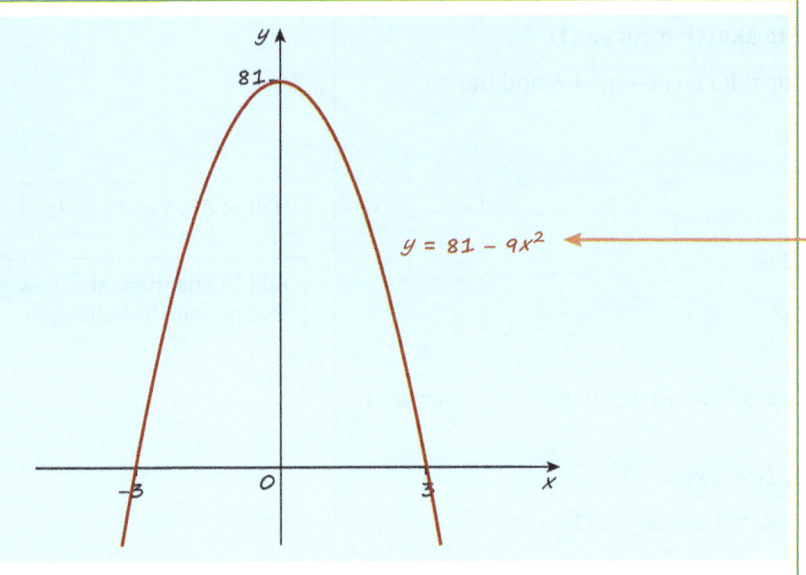

The coefficient of x^2 is negative so the curve is $\cap$-shaped.

3 Using quadratic equations

Solve:
  i $2x + 3\sqrt{x} - 2 = 0$
  ii $3 \times 2^{2x} - 10 \times 2^x - 8 = 0$

Hint: Some equations can be rewritten as a quadratic equation.

Solution

i Replacing $\sqrt{x}$ with z gives the quadratic equation
$2z^2 + 3z - 2 = 0$.

$\Rightarrow 2z^2 + 4z - z - 2 = 0$
$\Rightarrow 2z(z+2) - (z+2) = 0$
$\Rightarrow (2z-1)(z+2) = 0$
$\Rightarrow z = \frac{1}{2}$ or $z = -2$
$\Rightarrow \sqrt{x} = \frac{1}{2}$ or $\sqrt{x} = -2$ (no solution)

Squaring gives $x = \frac{1}{4}$

Remember that $(\sqrt{x})^2 = x$.

Split the middle term to help you factorise.

Note that $\sqrt{x} = -2$ has no solution as $\sqrt{x}$ means the positive square root of -2.

Common mistake: You need to solve the equation for x not z, so don't forget to work out what x is.

ii Replacing 2^x with z gives the quadratic equation
$3z^2 - 10z - 8 = 0$.

$\Rightarrow 3z^2 - 12z + 2z - 8 = 0$
$\Rightarrow 3z(z-4) + 2(z-4) = 0$
$\Rightarrow (3z+2)(z-4) = 0$
$\Rightarrow z = -\frac{2}{3}$ or $z = 4$
$\Rightarrow 2^x = -\frac{2}{3}$ or $2^x = 4$
So $x = 2$ (by inspection $2^2 = 4$)

Note that $2^x = -\frac{2}{3}$ has no solution as 2^x is positive for all real values of x.

Remember that $(2^x)^2 = 2^{2x}$.

Split the middle term to help you factorise.

Common mistake: Make sure you think carefully about which solutions are valid.

4 Using completing the square to sketch a curve (1)

Express $f(x) = x^2 + 8x + 10$ in the form $f(x) = (x+a)^2 + b$ and hence sketch the curve $y = f(x)$.

Solution

$$f(x) = x^2 + 8x + 10$$
$$= x^2 + 4x + 16 + 10 - 16$$
$$= (x+4)^2 + 10 - 16$$
$$= (x+4)^2 - 6$$

Half of 8 is 4 and $4^2 = 16$.

Add 16 and then take it away so that nothing has changed.

$y = (x+4)^2 - 6$ has the line of symmetry at $x = -4$ and a vertex at $(-4, -6)$.

Find where the curve cuts the axes.

When $x = 0$, $y = (0+4)^2 - 6 = 16 - 6 = 10$

When $y = 0$, $(x+4)^2 - 6 = 0$

$$\Rightarrow (x+4)^2 = 6$$
$$\Rightarrow x+4 = \pm\sqrt{6}$$
$$\Rightarrow x = -4 \pm \sqrt{6}$$

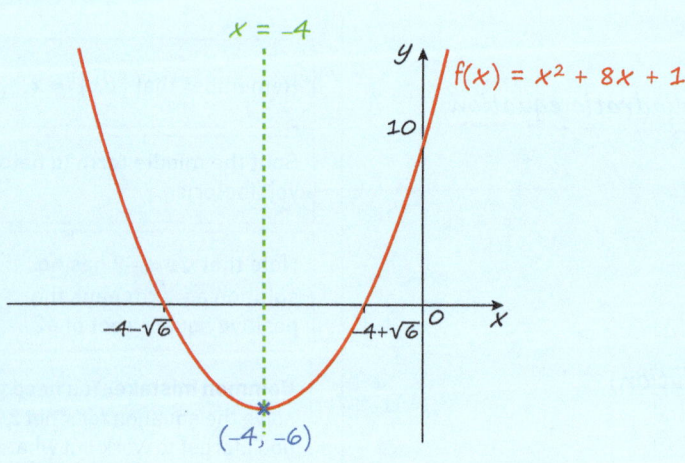

5 Using completing the square to sketch a curve (2)

Use completing the square to sketch the curve $y = -2x^2 + 4x + 6$.

Solution

$$y = -2x^2 + 4x + 6$$
$$= -2(x^2 - 2x) + 6$$
$$= -2(x^2 - 2x + 1 - 1) + 6$$
$$= -2(x^2 - 2x + 1) + 2 + 6$$
$$= -2(x-1)^2 + 8$$

Half of -2 is -1 and $(-1)^2 = 1$.

Add 1 and then take it away so that nothing has changed.

Take -2×-1 outside the bracket.

Common mistake: Do not divide by -2 because this would change the equation; instead you should factorise out -2 from the first two terms.

$y = -2(x-1)^2 + 8$ has the line of symmetry at $x = 1$ and a vertex at $(1, 8)$.

Find where the curve cuts the axes.

When $x = 0$, $y = -2(0-1)^2 + 8 = -2 + 8 = 6$

You can see this from the equation of the curve $y = -2x^2 + 4x + 6$.

When $y = 0$, $-2(x - 1)^2 + 8 = 0$
$\Rightarrow -2(x - 1)^2 = -8$
$\Rightarrow x - 1 = \pm\sqrt{4}$
$\Rightarrow x = 1 \pm 2$
$\Rightarrow x = -1$ or $x = 3$

> Divide both sides by –2 and then square root.

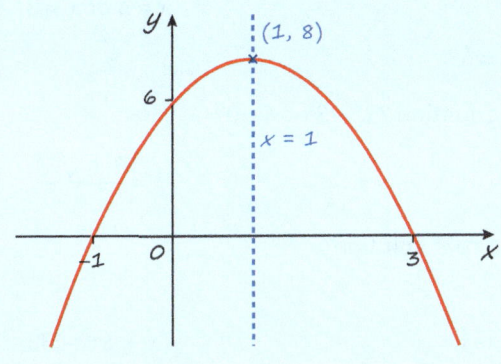

6 Using the quadratic formula

Solve the equation $3x^2 - 5x + 1 = 0$, giving your answers in exact form.

> **Hint: Exact** means leave your answer as a surd or a fraction – not a rounded decimal.

Solution

In this case, $a = 3$, $b = -5$ and $c = 1$.
Using the quadratic formula:

$x = \dfrac{-b \pm \sqrt{b^2 - 4ac}}{2a}$

> **Common mistake:** Take care with your signs!

$x = \dfrac{-(-5) \pm \sqrt{(-5)^2 - 4 \times 3 \times 1}}{2 \times 3}$

$= \dfrac{5 \pm \sqrt{13}}{6}$

> **Common mistake:** The rounded solutions are $x = 0.232$ or $x = 1.43$ (to 3.s.f.), but you aren't asked for these!

7 Using the discriminant

The equation $2x^2 + kx + 8 = 0$ has no real roots.

Find the possible values of k.

Solution

When there are no real roots the discriminant $b^2 - 4ac < 0$.
From the equation
$a = 2$, $b = k$ and $c = 8 \Rightarrow k^2 - 4 \times 2 \times 8 < 0$
$\Rightarrow k^2 - 64 < 0$
$\Rightarrow k^2 < 64$
$\Rightarrow -8 < k < 8$

> **Hint:** Use the discriminant for questions about number of roots.

> **Common mistake:** Take care with the inequality signs. Check values of k if you aren't sure which way round the signs should go.

Edexcel A Level Mathematics (Pure)

Test yourself

1 Factorise $6x^2 + 19x - 20$
   A $(x+4)(6x-5)$
   B $(3x+10)(2x-2)$
   C $(x+20)(6x-1)$
   D $(3x+4)(2x-5)$
   E $(6x+10)(x-2)$

2 Which of the following is the solution of the quadratic equation $2x^2 - 9x - 18 = 0$?
   A $x = \frac{3}{2}$ or $x = -6$
   B $x = \frac{9}{2}$ or $x = -2$
   C $x = 6$ or $x = 3$
   D $x = -\frac{9}{2}$ or $x = 2$
   E $x = -\frac{3}{2}$ or $x = 6$

3 Which of the following is the solution of the quadratic equation $2x^2 - 3x - 4 = 0$?
   A $x = \frac{-3 \pm \sqrt{41}}{4}$
   B $x = \frac{3 \pm \sqrt{41}}{4}$
   C $x = \frac{3 \pm \sqrt{23}}{4}$
   D $x = \frac{-3 \pm \sqrt{23}}{4}$
   E There are no real solutions

4 Write $x^2 - 12x + 3$ in completed square form.
   A $(x-6)^2 + 3$
   B $(x-12)^2 - 141$
   C $(x+6)^2 - 33$
   D $(x - 6 - \sqrt{33})(x - 6 + \sqrt{33})$
   E $(x-6)^2 - 33$

5 The curve $y = -2(x-5)^2 + 3$ meets the y-axis at A and has a maximum point at B. Find the coordinates of A and B.
   A A(0, 3) and B(5, 3)
   B A(0, −47) and B(5, 3)
   C A(0, −47) and B(5, −6)
   D A(0, 3) and B(−10, −47)
   E A(0, −47) and B(−10, 3)

6 Four of the following statements are true and one is false. Which one is false?
   A $3x^2 - 2x + 1 = 0$ has no real roots.
   B $2x^2 - 5x + 1 = 0$ has two distinct real roots.
   C $9x^2 - 6x + 1 = 0$ has one repeated real root.
   D $x^2 + 2x - 5 = 0$ has two distinct real roots.
   E $4x^2 - 9 = 0$ has one repeated real root.

Full worked solutions online

Exam-style question

You are given that $f(x) = 2x^2 + 12x + 10$.
  i Express $f(x)$ in the form $a(x+b)^2 + c$ where a, b and c are integers.
  ii The curve C with equation $y = f(x)$ meets the y-axis at P and has a minimum point at Q.
     a State the coordinates of P and Q.
     b Sketch the curve.

Short answers on page 215

Full worked solutions online

Simultaneous equations

REVISED

Key facts

1. A **linear equation** is an equation whose graph is a straight line. When you draw the graphs of two linear equations, then, unless the graphs are parallel, they intersect.
   The coordinates of the point of intersection give the solution to the two **linear simultaneous equations**.

   Linear equations do not involve terms in any powers of x or y, or terms like xy. Examples of linear equations are $2x - y = 8$ and $4x + 3y = 6$.

   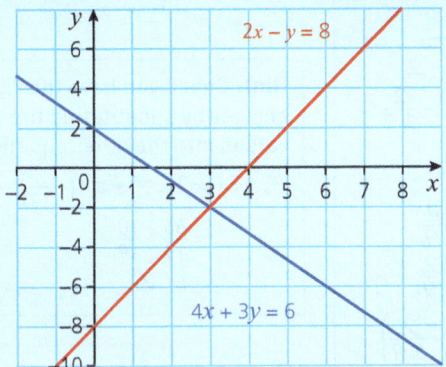

   The graphs intersect at the point $(3, -2)$…

   …so the solution of the two linear simultaneous equations $2x - y = 8$ and $4x + 3y = 6$ is $x = 3, y = -2$.

   Multiply one or both of the equations through by a constant, so that adding or subtracting the resulting equations eliminates one of the unknowns.

2. You should solve simultaneous equations algebraically. There are two different methods: the **elimination method** and the **substitution method**.

   Substitute one equation into the other so the resulting equation is in terms of just one unknown.

3. **Non-linear simultaneous equations** involve terms like x^2, y^2 or other powers.
   When you draw the graphs of a pair of equations where one is linear and one is quadratic, then there could be different solutions, as shown below.

   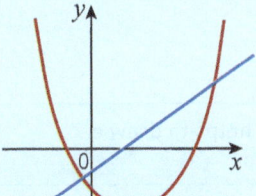 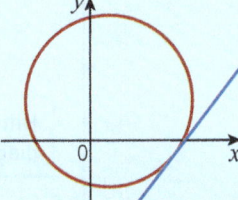 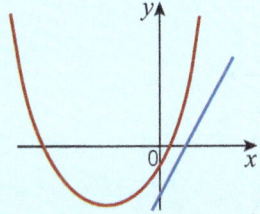

   Two points of intersection One point of intersection (a tangent) No points of intersection

4. Use the **substitution method** to solve one linear and one quadratic equation:
   - rearrange the linear equation if necessary so that one unknown is given in terms of the other
   - substitute the resulting equation into the quadratic equation
   - solve to find the value(s) of one of the unknowns
   - substitute back into the linear equation to find the value of the other unknown.

   There could be 2 solutions, 1 repeated solution or 0 solutions.

Edexcel A Level Mathematics (Pure)

Worked examples

1 Finding the intersection of two lines

Find the coordinates of the point where the lines $2x + 3y = 2$ and $3x - 5y = 4$ intersect.

Solution

(1): $2x + 3y = 2$ $\quad 3 \times (1):$ $\quad 6x + 9y = 6$
(2): $3x - 5y = 4$ $\quad 2 \times (2):$ $\quad 6x - 10y = 8$

Subtract:
$$19y = -2$$
$$\Rightarrow y = -\frac{2}{19}$$

Substitute $y = -\frac{2}{19}$ into (1): $\quad 2x + 3 \times \left(-\frac{2}{19}\right) = 2$
$$\Rightarrow 2x = \frac{44}{19}$$
$$\Rightarrow x = \frac{22}{19}$$

The coordinates of the point of intersection are $\left(\frac{22}{19}, -\frac{2}{19}\right)$.

Hint: Check you have the right answer by substituting the values into the other equation:
$$3x - 5y = 3 \times \frac{22}{19} - 5 \times \left(-\frac{2}{19}\right)$$
$$= \frac{66}{19} + \frac{10}{19}$$
$$= 4 \checkmark$$

Common mistake: Don't forget to state the coordinates. This is what the question asked for.

2 Finding the intersection of a line and a curve

Find the coordinates of the two points where the line $2x + y = 8$ intersects the circle $(x-2)^2 + (y+1)^2 = 10$.

Solution

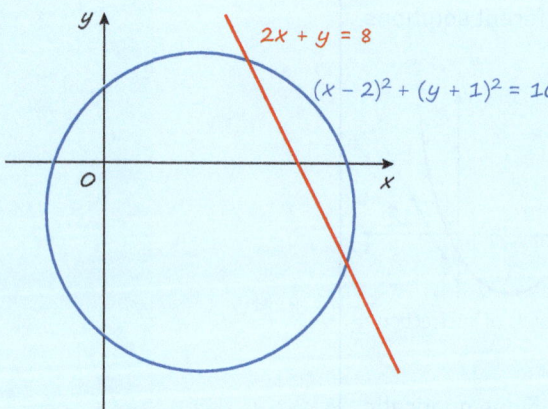

$2x + y = 8 \Rightarrow y = 8 - 2x$

Substitute $y = 8 - 2x$ into $(x-2)^2 + (y+1)^2 = 10$:
$$(x-2)^2 + (8 - 2x + 1)^2 = 10$$
$$\Rightarrow (x-2)^2 + (9 - 2x)^2 = 10$$
$$\Rightarrow x^2 - 4x + 4 + 81 - 36x + 4x^2 = 10$$
$$\Rightarrow 5x^2 - 40x + 75 = 0$$
$$\Rightarrow x^2 - 8x + 15 = 0$$
$$\Rightarrow (x-3)(x-5) = 0$$
$$\Rightarrow x = 3 \text{ or } x = 5$$

These values can then be substituted into $y = 8 - 2x$:
$y = 8 - 2 \times 3 = 2$ and $y = 8 - 2 \times 5 = -2$.
So the points of intersection are $(3, 2)$ and $(5, -2)$.

Hint: It helps to draw a diagram.

Hint: You should always substitute the linear equation into the equation for the curve.

Hint: You should use the linear equation to find the y-values once you have found the x-values.

Common mistake: Make sure you have the correct y-value paired-up with the correct x-value.

3 The tangent to a curve

The line $y = 5x + k$ is a tangent to the curve $y = x^2 - 3x - 6$.

Find the value of k and give the coordinates of the point where the line and curve meet.

Solution

Substituting $y = 5x + k$ into $y = x^2 - 3x - 6$:
$$5x + k = x^2 - 3x - 6$$
$$\Rightarrow x^2 - 8x - 6 - k = 0$$

The line $y = 5x + k$ forms a tangent when there is one repeated root $\Rightarrow b^2 - 4ac = 0$.

$a = 1$, $b = -8$ and $c = -(6 + k)$ ← Take care with your signs!

$$b^2 - 4ac = 0$$
$$\Rightarrow (-8)^2 + 4 \times 1 \times (6 + k) = 0$$
$$\Rightarrow 64 + 24 + 4k = 0$$
$$\Rightarrow 4k = -88$$
$$\Rightarrow k = -22$$

To find the x coordinates of the point of contact solve $x^2 - 8x - 6 - k = 0$ when $k = -22$.
$$\Rightarrow x^2 - 8x - 6 + 22 = 0$$
$$\Rightarrow x^2 - 8x + 16 = 0$$ ← This is a perfect square as it is in the form $x^2 - 2ax + a^2 = (x - a)^2$.
$$\Rightarrow (x - 4)^2 = 0$$
$$\Rightarrow x = 4$$

Substituting $x = 4$ and $k = -22$ into $y = 5x + k$ gives:
$$y = 5 \times 4 - 22 = -2$$

The coordinates of the point of intersection are $(4, -2)$.

There is a repeated (single) point of intersection and therefore the line is a tangent.

4 Finding the intersection of two curves

Find the coordinates of the points where $y = 2x^2 + 3x - 2$ and $y = x^2 + x + 1$ intersect.

Solution

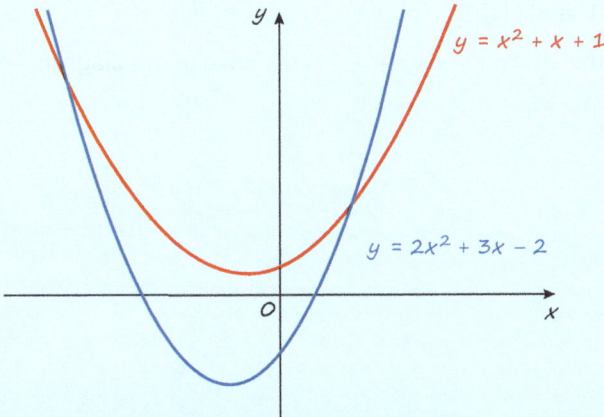

Equating the two expressions for y: ← This just means that you put the two expressions for y equal to each other.
$$2x^2 + 3x - 2 = x^2 + x + 1$$
$$\Rightarrow x^2 + 2x - 3 = 0$$
$$\Rightarrow (x + 3)(x - 1) = 0$$
$$\Rightarrow x = -3 \text{ or } x = 1$$

> Substituting the x coordinates into $y = x^2 + x + 1$:
> $x = 1 \Rightarrow y = 1 + 1 + 1 = 3$
> $x = -3 \Rightarrow y = 9 - 3 + 1 = 7$
> So the two points of intersection are at $(1, 3)$ and $(-3, 7)$.

Hint: You can check your answer is correct by substituting the x coordinates into $y = 2x^2 + 3x - 2$.

Test yourself

TESTED

1. Which one of the following is the correct x-value for the linear simultaneous equations $5x - 3y = 1$ and $3x - 4y = 4$?
   - A $x = \dfrac{16}{29}$
   - B $x = \dfrac{16}{11}$
   - C $x = -\dfrac{8}{11}$
   - D $x = -\dfrac{3}{11}$
   - E $x = \dfrac{5}{29}$

2. Simon is solving the simultaneous equations $y(1 - x) = 1$ and $2x + y = 3$. Simon's working is shown below.

Rearrange second equation:	$y = 2x - 3$	Line X
Substitute into first equation:	$(2x-3)(1-x) = 1$	Line Y
	$-2x^2 + 5x - 3 = 1$	
	$2x^2 - 5x + 4 = 0$	
Discriminant $= -5^2 - 4 \times 2 \times 4 = -25 - 32 = -57$		Line Z
There are no real solutions.		

   Simon knows that he must have made at least one mistake, as his teacher has told him that the equations do have real solutions. In which line(s) of the working has Simon made a mistake?
   - A Line X only
   - B Line Y only
   - C Lines Y and Z
   - D Line Z only
   - E Lines X and Z

3. Look at the simultaneous equations $3x^2 + 2y^2 = 5$ and $y - 2x = 1$. Which one of the following is the correct pair of x-values for the solution of these equations?
   - A $x = \dfrac{3}{11}$ or $x = -1$
   - B $x = \pm\sqrt{\dfrac{3}{11}}$
   - C $x = \dfrac{3}{7}$ or $x = -1$
   - D $x = -\dfrac{3}{11}$ or $x = 1$
   - E $x = -\dfrac{3}{7}$ or $x = 1$

4. Find the coordinates of the point where the lines $2x + 3y = 12$ and $3x - y = 7$ intersect.
   - A $\left(\dfrac{9}{7}, \dfrac{22}{7}\right)$
   - B $(3, 2)$
   - C $(9, 2)$
   - D $(3, 6)$
   - E $\left(\dfrac{33}{8}, \dfrac{5}{4}\right)$

5. Find the coordinates of the points where the line $y = 2x + 3$ intersects the curve $y = x^2 + 3x + 1$.
   - A $(-2, 5)$ and $(1, -1)$
   - B $(-4, -5)$ and $(-1, 1)$
   - C $(-2, -9)$ and $(1, 5)$
   - D $(-1, 1)$ and $(2, 7)$
   - E $(-2, -1)$ and $(1, 5)$

6. Which of the following lines does **not** intersect the circle $x^2 - 8x + y^2 - 9 = 0$ shown in the diagram?

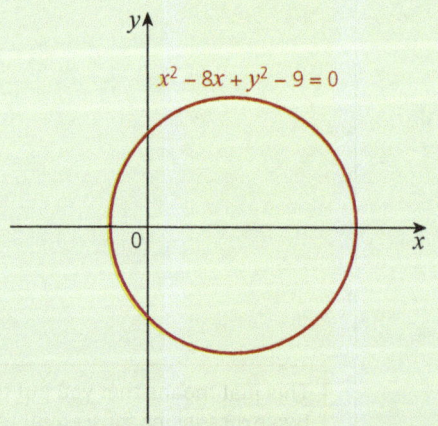

   - A $y = 2x + 2$
   - B $y = -5$
   - C $y = 2x + 4$
   - D $y = x - 1$
   - E $x = 7$

Full worked solutions online

CHECKED ANSWERS

Exam-style question

i The line $y - 3x = 3$ intersects the circle $(x + 3)^2 + (y - 2)^2 = 8$ at the points A and B.
   Find the coordinates of the points A and B.
ii The line $y - 3x = k$ forms a tangent to the curve $y = 4x^2 - x + 8$ at the point C.
   Find the value of k and the coordinates of C.

Short answers on page 215

Full worked solutions online

CHECKED ANSWERS

Inequalities

REVISED

Key facts

1 **Solving linear inequalities** is very similar to solving linear equations; however, you should bear in mind the following:
   - When swapping the sides of an inequality you should reverse the direction of the inequality sign. ← $6 < x$ so $x > 6$.
   - When multiplying or dividing both sides of an inequality by a negative number you should reverse the direction of the inequality. It is often best to avoid multiplying or dividing by a negative number if you can. ← $-2x < 6$ so $x > -3$.

2 **To solve a quadratic inequality**, you should solve the corresponding quadratic equation and then use a graph to determine the solution to the inequality.

The quadratic graph has a positive y-value here. So the range of values where the quadratic is > 0 is $x < x_1$ or $x > x_2$.

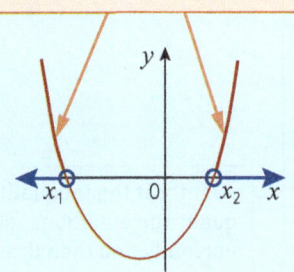

Use an open circle, O, to show the value of x_2 is not included.
Use a filled in circle, ●, to show the value of x_2 **is** included.

The quadratic graph has a negative y-value here. So the range of values where the quadratic is < 0 is $x_1 < x < x_2$.

3 You can write the solutions to inequalities using **set notation**.
   For example: $x < -2$ or $x \geqslant 5$ can be written as $\{x : x < -2\} \cup \{x : x \geqslant 5\}$.
   $-2 \leqslant x < 5$ can be written as $\{x : x \geqslant -2\} \cap \{x : x < 5\}$.

x belongs to the union of the set of numbers which are less than -2 and the set of the numbers which are greater or equal to 5.

4 You can represent inequalities graphically by sketching the line and then shading the relevant side of the line.

x belongs to the intersection of the set of numbers which are greater or equal to -2 and the set of the numbers which are less than 5.

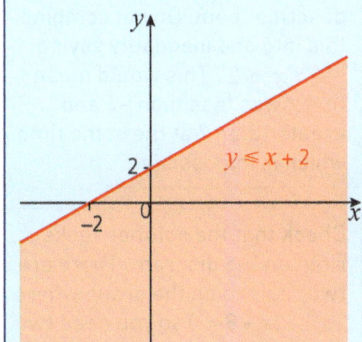

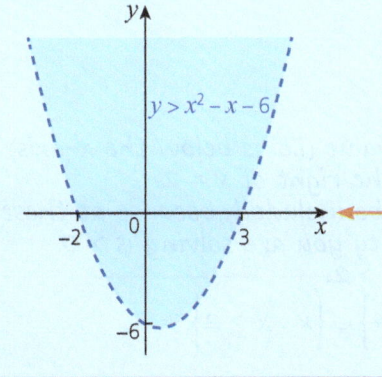

Use dashed line, - - -, to show that the line **is not** included.
Use a solid line, ——, to show that the line **is** included.

Edexcel A Level Mathematics (Pure) 35

Worked examples

1 Solving linear inequalities

Solve the inequality $1 - 2x > \frac{4-x}{3}$.

Solution

Method 1 – dividing by a negative number

$$1 - 2x > \frac{4-x}{3}$$

Multiply both sides by 3: $\quad 3 - 6x > 4 - x$

Add x to both sides: $\quad 3 - 5x > 4$

Subtract 3 from both sides: $\quad -5x > 1$

Divide both sides by -5: $\quad x < -\frac{1}{5}$

Using set notation: $\left\{x : x < -\frac{1}{5}\right\}$

Common mistake: Dividing by a negative number reverses the direction of the inequality sign.

Method 2 – a method that avoids dividing by a negative number

$$1 - 2x > \frac{4-x}{3}$$

Multiply both sides by 3: $\quad 3 - 6x > 4 - x$

Add $6x$ to both sides: $\quad 3 > 4 + 5x$

Subtract 4 from both sides: $\quad -1 > 5x$

Divide both sides by 5: $\quad -\frac{1}{5} > x$

Rewrite with x the subject: $\quad x < -\frac{1}{5}$

Hint: Inequalities should be written with x as the subject. Notice that '$-\frac{1}{5}$ is greater than or equal to x' means the same as 'x is less than or equal to $-\frac{1}{5}$'.

2 Solving quadratic inequalities

Solve the inequality $-x^2 - 2x + 8 < 0$.

Solution

The corresponding quadratic equation is $-x^2 - 2x + 8 = 0$.

Factorising: $(2 - x)(x + 4) = 0$.

So the graph of $y = -x^2 - 2x + 8$ intercepts the x-axis at $x = -4$ and $x = 2$.

Just treat the inequality as a quadratic equation: solve it normally and then draw the graph.

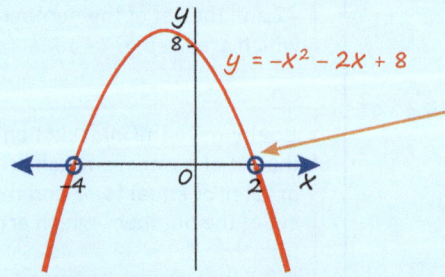

Empty circles are used to show that -4 and 2 are not included.

Common mistake: Two regions need two inequalities to describe them. Do not combine this into one inequality saying '$-4 > x > 2$'. This would mean that x was less than -4 and greater than 2 at the same time which is impossible!

The graph has a negative y-value (i.e. is below the x-axis) to the left of $x = -4$ and to the right of $x = 2$.
$x = -4$ and $x = 2$ should **not** be included, because at these points $y = 0$ and the inequality you are solving is > 0.
So the solution is $x < -4$ or $x > 2$.

Using set notation: $\{x : x < -4\} \cup \{x : x > 2\}$.

Check that the solution looks right on the diagram. There are two regions on the graph where $-x^2 - 2x + 8 < 0$ so you need two inequalities.

3 Representing inequalities graphically

Draw a graph to illustrate the inequality $2x^2 + 3x - 2 \leq 1 - 2x$.

Hence solve the inequality.

Solution

Start by sketching the graphs of $y = 2x^2 + 3x - 2$ and $y = 1 - 2x$.

Factorising $y = 2x^2 + 3x - 2$ gives $y = (2x - 1)(x + 2)$

So when $y = 0$, $x = \frac{1}{2}$ or $x = -2$,

and when $x = 0$, $y = -2$. ← *Find where the graph cuts the x- and the y-axes.*

The graph of $y = 1 - 2x$ passes through $(0, 1)$ and $\left(\frac{1}{2}, 0\right)$

The curve and line meet when $2x^2 + 3x - 2 = 1 - 2x$

$\Rightarrow 2x^2 + 5x - 3 = 0$

$\Rightarrow (2x - 1)(x + 3) = 0$

$\Rightarrow x = \frac{1}{2}$ or $x = -3$

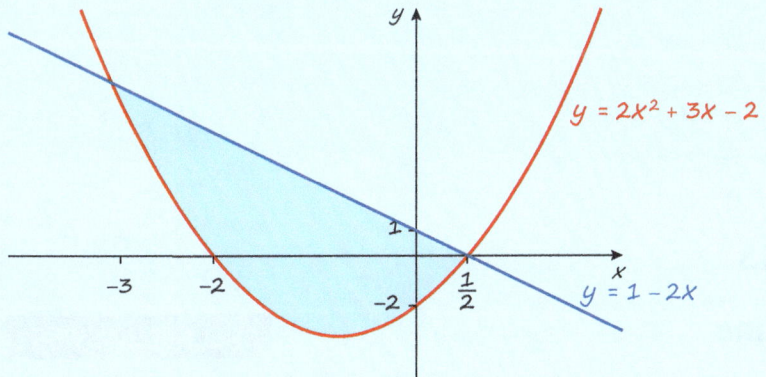

So the solution is $-3 \leq x \leq \frac{1}{2}$. ← *One region only needs one inequality to describe it.*

Using set notation: $\{x : -3 \leq x\} \cap \{x : x \leq \frac{1}{2}\}$.

Test yourself

1. Solve $x + 7 < 3x - 5$
   - A $x > 6$
   - B $x > 1$
   - C $x < 1$
   - D $x < 6$
   - E $x > 4$

2. Solve $\frac{2(2x+1)}{3} \geq 6$
   - A $x \geq 5$
   - B $x \geq \frac{3}{2}$
   - C $x > 4$
   - D $x \geq 4$
   - E $x > 5$

Edexcel A Level Mathematics (Pure)

3 The diagram shows the lines $y=3x-3$ and $y=-x+5$. For what values of x is $y=3x-3$ above $y=-x+5$?

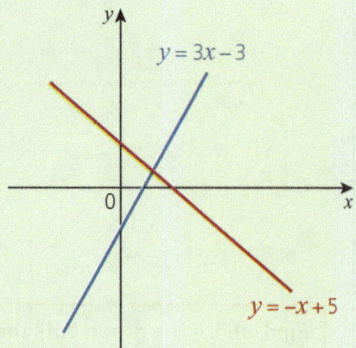

   A $x<4$ B $x<2$ C $x>\dfrac{1}{2}$ D $x>2$ E $x>4$

4 Solve the inequality $x^2+2x-15\leqslant 0$
   A $-5\leqslant x\leqslant 3$ B $-3\leqslant x\leqslant 5$ C $3\leqslant x\leqslant 5$ D $x=-5$ or $x=3$ E $x\leqslant -5$ or $x\geqslant 3$

5 Solve the inequality $6x-6<x^2-1$
   A $x<-1$ or $x>7$ B $1<x<5$ C $x<1$ or $x>7$ D $x<-5$ or $x>-1$ E $x<1$ or $x>5$

Full worked solutions online CHECKED ANSWERS

Exam-style question

Solve the following inequalities:

i $\dfrac{(3-5x)}{2}\leqslant x$

ii $2x^2>7x+4$.

Short answers on page 215

Full worked solutions online CHECKED ANSWERS

Working with polynomials and algebraic fractions

REVISED

Key facts

1 A **polynomial** consists of one or more terms where each term has a variable raised to a positive (or zero) integer power.

   The **order** of the polynomial is the highest power of the variable it contains. So $4-7x^5+3x^{12}$ has order 12.

For example: $3+x^2+3x^9$ or $2-x^3$.

2 When you **multiply** polynomials remember:
  - x means x^1
  - to multiply powers of x add the indices: $x^3\times x^5=x^{3+5}=x^8$
  - the rules for multiplying positive and negative numbers are
    $+\times +=+;\quad -\times -=+;\quad +\times -=-;\quad -\times +=-.$

3 You can divide one polynomial by another to find the **quotient**.

   The words you use for dividing two whole numbers are shown below

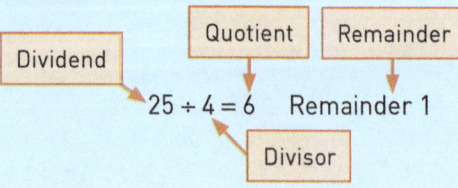

4. One way of **dividing polynomials** is to set them out in columns, rather like a long division using numbers.
   Reminder: long division, using numbers:

   $325 \div 25$

   $$\begin{array}{r} 13 \\ 25\overline{)325} \\ -25\downarrow \\ \overline{75} \\ -75 \\ \overline{0} \end{array}$$

   - 1×25
   - 3×25
   - No remainder.

5. Another way of dividing (when there is no remainder) is to set the question out as a multiplication and then compare the coefficients to find the quotient.

   See Worked example 3.

6. A **rational expression** is a fraction where the numerator and/or denominator is a polynomial. This can also be referred to as an **algebraic fraction**.

   In an algebraic fraction, the denominator must contain a variable. So $\frac{x}{3}$ is not an algebraic fraction, but $\frac{3}{x}$ is an algebraic fraction.

7. You can **simplify** algebraic fractions in the same way as you do ordinary fractions.

   - $\dfrac{2x}{x^2+1} + \dfrac{3x}{x^2+1} = \dfrac{5x}{x^2+1}$

   - $\dfrac{3(x+5)(x+1)}{(x+1)(x-2)} = \dfrac{3(x+5)}{(x-2)}$

   To add or subtract algebraic fractions you must have a common denominator.

   You can cancel common factors.

Worked examples

1 Adding and subtracting polynomials

You are given that $f(x) = 4x^3 - 3x^2 - 2x + 1$ and $g(x) = x^3 - x + 3$.

Find

i $f(x) + g(x)$

ii $f(x) - g(x)$.

Solution

i **Collecting like terms**

$(4x^3 - 3x^2 - 2x + 1) + (x^3 - x + 3)$
$= 4x^3 + x^3 - 3x^2 - 2x - x + 1 + 3$
$= 5x^3 - 3x^2 - 3x + 4$

Using columns

$$\begin{array}{r} 4x^3 - 3x^2 - 2x + 1 \\ + x^3 - x + 3 \\ \hline 5x^3 - 3x^2 - 3x + 4 \end{array}$$

Remember that x^3 means $1x^3$ so the coefficient of x^3 is 1.

Take care: there is no term of x^2 in the second polynomial, so make sure you leave a space.

So $f(x) + g(x) = 5x^3 - 3x^2 - 3x + 4$

ii **Collecting like terms**

$(4x^3 - 3x^2 - 2x + 1) - (x^3 - x + 3)$
$= 4x^3 - 3x^2 - 2x + 1 - x^3 + x - 3$
$= 4x^3 - x^3 - 3x^2 - 2x + x + 1 - 3$
$= 3x^3 - 3x^2 - x - 2$

So $f(x) - g(x) = 3x^3 - 3x^2 - x - 2$

Using columns

$$\begin{array}{r} 4x^3 - 3x^2 - 2x + 1 \\ - x^3 - x + 3 \\ \hline 3x^3 - 3x^2 - x - 2 \end{array}$$

Change the signs first.

Common mistake: A '−' sign in front of the brackets means that you have to multiply each term of the brackets by −1 or change the sign in front of each term in the brackets. Remember that $-2x - (-x) = -2x + x = -x$.

Edexcel A Level Mathematics (Pure)

2 Multiplying polynomials

You are given that $f(x) = x^2 - 5x + 1$ and $g(x) = 2x - 3$.

Find $f(x) \times g(x)$.

Solution

$$(2x - 3)(x^2 - 5x + 1) = 2x(x^2 - 5x + 1) - 3(x^2 - 5x + 1)$$
$$= 2x^3 - 10x^2 + 2x - 3x^2 + 15x - 3$$
$$= 2x^3 - 13x^2 + 17x - 3$$

Common mistake: Check the signs in this line carefully.

3 Dividing polynomials

Simplify $\dfrac{(6x^3 - 11x^2 + 10x - 3)}{(2x - 1)}$.

Solution

Hint: Lay it out in columns.

Method 1: Using long division to find a quotient

$$
\begin{array}{r}
3x^2 \\
2x-1 \overline{\smash{)}6x^3 - 11x^2 + 10x - 3} \\
\underline{-6x^3 - 3x^2} \\
-8x^2 + 10x
\end{array}
$$

- Divide $6x^3$ by $2x$ to get $3x^2$.
- Multiply $(2x - 1)$ by $3x^2$.
- Subtract $(6x^3 - 3x^2)$ from $6x^3 - 11x^2$ which gives $-8x^2$.
- Bring down $10x$.

$$
\begin{array}{r}
3x^2 - 4x \\
2x-1 \overline{\smash{)}6x^3 - 11x^2 + 10x - 3} \\
\underline{-6x^3 - 3x^2} \\
-8x^2 + 10x \\
\underline{--8x^2 + 4x} \\
6x - 3
\end{array}
$$

- Divide $-8x^2$ by $2x$ to get $-4x$ and write the answer here.
- Multiply $(2x - 1)$ by $4x$ and write the result down.
- Subtract which gives $6x$.
- Bring down -3.

$$
\begin{array}{r}
3x^2 - 4x + 3 \\
2x-1 \overline{\smash{)}6x^3 - 11x^2 + 10x - 3} \\
\underline{-6x^3 - 3x^2} \\
-8x^2 + 10x \\
\underline{--8x^2 + 4x} \\
6x - 3 \\
\underline{-6x - 3} \\
0
\end{array}
$$

- Divide $6x$ by $2x$ and write the answer here.
- Multiply $(2x - 1)$ by $+3$.
- Subtract. The answer is zero showing that there is no remainder.

So $(6x^3 - 11x^2 + 10x - 3) \div (2x - 1) = 3x^2 - 4x + 3$.

Method 2: Comparing coefficients to find a quotient

Hint: This method is suitable only when the remainder is 0.

$$\frac{(6x^3 - 11x^2 + 10x - 3)}{(2x - 1)} = ax^2 + bx + c$$

$$\Rightarrow 6x^3 - 11x^2 + 10x - 3 = (2x - 1)(ax^2 + bx + c)$$

Since $6x^3 \div 2x = 3x^2$ the first term in the right-hand bracket should be $3x^2$.

Step 1: Look at the term in x^3: $6x^3 = 2x \times ax^2 \Rightarrow a = 3$.

So $6x^3 - 11x^2 + 10x - 3 = (2x - 1)(3x^2 + bx + c)$

Step 2: Look at the constant term: $-3 = -1 \times c \Rightarrow c = 3$.

So $6x^3 - 11x^2 + 10x - 3 = (2x - 1)(3x^2 + bx + 3)$

The constant term on the left is -3. To get -3 when you multiply out the brackets you need '+3' in the last bracket.

Step 3: Look at the term in x: $+10x = 2x \times 3 + (-1) \times bx$
$\Rightarrow \qquad 10x = 6x - bx$.

Comparing coefficients of x gives: $10 = 6 - b \Rightarrow b = -4$

So $6x^3 - 11x^2 + 10x - 3 = (2x - 1)(3x^2 - 4x + 3)$.

Alternative Step 3:

Look at the coefficients of x^2 to find b:
$-11x^2 = 2x \times bx + (-1) \times 3x^2$
$\Rightarrow -11x^2 = 2bx^2 - 3x^2$

Comparing coefficients of x^2 gives: $-11 = 2b - 3$
$\Rightarrow \qquad\qquad\qquad\qquad\qquad b = -4$

So $6x^3 - 11x^2 + 10x - 3 = (2x - 1)(3x^2 - 4x + 3)$, as before.

> Look for the pairs of terms that multiply together to give a term in x.

> On the left side you have $10x$ and on the right you have $(6 - b)x$.

> Look for the pairs of terms that multiply together to give a term in x^2.

Hint: Check your result by multiplying out the brackets on the right-hand side.
$(2x - 1)(3x^2 - 4x + 3)$
$= 2x(3x^2 - 4x + 3) - 1 \times (3x^2 - 4x + 3)$
$= 6x^3 - 8x^2 + 6x - 3x^2 + 4x - 3$
$= 6x^3 - 11x^2 + 10x - 3$ ✓

4 Simplifying rational expressions

Simplify

i $\quad \dfrac{2x^2 + x - 15}{(2x - 5)(x + 1)}$

ii $\quad \dfrac{2}{(2x-1)} + \dfrac{3x}{(2x-1)^2} - \dfrac{1}{x}$.

Solution

i $\quad \dfrac{2x^2 + x - 15}{(x+1)(2x-5)} = \dfrac{(2x-5)(x+3)}{(x+1)(2x-5)}$

$\qquad\qquad\qquad\quad = \dfrac{x+3}{x+1}$

> Factorise and then cancel any common factors.

ii $\quad \dfrac{2}{(2x-1)} + \dfrac{3x}{(2x-1)^2} - \dfrac{1}{x}$

$= \dfrac{2}{(2x-1)} \times \dfrac{x(2x-1)}{x(2x-1)} + \dfrac{3x}{(2x-1)^2} \times \dfrac{x}{x} - \dfrac{1}{x} \times \dfrac{(2x-1)^2}{(2x-1)^2}$

$= \dfrac{2x(2x-1)}{x(2x-1)^2} + \dfrac{3x^2}{x(2x-1)^2} - \dfrac{(2x-1)^2}{x(2x-1)^2}$

$= \dfrac{2x(2x-1) + 3x^2 - (2x-1)^2}{x(2x-1)^2}$

$= \dfrac{4x^2 - 2x + 3x^2 - (4x^2 - 4x + 1)}{x(2x-1)^2}$

$= \dfrac{3x^2 + 2x - 1}{x(2x-1)^2}$

$= \dfrac{(3x-1)(x+1)}{x(2x-1)^2}$

> Multiply the top and bottom lines of each fraction so that all fractions have a common denominator.

Test yourself

1. When $(3x^4 - 2x^3 - x + 1)$ is divided by $(x - 1)$ the answer is $(3x^3 + x^2 + x)$ and the remainder is $+1$. Four of the statements below are true and one is false. Which one is false?
   - A $3x^4 - 2x^3 - x + 1 = (x - 1)(3x^3 + x^2 + x) + 1$.
   - B The quotient is a polynomial of order 3.
   - C The remainder is a polynomial of order 1.
   - D There is no remainder when $3x^4 - 2x^3 - x$ is divided by $x - 1$.
   - E The remainder is a constant.

2. Simplify $\dfrac{3x^3 - 5x^2 + 6x - 4}{x - 1}$.
   - A $3x^2 + 2x + 4$
   - B $-3x^2 - 2x + 4$
   - C $3x^2 - 2x - 4$
   - D $3x^2 - 2x + 4$
   - E $3x^3 - 1$

3. You are given that $2x^3 + x^2 - 7x - 6 = (x - 2)(2x^2 + bx + 3)$. The value of b is:
   - A -3
   - B 5
   - C 3
   - D -5
   - E $-\dfrac{1}{4}$

4. You are given that $6x^3 + 13x^2 - 2x - 12 = (2x + 3)(ax^2 + bx + c)$. The quadratic factor $(ax^2 + bx + c)$ is
   - A $3x^2 + 2x + 4$
   - B $3x^2 + 2x - 4$
   - C $3x^2 - \dfrac{10}{3}x + 4$
   - D $3x^2 + \dfrac{10}{3}x - 4$
   - E $3x^2 - 2x - 4$

5. When $2x^3 - x^2 - x - 1$ is divided by $x - 2$ the remainder is:
   - A -23
   - B 9
   - C -11
   - D -15
   - E 13

6. Simplify the expression $\dfrac{x^2 - 9}{x^2 - x - 12}$ as far as possible.
   - A $\dfrac{9}{x - 12}$
   - B $\dfrac{3}{4}$
   - C $\dfrac{x + 3}{x - 4}$
   - D $\dfrac{x - 3}{x - 4}$
   - E $\dfrac{x + 3}{x + 4}$

Full worked solutions online

Exam-style question

i Simplify

   a $\dfrac{x^2 - 4}{x^2 + 5x + 6}$

   b $\dfrac{1}{(x+2)} + \dfrac{1}{(x-2)} - \dfrac{1}{x}$

ii Given $\dfrac{6x^3 - 7x^2 + 14x - 8}{3x - 2} = ax^2 + bx + c$, find the values of a, b and c.

Short answers on page 215

Full worked solutions online

The factor theorem and curve sketching

Key facts

1. The **factor theorem** says:
   - If $(x - a)$ is a factor of $f(x)$ then $f(a) = 0$ and $x = a$ is a root of the equation $f(x) = 0$.
   - Conversely, if $f(a) = 0$ then $(x - a)$ is a factor of $f(x)$.
   - When $(ax + b)$ is a factor of a polynomial $f(x)$ then $f\left(-\frac{b}{a}\right) = 0$ and $x = -\frac{b}{a}$ is a root of $f(x) = 0$.

   > To find $f(a)$, substitute $x = a$ into the polynomial. For example, given $f(x) = 3x - 1$ then $f(2) = 3 \times 2 - 1 = 5$.

2. You can use the factor theorem to test for factors or roots of a polynomial.
   Start by looking at the constant term.
   For example, to find factors of $f(x) = 2x^3 + x^2 - 13x + 6$, look at the constant term '+6'. 6 is divisible by $\pm 1, \pm 2, \pm 3, \pm 6$.
   $f(1) = 2 \times 1^3 + 1^2 - 13 \times 1 + 6 = 2 + 1 - 13 + 6 = -4$ ✗
   $f(2) = 2 \times 2^3 + 2^2 - 13 \times 2 + 6 = 16 + 4 - 26 + 6 = 0$ ✓
   $\Rightarrow x = 2$ is a root and $(x - 2)$ is a factor of $f(x) = 2x^3 + x^2 - 13x + 6$.
   Once you have found one factor, you can use long division or the comparing coefficients method (see page 40) to find any other factors.

   > Any integer roots will be factors of the constants term.

3. Polynomial curves have **turning points**.

   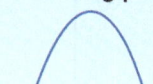

   Minimum point   Maximum point

4. The number of turning points depends on the **order** of a polynomial.
   A polynomial of degree n has at most $(n - 1)$ turning points.

   > For example, a cubic equation (order 3) has at most 2 turning points.

5. To **sketch** the curve of a polynomial:
   Step 1: Decide on the shape of the curve by looking at the highest power of x, ax^n.

                    n even n odd

   a positive

   a negative

   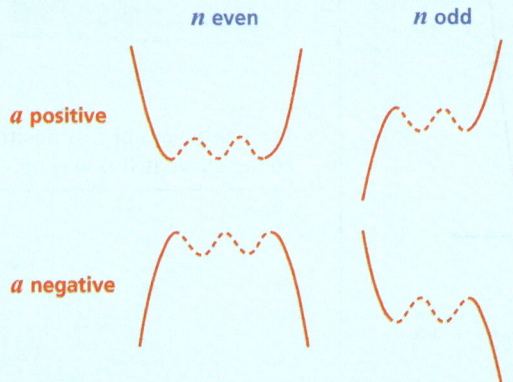

   Step 2: Show the turning points. Sometimes you will be asked to write the coordinates of the turning points.
   Step 3: Give the coordinates of the points where the curve crosses (intersects) the x-axis and the y-axis.

   > Remember that for a polynomial of degree n there are at most $(n - 1)$ turning points.

6. To **plot** the curve of a polynomial you have to be more accurate. Calculate the values of y for suitable values of x, plot these points and join them as a smooth curve.

Worked examples

1 Using the factor theorem to sketch curves

You are given that $f(x) = x^4 - x^3 - 11x^2 + 9x + 18$.

i Find the values of f(1), f(–1), f(2), f(–2), f(3) and f(–3). Hence factorise the polynomial.

ii Sketch the graph of $y = f(x)$.

Solution

i $f(x) = x^4 - x^3 - 11x^2 + 9x + 18$

$f(1) = 1^4 - 1^3 - 11 \times 1^2 + 9 \times 1 + 18 = 16$
$\Rightarrow (x - 1)$ is not a factor

> The highest power of x is 4 so you expect at most four linear factors.

$f(-1) = (-1)^4 - (-1)^3 - 11 \times (-1)^2 + 9 \times (-1) + 18 = 0$
$\Rightarrow (x + 1)$ is a factor. ✓

$f(2) = 2^4 - 2^3 - 11 \times 2^2 + 9 \times 2 + 18 = 0$
$\Rightarrow (x - 2)$ is a factor. ✓

$f(-2) = (-2)^4 - (-2)^3 - 11 \times (-2)^2 + 9 \times (-2) + 18 = -20$
$\Rightarrow (x + 2)$ is not a factor.

$f(3) = 3^4 - 3^3 - 11 \times 3^2 + 9 \times 3 + 18 = 0$
$\Rightarrow (x - 3)$ is a factor. ✓

$f(-3) = (-3)^4 - (-3)^3 - 11 \times (-3)^2 + 9 \times (-3) + 18 = 0$
$\Rightarrow (x + 3)$ is a factor. ✓

So $f(x) = (x + 1)(x - 2)(x + 3)(x - 3)$.

ii The curve cuts the x-axis when $f(x) = 0$

So $x = -1, x = -3, x = 2$ and $x = 3$ are roots of $f(x) = 0$.

When $x = 0$, $f(x) = 18$.

> $f(a) = 0 \Rightarrow x = a$ is a root of $f(x) = 0$.

> Substitute $x = 0$ into $f(x) = x^4 - x^3 - 11x^2 + 9x + 18$.

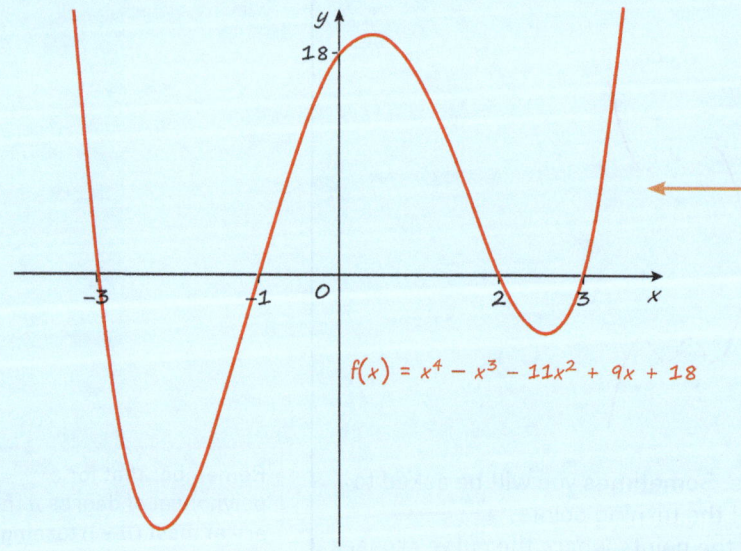

> The coefficient of x^4 is positive so the curve is this way up.

$f(x) = x^4 - x^3 - 11x^2 + 9x + 18$

2 Using the factor theorem to solve equations

  i Show that $(2x + 1)$ is a factor of $f(x) = 2x^3 + 3x^2 - 11x - 6$ and hence factorise $f(x)$ fully.
  ii Solve the equation $f(x) = 0$.
  iii Sketch the graph of $y = f(x)$.

Hint: Factorise completely or factorise fully means keep going until there are no further factors to be found.

Solution

  i Using the factor theorem:
  if $(2x + 1)$ is a factor of $f(x)$ then $f\left(-\tfrac{1}{2}\right) = 0$

  $f\left(-\tfrac{1}{2}\right) = 2 \times \left(-\tfrac{1}{2}\right)^3 + 3 \times \left(-\tfrac{1}{2}\right)^2 - 11 \times \left(-\tfrac{1}{2}\right) - 6$

  $= -\tfrac{2}{8} + \tfrac{3}{4} + \tfrac{11}{2} - 6$

  $= 0$

  So $(2x + 1)$ is a factor of $2x^3 + 3x^2 - 11x - 6$.

  $2x^3 + 3x^2 - 11x - 6 = (2x + 1)(ax^2 + bx + c)$ ◄

  Hint: When the question says 'show', you must show **all** your working.

  You can also use long division to find the quadratic factor.

  Step 1: Look at the term in x^3: $2x^3 = 2x \times ax^2 \Rightarrow a = 1$.
  Step 2: Look at the constant term: $-6 = 1 \times c \Rightarrow c = -6$.
  So $2x^3 + 3x^2 - 11x - 6 = (2x + 1)(x^2 + bx - 6)$.
  Step 3: Look at the term in x: $-11x = -12x + bx \Rightarrow b = 1$. ◄

  $2x \times (-6) + 1 \times bx$.

  So, $2x^3 + 3x^2 - 11x - 6 = (2x + 1)(x^2 + x - 6)$
  $= (2x + 1)(x + 3)(x - 2)$ ◄

  Factorise the quadratic.

  ii $f(x) = 0 \Rightarrow x = -\tfrac{1}{2}, x = -3$ or $x = 2$. ◄

  When $(ax + b)$ is a factor of a polynomial $f(x)$, then $x = -\tfrac{b}{a}$ is a root of $f(x) = 0$.

  iii The curve $y = 2x^3 + 3x^2 - 11x - 6$ crosses the x-axis at $(-3, 0)$, $\left(-\tfrac{1}{2}, 0\right)$ and $(2, 0)$.

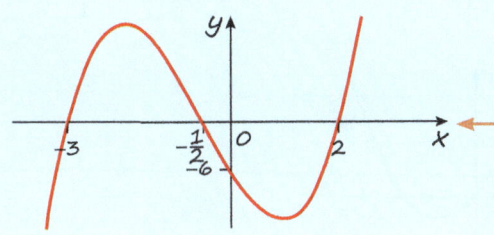

When $x = 0$, $f(x) = -6$ so the curve cuts the y-axis at $(0, -6)$.

Test yourself

TESTED

1 $(x - 2)$ is a factor of one of these polynomials. Which one is it?
   A $x^3 - 2x^2 - 3x + 1$
   B $x^4 - 2x^3$
   C $2x^3 - x^2 - 4$
   D $x^2 + 4x - 16$
   E $2x^3 - 4x^2 - 3x + 2$

2 $f(x) = 3x^3 - 2x^2 - 3x + 2$.
   Four of the statements below are true and one is false. Which one is false?
   A $(x - 1)$ is a factor of the polynomial $f(x)$.
   B When the polynomial $3x^3 - 2x^2 - 3x + 2$ is divided by $(x + 1)$ there is no remainder.
   C $(x + 2)$ is not a factor of polynomial $f(x)$.
   D $(3x + 2)$ is a factor of polynomial $f(x)$.
   E There are three different values of x for which $f(x) = 0$.

3 A polynomial $f(x)$ given by $x^3 - 3x^2 - 7x - 15$. Which one of these is a factor of $f(x)$?
   A $(x - 1)$
   B $(x - 3)$
   C $(x - 5)$
   D $(x + 1)$
   E $(x + 3)$

Edexcel A Level Mathematics (Pure)

4 A polynomial is given by $f(x) = x^4 + 2x^3 + 3x^2 + 4x + 2$. Four of the following statements are false and one is true. Which one is true?
  A The four roots of $x^4 + 2x^3 + 3x^2 + 4x + 2 = 0$ are $x = 1$, $x = -1$, $x = 2$ and $x = -2$.
  B There are two different roots.
  C The curve $y = f(x)$ touches the x-axis.
  D $(x+1)$ is a factor of $f(x)$ but $(x+1)^2$ is not a factor.
  E $(x+2)$ is a factor of $f(x)$.

5 Which of the following is the equation of the curve?

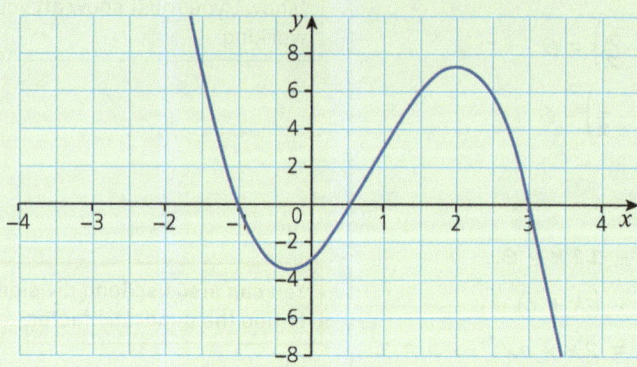

  A $y = 2x^4 + x^3 - 19x^2 - 9x + 9$
  B $y = 2x^3 - 5x^2 - 4x + 3$
  C $y = x^2 - 2x + 3$
  D $y = -2x^2 - x + 1$
  E $y = -2x^3 + 5x^2 + 4x - 3$

6 Which of the following is the graph of the curve $y = (x^2 - 1)(x + 1)$?

A
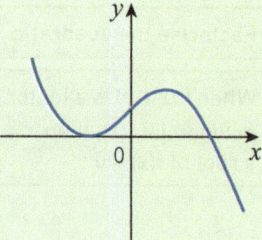

B
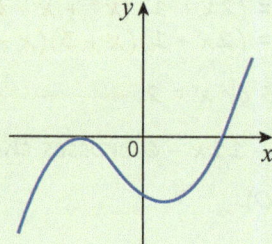

C
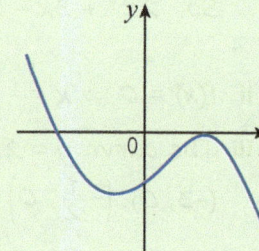

D
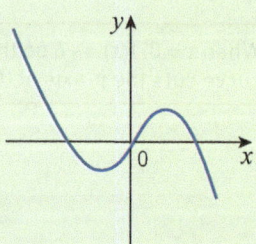

E
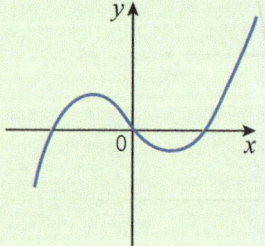

Full worked solutions online   CHECKED ANSWERS

Exam-style question

Given that $(2x - 1)$ and $(x + 4)$ are factors of $2x^3 - x^2 + ax + b$, find the values of a and b.

Short answers on page 215

Full worked solutions online   CHECKED ANSWERS

Chapter 4 Coordinate geometry

About this topic

In this chapter you will solve problems involving coordinates, straight lines and circles. To find the equation of a straight line, you need to know either two points on the line, or one point and the direction of the line. To define a circle, you need the centre and radius.

Before you start, remember

- Pythagoras' theorem
- how to solve simultaneous equations – see Chapter 3
- how to solve linear and quadratic equations – see Chapter 3.

Straight lines

REVISED

Key facts

1 The diagram shows the line joining $A(x_1, y_1)$ and $B(x_2, y_2)$.

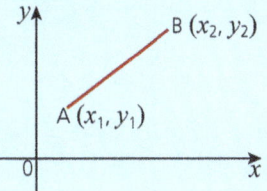

- Gradient of AB = $\dfrac{y_2 - y_1}{x_2 - x_1}$. ← This is $\dfrac{\text{difference in } y}{\text{difference in } x}$.

- Length AB = $\sqrt{(x_2 - x_1)^2 + (y_2 - y_1)^2}$. ← Length of a line just uses Pythagoras' theorem.

- Midpoint of AB = $\left(\dfrac{x_1 + x_2}{2}, \dfrac{y_1 + y_2}{2}\right)$. ← The midpoint has x-value half-way between the x-values of the two points (and the same for y). This is the mean of the x coordinates and the mean of the y coordinates.

2 Parallel lines have the same gradient, i.e. $m_1 = m_2$.

gradient = m_1
gradient = m_2

3 Perpendicular lines have gradients such that $m_1 m_2 = -1$. This is sometimes written as: $m_2 = -\dfrac{1}{m_1}$.

gradient = m_1
gradient = m_2

4 In the equation of the line $y = mx + c$:
m is the gradient
c is the intercept with the y-axis.

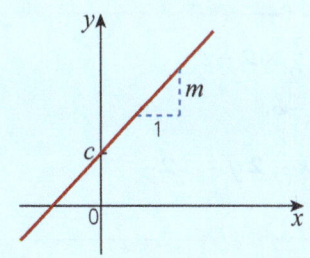

Edexcel A Level Mathematics (Pure)

5. The equation of the line with gradient m passing through the point (x_1, y_1) is $y - y_1 = m(x - x_1)$.
6. The equation of the line passing through the points (x_1, y_1) and (x_2, y_2) is $\dfrac{y - y_1}{y_2 - y_1} = \dfrac{x - x_1}{x_2 - x_1}$.
7. The equation of a line can also be written in the form $ax + by + c = 0$.

> Usually this is written so that a, b and c are integers.

8. Vertical lines have equation $x = a$.

> These lines are parallel to the y-axis.

9. Horizontal lines have equation $y = b$.

> These lines are parallel to the x-axis.

10. There are several different ways to find the equation of a line.
    - By:
        - obtaining its gradient,
        - substituting this value for m into $y = mx + c$
        - and then finding c by substituting a point into the resulting equation.
    - If the gradient of the line m and a point (x_1, y_1) are known, then the equation of the line can be found with
    $y - y_1 = m(x - x_1)$.
    - If two points on the line are known, you can use
    $\dfrac{y - y_1}{y_2 - y_1} = \dfrac{x - x_1}{x_2 - x_1}$.

Worked examples

1 Finding the equation of a line parallel to another line

Find the equation of the line parallel to $2y - 3x = 2$ through $(2, -3)$.
Give your answer in the form $ax + by = c$, where a, b, and c are integers.

Solution

$2y - 3x = 2 \Rightarrow y = \dfrac{3}{2}x + 1$

> Rearrange $2y - 3x = 2$ into the form $y = mx + c$.

The gradient of the line is $\dfrac{3}{2}$.

Method 1:

Equation is: $y = \dfrac{3}{2}x + c$.

> Parallel lines have the same gradient.

Substituting $(2, -3)$: $-3 = \dfrac{3}{2} \times 2 + c$
$\Rightarrow c = -6$.

Equation: $y = \dfrac{3}{2}x - 6 \Rightarrow 3x - 2y = 12$.

> **Common mistake:** Make sure you give your answer in the right form.

Method 2:

Equation: $y - (-3) = \frac{3}{2}(x - 2)$ ◀ — Using $y - y_1 = m(x - x_1)$, where $m = \frac{3}{2}$, $x_1 = 2$ and $y_1 = -3$.

$\Rightarrow 2y + 6 = 3x - 6$

$\Rightarrow 3x - 2y = 12$

2 Solving problems involving perpendicular lines

The point $(6, k)$ lies on the perpendicular bisector of $(1, -2)$ and $(5, 6)$. Find the value of k.

Solution

The gradient of the line joining $(1, -2)$ and $(5, 6)$ is

$\frac{y_2 - y_1}{x_2 - x_1} = \frac{6 - (-2)}{5 - 1} = \frac{8}{4} = 2.$ ◀ — difference in y / difference in x

So the gradient of the perpendicular line is $-\frac{1}{2}$. ◀ — When two lines are perpendicular lines, their gradients are the negative reciprocals of each other. $m_1 m_2 = -1 \Rightarrow m_2 = -\frac{1}{m_1}$

Midpoint, $M = \left(\frac{x_1 + x_2}{2}, \frac{y_1 + y_2}{2}\right)$

$= \left(\frac{1 + 5}{2}, \frac{(-2) + 6}{2}\right)$

$= (3, 2).$ ◀ — The midpoint is the mean of the coordinates.

The perpendicular bisector has gradient $-\frac{1}{2}$ and passes through $(3, 2)$ ◀ — Bisect means cut exactly in half.

Equation: $y - 2 = -\frac{1}{2}(x - 3)$ ◀ — Using $y - y_1 = m(x - x_1)$, where $m = -\frac{1}{2}$, $x_1 = 3$ and $y_1 = 2$.

$\Rightarrow 2y - 4 = 3 - x$

$\Rightarrow 2y = 7 - x$

The point $(6, k)$ lies on the line $2y = 7 - x$

$\Rightarrow 2k = 7 - 6$

$\Rightarrow k = \frac{1}{2}$

3 Solving geometric problems

A quadrilateral has vertices $A(-5, -2)$, $B(3, 2)$, $C(0, 4)$ and $D(-4, 2)$.
  i Find the length of AD and of BC.
  ii Show that AB and DC are parallel.
    What type of quadrilateral is ABCD?
  iii The diagonals of ABCD meet at M. Find the coordinates of M.
  iv Show that neither diagonal bisects the other.

Edexcel A Level Mathematics (Pure)

Solution

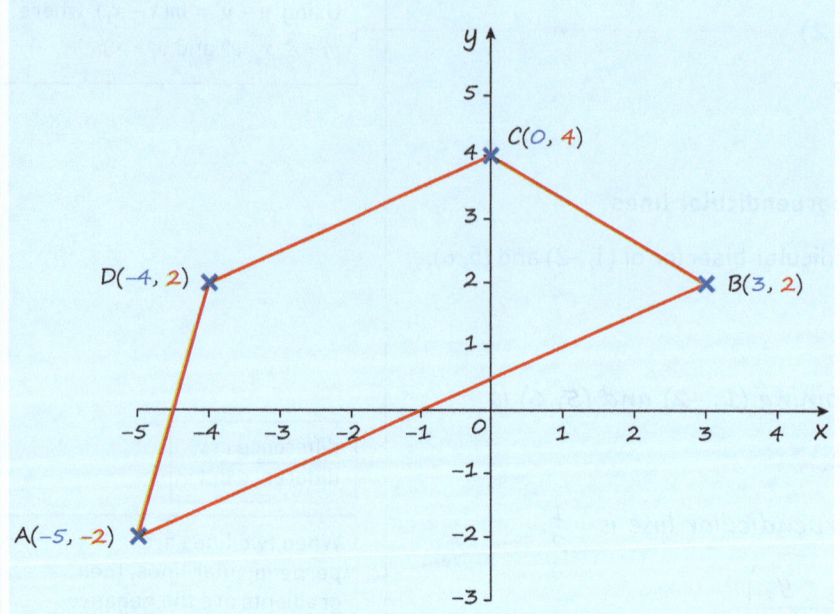

Hint: It helps to draw a diagram, even if the question does not ask for it.

i Length AD $= \sqrt{(x_2 - x_1)^2 + (y_2 - y_1)^2}$
$= \sqrt{(-4-(-5))^2 + (2-(-2))^2}$ ← A(−5,−2) and D(−4, 2).
$= \sqrt{1^2 + 4^2}$
$= \sqrt{17}$.

Common mistake: Make sure you use brackets.

Length BC $= \sqrt{(x_2 - x_1)^2 + (y_2 - y_1)^2}$
$= \sqrt{(3-0)^2 + (2-4)^2}$ ← B(3, 2) and C(0, 4).
$= \sqrt{3^2 + (-2)^2}$
$= \sqrt{13}$.

ii Gradient AB $= \dfrac{2-(-2)}{3-(-5)}$ Gradient DC $= \dfrac{2-4}{-4-0}$

$= \dfrac{4}{8}$ $= \dfrac{-2}{-4}$

$= \dfrac{1}{2}$ $= \dfrac{1}{2}$

Common mistake: Use gradient $= \dfrac{y_2 - y_1}{x_2 - x_1}$. It doesn't matter which point you choose to be (x_1, y_1) or (x_2, y_2) but you must be consistent!

The gradients are the same so the lines are parallel. ABCD has one pair of parallel sides ⇒ ABCD is a trapezium.

iii The diagonals are BD and AC.

Gradient BD $= \dfrac{2-2}{-4-2}$ ← Take care! BD is horizontal so its gradient is 0.

$= 0$

Equation of BD: $y = 2$

Gradient AC $= \dfrac{4-(-2)}{0-(-5)}$

$= \dfrac{6}{5}$

Equation of AC: $y - 4 = \dfrac{6}{5}x \Rightarrow 5y - 20 = 6x$.

Substitute $y = 2$ into equation of AC:

$5 \times 2 - 20 = 6x \Rightarrow x = -\frac{5}{3}$.

So the coordinates of M are $\left(-\frac{5}{3}, 2\right)$.

Find where the two lines cross by solving the equations of the lines simultaneously.

iv If one diagonal bisects the other then M is at the midpoint of either AC or BD.

Midpoint of AC is $\left(\frac{-5+0}{2}, \frac{(-2)+4}{2}\right) = \left(-\frac{5}{2}, 1\right)$.

Midpoint of BD is $\left(\frac{-4+3}{2}, \frac{2+2}{2}\right) = \left(-\frac{1}{2}, 2\right)$.

Since M is neither midpoint, the diagonals do not bisect each other.

Test yourself

TESTED

1. The points A and B have coordinates (−3, 1) and (2, 5). Find the length of the line AB.
   - A 3
   - B $\sqrt{41}$
   - C $\sqrt{17}$
   - D 41
   - E 5

2. The points A, B, C and D have coordinates (1, 5), (3, −1), (1, 2) and (x, y). CD is perpendicular to AB. Which of the following could be the coordinates of D?
   - A (7, 4)
   - B (2, −1)
   - C (2, 5)
   - D (3, 8)
   - E (7, 0)

3. Find the equation of the line through (−1, 3) and (2, −3).
   - A $y = -6x - 3$
   - B $y = -2x + 5$
   - C $y = -2x + 1$
   - D $y = 2x + 5$
   - E $y = -6x + 17$

4. Which of these is the equation of the line in the diagram?
   - A $x + 2y = 6$
   - B $y = -2x + 3$
   - C $2x + y = 6$
   - D $x + 2y = 3$
   - E $2y - x = 6$

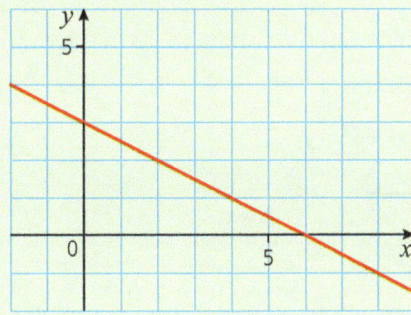

5. Find the equation of the line perpendicular to $y = -4x + 1$ and passing through (2, 1).
   - A $y = -\frac{1}{4}x + \frac{3}{2}$
   - B $y = \frac{1}{4}x + \frac{3}{2}$
   - C $y = \frac{1}{4}x + \frac{1}{2}$
   - D $y = 4x - 7$
   - E $y = \frac{1}{4}x + \frac{7}{4}$

6. The line L is parallel to $y = 3x - 2$ and passes through the point (−2, −1). Find the coordinates of the point of intersection with the x-axis.
   - A $\left(-\frac{1}{3}, 0\right)$
   - B $\left(-\frac{5}{3}, 0\right)$
   - C (0, 5)
   - D $\left(\frac{7}{3}, 0\right)$
   - E $\left(\frac{2}{3}, 0\right)$

Full worked solutions online

CHECKED ANSWERS

Exam-style question

A and B are points with coordinates (−4, 9) and (6, −3) respectively.

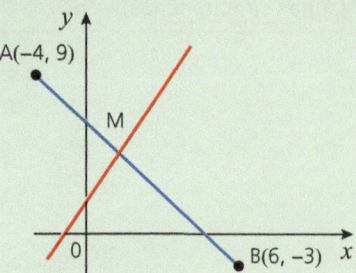

i   Find the coordinates of the midpoint, M, of AB. Show also that the equation of the perpendicular bisector of AB is $6y - 5x = 13$.

ii Find the area of the triangle bounded by the perpendicular bisector, the *x*-axis and the line MB.

Short answers on page 215

Full worked solutions online   CHECKED ANSWERS

Circles   REVISED

Key facts

1 The equation of a circle with radius *r* and centre at the origin is:
$x^2 + y^2 = r^2$.

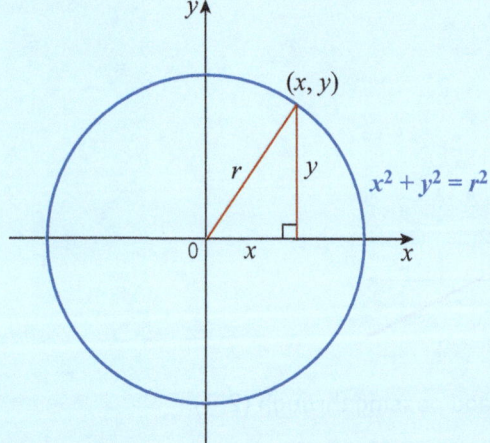

2 The equation of a circle with radius *r* and centre at (a, b) is:
$(x - a)^2 + (y - b)^2 = r^2$.

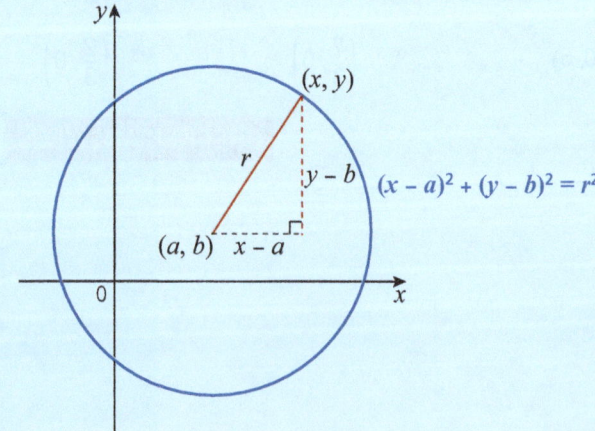

3 The equation of the circle can be rearranged and written in the form
$$x^2 + y^2 - 2ax - 2by + (a^2 + b^2 - r^2) = 0.$$

Note that
- there is no xy term
- the coefficients of x^2 and y^2 are equal.

4 You need to know the following **circle theorems** to help you solve problems involving circles.
- The angle in a semi-circle is a right-angle.

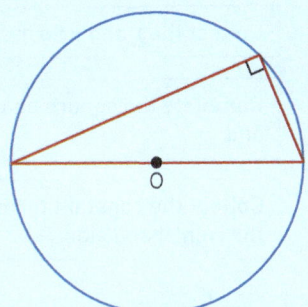

- The perpendicular from the centre of a circle to a chord bisects the chord.

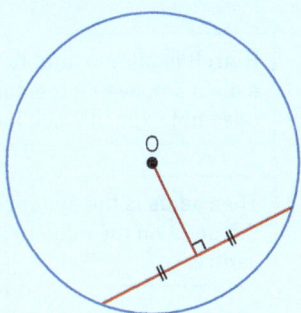

- The tangent to a circle at a point is perpendicular to the radius through that point.

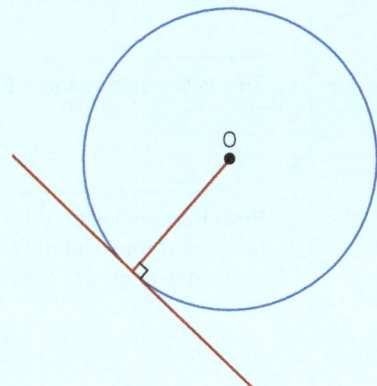

Worked examples

1 Finding the centre and radius of a circle given its equation

Give the centre and radius of the circle with equation $(x - 3)^2 + (y + 1)^2 = 36$.

Solution

The centre is at $(3, -1)$.

$r^2 = 36 \Rightarrow r = 6$ so the radius is 6.

$(y+1)^2$ is the same as $(y-(-1))^2$ so the y coordinate of the centre is -1.

Common mistake: Don't write that the radius is ±6 because the radius is a length and so is positive.

2 Working with the form $x^2 + y^2 - 2ax - 2by + (a^2 + b^2 - r^2) = 0$

The equation of a circle is $x^2 + y^2 - 4x + 6y + 4 = 0$.

Find the centre and radius of the circle.

Solution

$$x^2 + y^2 - 4x + 6y + 4 = 0$$
$$\Rightarrow x^2 - 4x + y^2 + 6y + 4 = 0$$ ← Collect the x and y terms.
$$\Rightarrow (x-2)^2 - 4 + (y+3)^2 - 9 + 4 = 0$$ ← Complete the square on both x and y.
$$\Rightarrow (x-2)^2 + (y+3)^2 = 9$$ ← Collect the constant terms on the right-hand side.

Therefore the circle has centre $(2, -3)$ and radius 3.

3 Finding the equation of a circle given the centre and a point on the circle

A circle has centre $(5, 0)$ and the point $(-1, 6)$ lies on the circumference of the circle.

Find the equation of the circle.

Solution

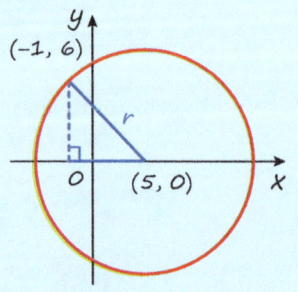

Hint: It is often useful to draw a diagram, even if the question does not explicitly ask for it.

The radius is the distance from the point on the edge to the centre.

$$r^2 = (-1-5)^2 + (6-0)^2$$
$$\Rightarrow r^2 = 6^2 + 6^2$$
$$\Rightarrow r^2 = 72$$

Common mistake: You don't need to square root 72 as the equation of a circle has r^2 in it.

Therefore the circle has equation: $(x-5)^2 + y^2 = 72$

This is the same as $(y-0)^2$.

4 Finding the equation of the tangent to the circle

Find the equation of the tangent to the circle $(x-2)^2 + (y+4)^2 = 10$ at the point $(5, -3)$.

Hint: Remember that the tangent at the point of contact is at right-angles to the radius.

Solution

The equation of the line from centre $(2, -4)$ to $(5, -3)$ is

$$\frac{y_2 - y_1}{x_2 - x_1} = \frac{(-3)-(-4)}{5-2} = \frac{1}{3}.$$

The tangent is at right-angles to the radius, so start by finding the gradient of the radius.

So the gradient of the tangent is -3.

So the equation of the tangent is

$$y - (-3) = -3(x - 5)$$
$$y + 3 = -3x + 15$$
$$y = -3x + 12.$$

Remember that $m_1 m_2 = -1$ for lines at right-angles, so their gradients are negative reciprocals of each other.

Use $y - y_1 = m(x - x_1)$ to find the equation of the line through $(5, -3)$ with gradient -3.

5 Finding the equation of a circle given three points on the circle

A(2, 3), B(0, 7) and C(8, 11) are three points.
i   Show that AB and BC are perpendicular.
ii Find the equation of the circle with AC as diameter and show that B lies on this circle.
iii Find the coordinates of D such that BD is a diameter.

Solution

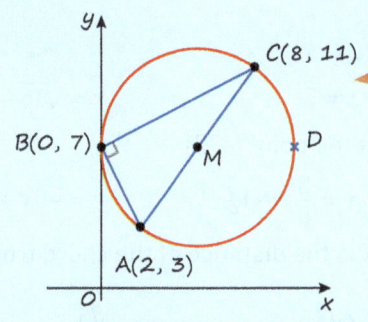

The circle ABC is called the circumcircle of triangle ABC.

i   Gradient of AB $= \dfrac{7-3}{0-2}$

$= -2$

Gradient of BC $= \dfrac{7-11}{0-8}$

$= \dfrac{-4}{-8}$

$= \dfrac{1}{2}$

$-2 \times \left(\dfrac{1}{2}\right) = -1$

Using $m_1 m_2 = -1$ for perpendicular lines.

therefore AB and BC are perpendicular.

Notice that angle ABC is an angle in a semi-circle and so is 90°.

ii Midpoint, M, of AC is at

$\left(\dfrac{2+8}{2}, \dfrac{3+11}{2}\right) = (5, 7)$.

The centre of the circle is at the midpoint of AC because AC is a diameter.

AM is the radius, r, of the circle.

$r^2 = (5-2)^2 + (7-3)^2$

$= 3^2 + 4^2$

$= 25$

$r^2 = AM^2 = (x_2 - x_1)^2 + (y_2 - y_1)^2$.

Equation of the circle is $(x-5)^2 + (y-7)^2 = 25$.

The radius of the circle is $\sqrt{25} = 5$.

At B: $(0-5)^2 + (7-7)^2 = 25$, therefore B is on the circle.

iii M (5, 7) is also the midpoint of B (0, 7) and D so the coordinates of D are (10, 7).

Common mistake: Don't forget this part of the question, it's easy to miss.

Test yourself

1. What is the equation of the circle with centre (1, −3) and radius 5?
   A $(x+1)^2 + (y-3)^2 = 25$
   B $(x-1)^2 + (y+3)^2 = 25$
   C $(x+1)^2 + (y-3)^2 = 5$
   D $(x-1)^2 + (y+3)^2 = 5$
   E $x^2 + y^2 = 25$

2. Give the centre and the radius of the circle with equation $x^2 + y^2 + 6x - 4y - 36 = 0$.
   A Centre (−3, 2), radius 7
   B Centre (−6, 4), radius 6
   C Centre (−3, 2), radius 6
   D Centre (3, −2), radius 7
   E Centre (3, −2), radius 6

3. Find where the circle $(x-1)^2 + (y+2)^2 = 16$ crosses the positive x-axis.
   A $x = 1 + 4\sqrt{3}$
   B $x = 3$
   C $x = 1 + 2\sqrt{3}$
   D $x = 5$
   E $x = \sqrt{15} - 2$

4. Find the equation of the tangent to the circle $(x-1)^2 + (y+1)^2 = 34$ at the point (6, 2).
   A $y = -7x + 44$
   B $y = -\frac{5}{3}x + 12$
   C $y = -\frac{5}{3}x + 9\frac{1}{3}$
   D $y = \frac{3}{5}x - 1\frac{3}{5}$
   E $y = -5x + 32$

5. A(2, −1) and B(4, 3) are two points on a circle with centre (1, 2). What is the distance of the chord from the centre of the circle?
   A $2\sqrt{5}$
   B 5
   C $\sqrt{10}$
   D $2\sqrt{2}$
   E $\sqrt{5}$

Full worked solutions online

Exam-style question

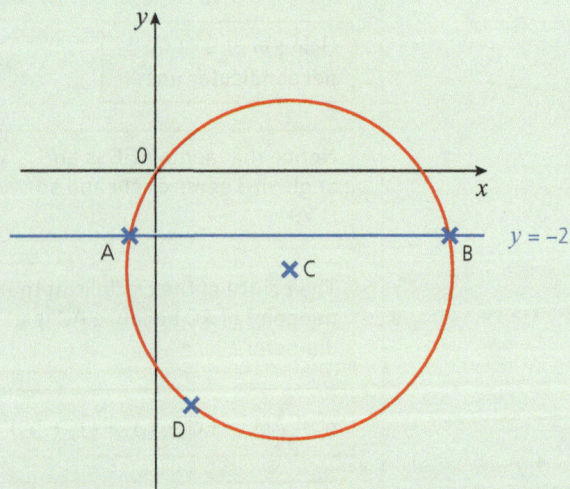

The diagram shows a circle with equation $x^2 + y^2 - 8x + 6y = 0$.
i Find the centre, C, and radius of the circle.
ii The circle meets the line $y = -2$ at the points A and B.
    Find the exact coordinates of A and B.
iii Verify that the point D(1, −7) lies on the circle.
    Find the equation of the tangent to the circle at D in the form $ax + by + c = 0$.

Short answers on page 215

Full worked solutions online

Chapter 5 Functions

About this topic

A mapping is any rule which associates two sets of items. In mathematics, a mapping often transforms one set of numbers into another set of numbers. A function is a type of mapping where each input has exactly one output.

Before you start, remember

- how to solve quadratic and linear equations
- how to sketch the curves of a variety of functions including trigonometric functions – see also Chapter 6.

Functions

REVISED

Key facts

1. Types of number

   Natural numbers $\mathbb{N}$ (e.g. 1, 2, 3, …)

   Integers $\mathbb{Z}$ – the natural numbers, zero and all the negative whole numbers (e.g. … –3, –2, –1, 0, 1, 2, 3, …)

   Rational numbers $\mathbb{Q}$ – all the integers and all the fractions $\left(\text{e.g.} \frac{2}{3}, -\frac{11}{5}\right)$

   Real numbers $\mathbb{R}$ – all rational and irrational numbers (e.g. $\sqrt{2}$, π)

2. A **mapping** is a rule which links items in one set (the **inputs** or **objects**) to items in another set (the **outputs** or **images**).

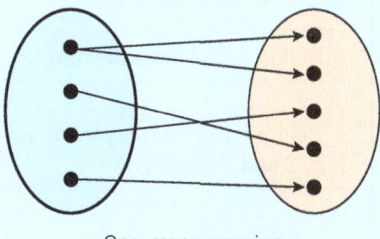

One–many mapping

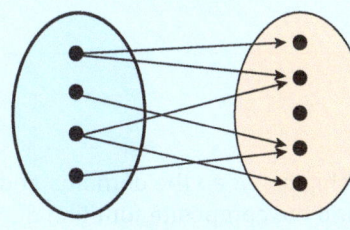

Many–many mapping

These mappings are **not** functions.

Edexcel A Level Mathematics (Pure)

3. In a **one–one mapping**, each input has exactly one output and each output has exactly one input.

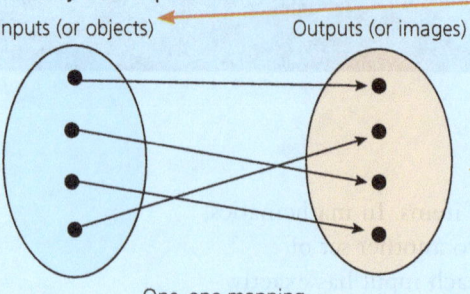

One–one mapping

> The set of inputs is called the **domain**.

> The set of outputs is called the **range**.

4. In a **many–one** mapping, each input maps to just one output, but some points in the range are the image of more than one point in the domain.

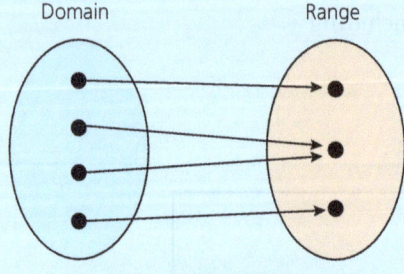

Many–one mapping

> Some inputs map to the same output.

5. A **function** is a mapping in which each point in the domain maps to just one image point. So a function is a **many–one** or a **one–one mapping**.

6. Functions are often written using the letters f, g and h.
   For example, $f(x) = 2x + 3$ or $g: x \to x^3$.

> You say 'g of x maps to x cubed'.
> $g(x) = 8 \Rightarrow x^3 = 8 \Rightarrow x = 2$.

> You say 'f of x equals two x plus 3'.
> $f(8) = 2 \times 8 + 3 = 19$

7. If you apply a function f to x to give $f(x)$, and then apply another function g to the result, giving $g[f(x)]$ or $gf(x)$, then you have applied the **composite function** gf to x.

   $x \xrightarrow{f} f(x) \xrightarrow{g} gf(x)$

   $gf(2)$ means substitute $x = 2$ into the function $f(x)$ and the result into $g(x)$.
   For example, $f(x) = 2x + 3$ and $g: x \to x^3$
   $2 \xrightarrow{f} 7 \xrightarrow{g} 343$
   $x \xrightarrow{f} 2x+3 \xrightarrow{g} (2x+3)^3$

   To find $fg(x)$, you apply g first.
   $x \xrightarrow{g} x^3 \xrightarrow{f} 2x^3 + 3$

> In general, $gf(x)$ is not the same as $fg(x)$. so $fg(2)$ is found by:
> $2 \xrightarrow{g} 8 \xrightarrow{f} 19$.

> To find g, you cube the input.

> To find f, you multiply the input by 2 and then add 3.

8. The diagram shows the relationship between the domains and ranges of the functions f and g and the composite function gf.

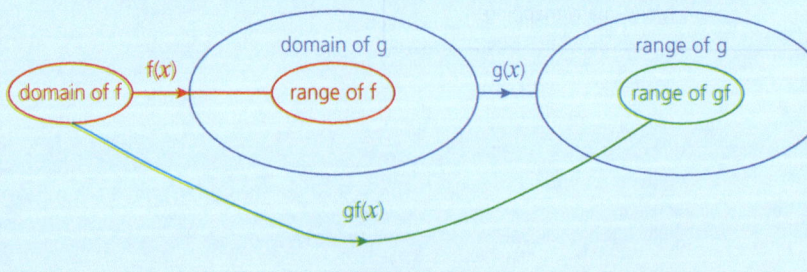

> Note that the range of f must be completely contained within the domain of g – otherwise some inputs won't be valid when you try to apply g to the outputs of f.

Worked examples

1 Recognising functions

Which of the following mappings are functions?

Write down the domain and range of each function.

i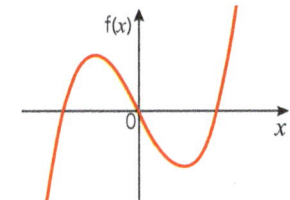

ii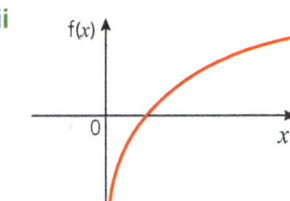

iii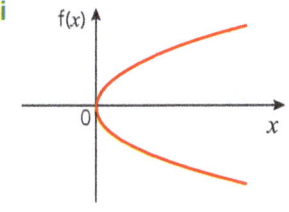

Solution

i This is a many–one function, with domain $x \in \mathbb{R}$ and range $f(x) \in \mathbb{R}$.

You can input any real number.

The set of outputs are real numbers.

ii This is a one–one function, with domain $x > 0$ and range $f(x) \in \mathbb{R}$.

You can input any positive number.

iii This is **not** a function. It is a one–many mapping, with domain $x \geq 0$ and range $f(x) \in \mathbb{R}$.

You can input any positive number or zero.

2 Finding the domain and range of a function

You are given that $g(x) = 1 - x^2$ for $-1 \leq x \leq 1$.

i Sketch the graph of $y = g(x)$.

ii State the range of the function g.

Solution

i

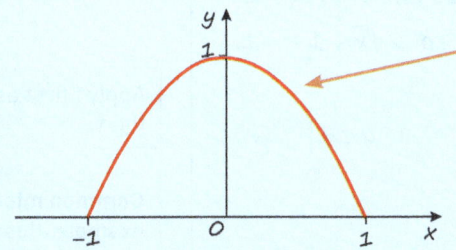

Reflect the graph of $y = x^2$ in the x-axis to get $y = -x^2$, then translate it 1 unit vertically to get $y = -x^2 + 1 = 1 - x^2$. See page 62 for a reminder of transformations.

ii The range of g is $0 \leq g(x) \leq 1$.

3 Solving equations involving functions

The function f is defined by $f(x) = (x-2)^2 - 1$ for $x \in \mathbb{R}$.
  i Sketch the graph of $y = f(x)$ and write down the range of f.
  ii Find $f(-3)$.
  iii Find the values of x for which $f(x) = 3$.

Hint: You can think of the range as your 'y-values' – look at the y-axis: the curve runs from $y = 0$ to $y = 1$.

Solution

  i This is a quadratic function, with minimum point $(2, -1)$. The range of f is $f(x) \geq -1$.

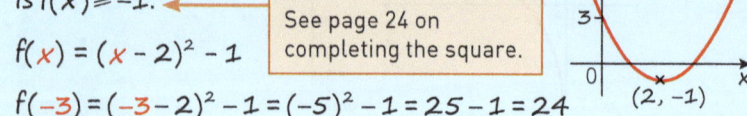

See page 24 on completing the square.

  ii $f(x) = (x-2)^2 - 1$
     $f(-3) = (-3-2)^2 - 1 = (-5)^2 - 1 = 25 - 1 = 24$

  iii $f(x) = 3$
     $\Rightarrow (x-2)^2 - 1 = 3$
     $\Rightarrow (x-2)^2 = 4$
     $\Rightarrow x - 2 = \pm\sqrt{4}$
     $\Rightarrow x = 2 \pm 2$
     So $x = 0$ or $x = 4$.

Hint: It is easier **not** to expand the brackets.

4 Finding composite functions

The functions f and g are defined as follows.

$f(x) = 2x - 1 \quad x \in \mathbb{R}$
$g(x) = \sqrt{x} \quad x \geq 0, \; x \in \mathbb{R}$

  i Find **a** $fg(x)$
          b $gf(x)$
     giving the domain and range of the composite function in each case.
  ii Solve the equation $ff(x) = x$.

Hint: Remember that $\sqrt{x}$ means the positive square root of x.

$ff(x)$ is sometimes written $f^2(x)$.

Solution

  i **a**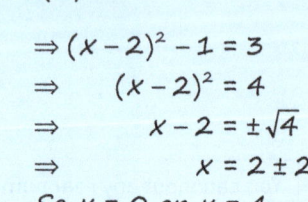

     $fg(x) = 2\sqrt{x} - 1$

     The domain is $x \geq 0$ since $\sqrt{x}$ does not exist for $x < 0$.
     The smallest possible value of $\sqrt{x}$ is 0, so $2\sqrt{x} - 1 \geq -1$.
     So the range is $fg(x) \geq -1$.
     Using set notation, the range is $\{y : y \geq -1, \; y \in \mathbb{R}\}$ where $y = fg(x)$.

Apply the function closest to the 'x' first.

Apply f first as it is closest to the x.

  b $x \xrightarrow{f} 2x - 1 \xrightarrow{g} \sqrt{2x-1}$

     $gf(x) = \sqrt{2x-1}$

     Since $\sqrt{2x-1}$ does not exist for $2x - 1 < 0$, the domain is $x \geq \frac{1}{2}, \; x \in \mathbb{R}$.
     Using set notation the domain is $\{x : x \geq \frac{1}{2}, \; x \in \mathbb{R}\}$.

     The smallest possible value of $\sqrt{2x-1}$ is 0, and so the range is $gf(x) \geq 0, \; x \in \mathbb{R}$.

     Using set notation, the range is $\{y : y \geq 0, y \in \mathbb{R}\}$ where $y = gf(x)$.

Common mistakes: This example illustrates some important points:
- $fg(x)$ is not the same as $gf(x)$. This is the case for most composite functions, although not in every case.
- You need to be careful with domains and ranges for composite functions. The domain for a composite function may not be the same as the domain for either of the functions involved.

ii $x \xrightarrow{f} 2x-1 \xrightarrow{f} 2(2x-1)-1$ ← Apply f to x and then f to the result.

$ff(x) = 2(2x-1)-1$
$= 4x-3$
$ff(x) = x \Rightarrow 4x-3 = x$
$\Rightarrow 3x = 3$
$\Rightarrow x = 1$ ← Check: $1 \xrightarrow{f} 1 \xrightarrow{f} 1$.

Test yourself

1 The mapping $x \to \sqrt[3]{x^2-1}$, $x \in \mathbb{R}$, can be described as
  A one–one   B one–many   C many–one   D many–many

2 The function f is defined as f: $x \to \sqrt{2x-3}$. What is the domain of f?
  A $x \in \mathbb{R}$ B $x \geq 0$ C $x \geq \frac{3}{2}$ D $x \geq 3$

3 The function g is defined as $g(x) = x^2 - 2x - 1$ for $-2 \leq x \leq 2$. What is the range of the function?
  A $-1 \leq g(x) \leq 7$ B $g(x) \leq 7$ C $g(x) \geq -2$ D $-2 \leq g(x) \leq 7$

4 The functions f and g are defined for all real numbers x as
  $f(x) = x^2 - 2$, $g(x) = 3 - 2x$
  Find an expression for the function fg(x) (for all real numbers x).
  A $fg(x) = 7 - 2x^2$ B $fg(x) = -2x^3 + 3x^2 + 4x - 6$ C $fg(x) = 4x^2 - 12x + 7$
  D $fg(x) = 7 + 4x^2$ E $fg(x) = 1 - 2x^2$

5 The functions p and q are defined as follows:
  $p(x) = \frac{1}{x}$ $x \neq 0$, $q(x) = 2x - 1$ $x \in \mathbb{R}$
  Find an expression for the function pq.
  A $pq(x) = \frac{2}{x} - 1$ $x \neq 0$ B $pq(x) = \frac{2}{x} - 1$ $x \in \mathbb{R}$ C $pq(x) = \frac{1}{2x-1}$ $x \neq 0$
  D $pq(x) = \frac{1}{2x-1}$ $x \in \mathbb{R}$ E $pq(x) = \frac{1}{2x-1}$ $x \neq \frac{1}{2}$

Full worked solutions online   CHECKED ANSWERS

Exam-style question

The function f(x) is defined by $f(x) = \frac{x^3}{(x+3)(x-3)}$ for $-2 \leq x \leq 2$.
The diagram shows a sketch of the graph of $y = f(x)$.

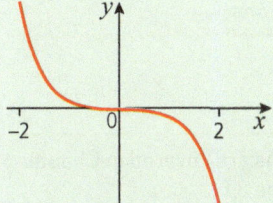

i Find the range of the function $y = f(x)$.
ii Show algebraically that $f(-x) = -f(x)$.
  State how this property relates to the shape of the curve.

The function g(x) is defined by $g(x) = x - 1$ for $-1 \leq x \leq 1$.
iii Find $y = fg(x)$ and give its domain and range.

Short answers on page 215

Full worked solutions online   CHECKED ANSWERS

Edexcel A Level Mathematics (Pure)

Graphs and transformations

REVISED

Key facts

1. Make sure you know the shapes of these common curves.

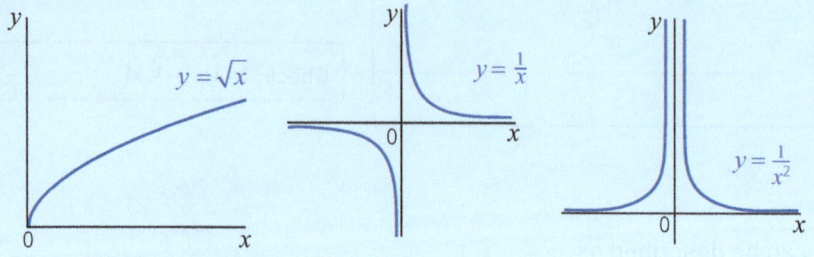

2. The symbol ∝ means 'proportional to'.
   - When y is **directly proportional** to x you write $y \propto x$.
     You can write this as an equation: $y = kx$, where k is a constant.
     The graph of $y = kx$ is a straight line through the origin.

     > For example, when y is **directly** proportional to the square of x, then $y \propto x^2 \Rightarrow y = kx^2$.

   - When y is **inversely proportional** to x you write $y \propto \dfrac{1}{x}$.
     You can write this as an equation: $y = \dfrac{k}{x}$, where k is a constant.

     > For example, when y is **inversely** proportional to the cube root of x, then $y \propto \dfrac{1}{\sqrt[3]{x}} \Rightarrow y = \dfrac{k}{\sqrt[3]{x}}$.

   You can use given values of y and x to find the value of k.

3. The following graphs can be obtained by **transformations** of the graph of $y = f(x)$.

$f(x - a) + b$	Translation through $\begin{pmatrix} a \\ b \end{pmatrix}$
$af(x)$	Stretch parallel to the y-axis, scale factor a
$f(ax)$	Stretch parallel to the x-axis, scale factor $\dfrac{1}{a}$
$f(-x)$	Reflection in the y-axis
$-f(x)$	Reflection in the x-axis

> Reflections are really just a special case of stretches, with scale factor −1.

4. Take care when combining transformations. When the transformations involve the same direction or are a mix of vertical and horizontal transformations then the order matters.

   You should deal with transformations of $f(x)$ to $af(bx + c) + d$ in the following order:
   1. Horizontal translation — bracket first
   2. Stretching or reflecting — multiplication/division
   3. Vertical translation — addition/subtraction

   > It doesn't matter which order you carry out the stretches in as they are in different directions.

   $$f(x) \xrightarrow[c \text{ units left}]{\text{translate}} f(x + c) \xrightarrow[\text{two way stretch}]{\text{stretch}} af(bx + c) \xrightarrow[d \text{ units up}]{\text{translate}} af(bx + c) + d$$

5. When you transform the graph of a trigonometric function make sure you know whether you are working in degrees or radians. Remember $360° = 2\pi$ radians.

   > Radians are covered in Chapter 6 on page 87.

> **Worked examples**

1 Finding the equation of a graph after successive transformations

Find the equation of the new graph when the following transformations are applied to the graph of $y = \sin x$.

  i A stretch, scale factor 2, parallel to the y-axis, followed by a translation of 1 unit vertically downwards.

  ii A translation of 1 unit vertically downwards, followed by a stretch, scale factor 2, parallel to the y-axis.

Sketch the new graph in each case.

Solution

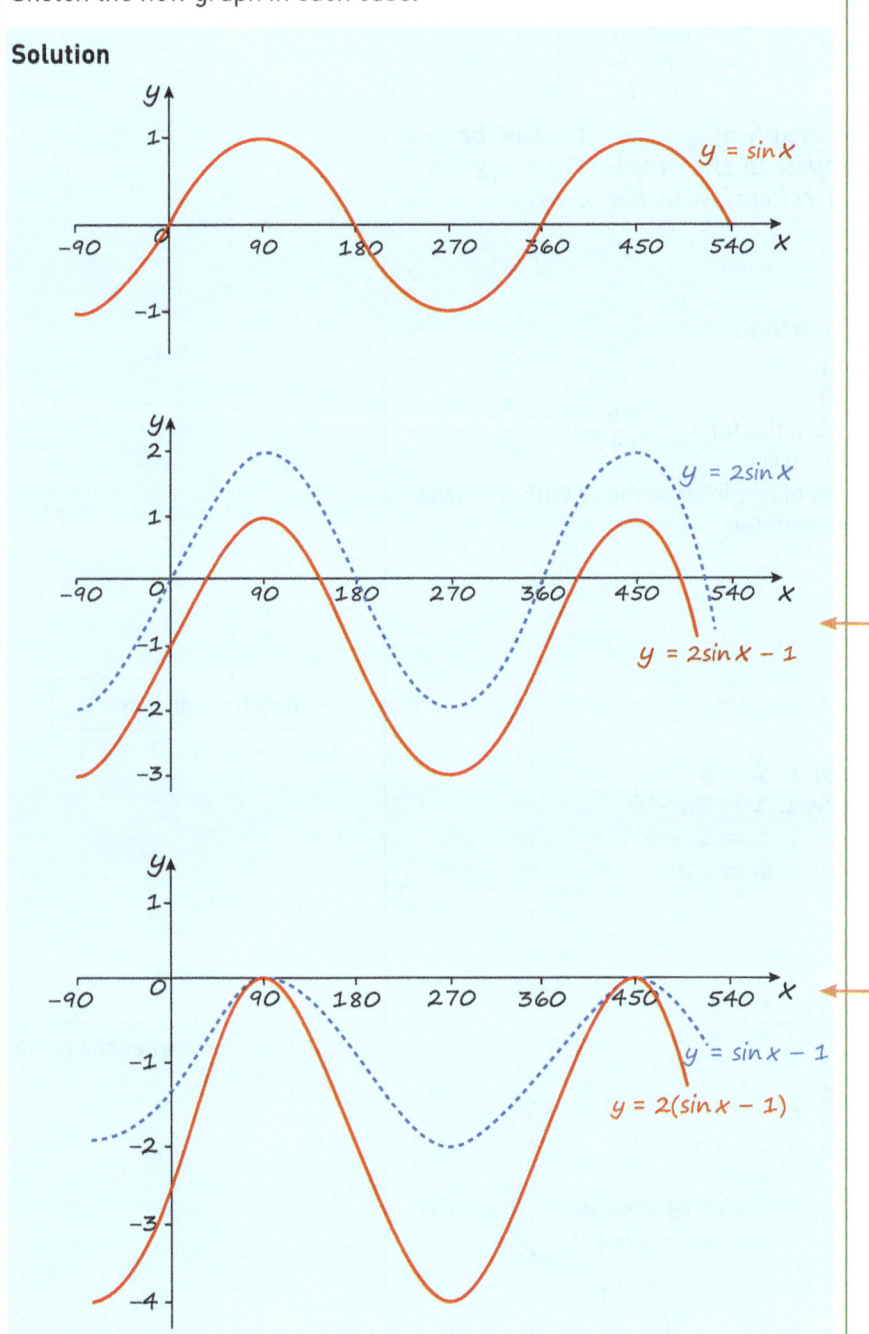

A stretch, scale factor 2, parallel to the y-axis, maps the graph of $y = \sin x$ to the graph of $y = 2 \sin x$. A translation of 1 unit vertically downwards maps the graph of $y = 2 \sin x$ to the graph of $y = 2 \sin x - 1$.

A translation of 1 unit vertically downwards maps the graph of $y = \sin x$ to the graph of $y = \sin x - 1$. A stretch, scale factor 2, parallel to the y-axis, maps the graph of $y = \sin x - 1$ to the graph of $y = 2(\sin x - 1)$, or $y = 2 \sin x - 2$.

2 Identify the transformations that map one graph onto another

Show how the graph of $y = -(x+1)^3$ can be obtained from the graph of $y = x^3$ using successive transformations, and hence sketch the graph of $y = -(x+1)^3$.

Solution

The graph of $y = x^3$ can be mapped to the graph of $y = (x+1)^3$ by a translation of 1 unit to the left.

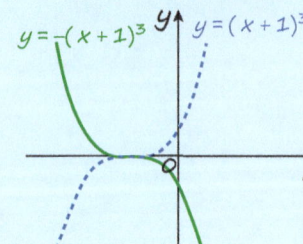

The graph of $y = (x+1)^3$ can be mapped to the graph of $y = -(x+1)^3$ by a reflection in the x-axis.

3 Sketching graphs of rational functions

i You are given that $f(x) = \dfrac{x+1}{x+3}$.

Show that $f(x)$ can be written in the form $a + \dfrac{b}{x+3}$.

ii Hence sketch the graph of $y = f(x)$.
Label clearly the coordinates of any intersections with the axes and the equations of any asymptotes.

Solution

i $\dfrac{x+1}{x+3} = a + \dfrac{b}{x+3}$

$\Rightarrow x + 1 = a(x+3) + b$ ← Multiply through by $x+3$.

$\Rightarrow x + 1 = ax + 3a + b$

Comparing coefficients of x: $1 = a$
Comparing constant terms: $1 = 3a + b$
$\Rightarrow \qquad\qquad\qquad\qquad 1 = 3 + b$
$\Rightarrow \qquad\qquad\qquad\qquad b = -2$

So $f(x) = 1 - \dfrac{2}{x+3}$.

ii $f(x) = \dfrac{x+1}{x+3}$

When $x = 0$ then $f(0) = \dfrac{0+1}{0+3} = \dfrac{1}{3}$.

← Start by finding when the curve cuts the axes.

When $f(x) = 0$ then $\dfrac{x+1}{x+3} = 0 \Rightarrow x + 1 = 0 \Rightarrow x = -1$.

Start with the curve of $y = \dfrac{1}{x}$.

Translate 3 units left to get $y = \dfrac{1}{x+3}$

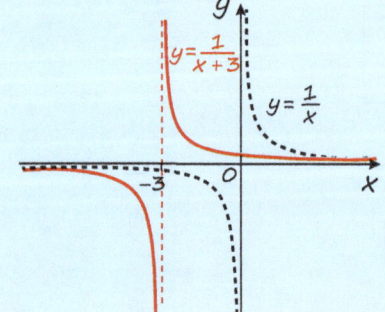

Stretch by scale factor −2 parallel to y-axis to get $y = -\dfrac{2}{x+3}$

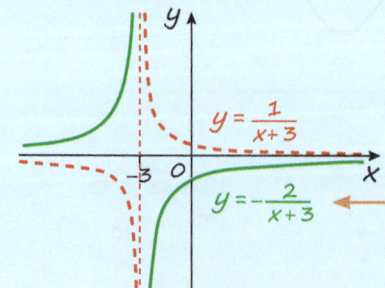

← This is the same as a vertical stretch scale factor 2 followed by a reflection in the x-axis.

Translate 1 unit vertically to get
$y = 1 - \dfrac{2}{x+3}$

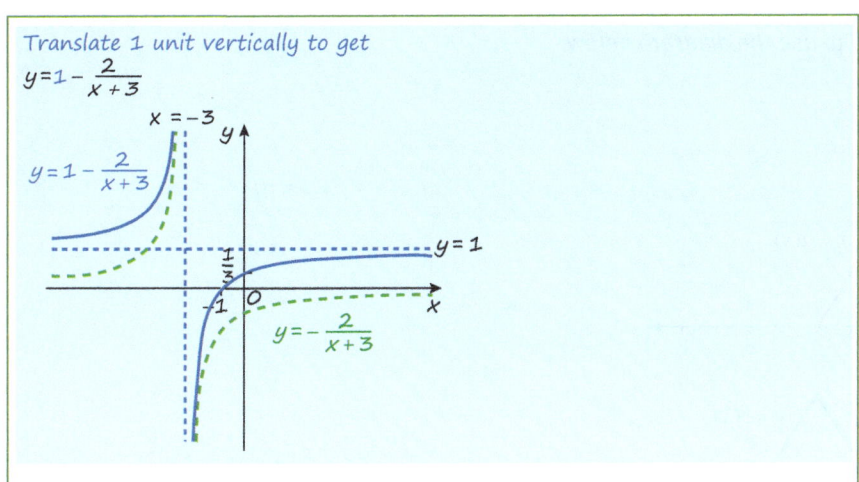

Common mistakes: Don't forget to label where the curve cuts the axes and give the equations of the asymptotes.

Test yourself

1. Which of these graphs show that
   a y is directly proportional to x and
   b y is inversely proportional to x?

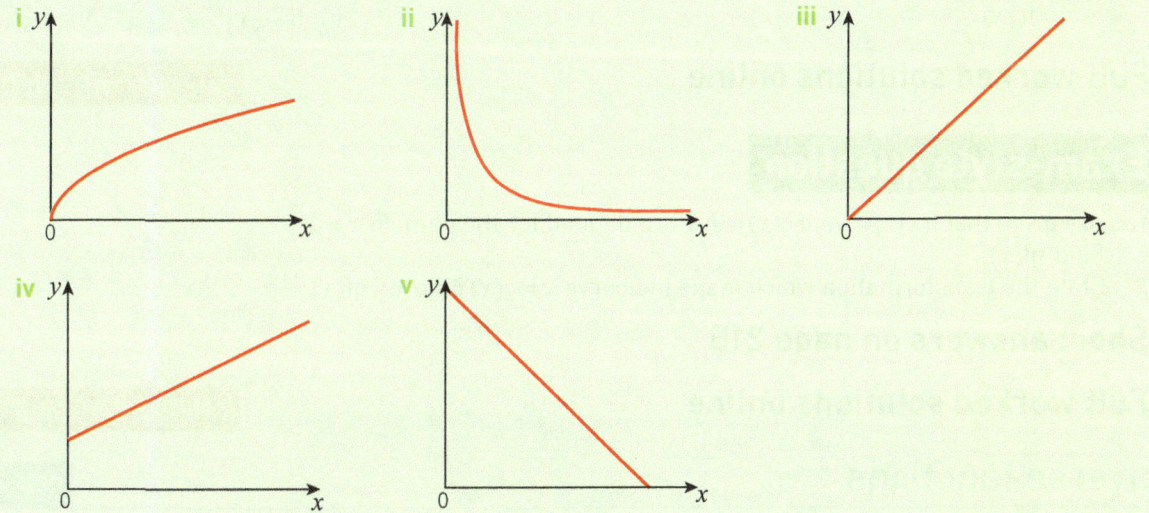

   A a iii and b v   B a iv and b ii   C a i and b v   D a i and b ii   E a iii and b ii

2. The curve $y = x^2 - 4x$ is translated and the equation of the new curve is $y = (x-1)^2 - 4(x-1) + 2$. What are the coordinates of the vertex of the new curve?
   A (1, 2)   B (1, −2)   C (3, −2)   D (1, 5)   E (1, −6)

3. The curve $y = x^2 - 2x + 3$ is translated through $\begin{pmatrix} 2 \\ -4 \end{pmatrix}$. What is the equation of the new curve?

   A $y = x^2 + 2x - 1$ B $y = x^2 - 6x + 11$ C $y = x^2 - 6x + 15$ D $y = 4x^2 - 4x - 1$ E $y = x^2 - 6x + 7$

4. The graph of $y = x^2$ is first translated 2 units to the right and 1 unit vertically upwards, and then reflected in the y-axis. Which of the following is the equation of the new graph?
   A $y = x^2 + 4x + 5$ B $y = -x^2 + 4x - 5$ C $y = x^2 - 4x + 5$ D $y = -x^2 - 4x - 5$

5. The graph of $y = f(x)$ has a maximum point at $(-3, 2)$. Which of the following is the maximum point of the graph of $y = 2 + 3f(x)$?
   A (−3, 12)   B (−7, 8)   C (−1, 4)   D (−3, 8)

Edexcel A Level Mathematics (Pure)

For questions 6 and 7 you will need to use the diagram below.

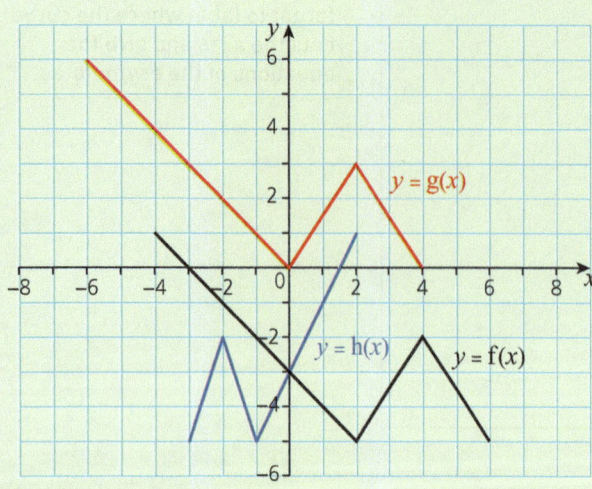

6 The diagram shows the graph of $y = f(x)$ and its image $y = g(x)$ after a transformation. What is the equation of the image?
   A $y = f(x+2) - 5$ B $y = f(x-2) + 5$ C $y = f(x-2) - 5$ D $y = f(x+2) + 5$ E $y = f(x+5) + 2$

7 The diagram shows the graph of $y = f(x)$ and its image $y = h(x)$ after a transformation. What is the equation of the image?
   A $y = f(2x)$ B $y = f(\tfrac{1}{2}x)$ C $y = f(-2x)$ D $y = f\left(-\tfrac{1}{2}x\right)$ E $y = -\tfrac{1}{2}f(x)$

Full worked solutions online   CHECKED ANSWERS

Exam-style question

You are given that $f(x) = x^2$ and $g(x) = \ln x$ are defined for the domain $x > 0$.
  i Find $gf(x)$.
  ii State the transformation which maps the curve $y = g(x)$ onto $y = gf(x)$.

Short answers on page 215

Full worked solutions online   CHECKED ANSWERS

Inverse functions   REVISED

Key facts

1 The **inverse** of a function 'undoes' the effect of a function.
   The inverse of a function $f(x)$ is written as $f^{-1}(x)$.
   Since an inverse 'undoes' a function then $ff^{-1}(x) = x$ and $f^{-1}f(x) = x$.

   For example, the inverse of the function $f(x) = x + 2$ is $f^{-1}(x) = x - 2$, because subtracting 2 undoes the effect of adding 2.

2 For a function f to have an inverse function f^{-1}, then f must be a one–one function over the domain of f.

   Sometimes you need to restrict the domain of f to ensure the function is one–one over its domain.

   The range of f is the same as the domain of f^{-1}, and the domain of f is the same as the range of f^{-1}.
   For example: $f : x \to 2x + 3$, $x = 1, 2, 3$ and $f^{-1} : x \to \dfrac{x-3}{2}$, $x = 5, 7, 9$.

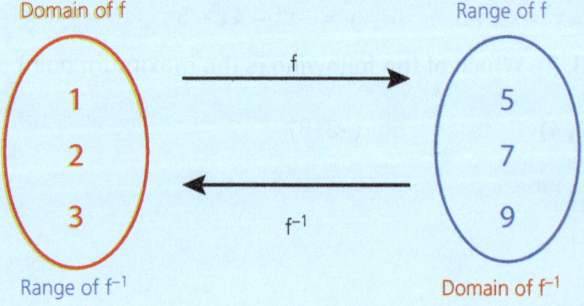

Chapter 5 Functions

3. The **graphs** of a function and its inverse function are reflections of each other in the line $y = x$.
   For example:

   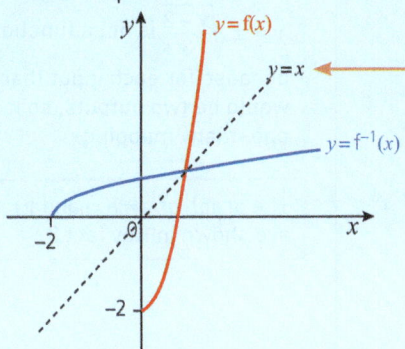

   > Remember that for the line $y = x$ to be at 45° to the axes, the x- and y- scales must be the same.

4. To find the inverse of a function:
   - write the function in the form $y = f(x)$
   - interchange y and x, to give $x = f(y)$
   - rearrange to make y the subject.

5. The trigonometric functions sin, cos and tan are not one–one functions, so they do not have inverse functions. However, if the domain of the trigonometric functions is restricted so that they are one–one over this restricted domain, they do have inverse functions, which are called **arcsin**, **arccos** and **arctan**.

   > These inverse trigonometric functions can also be written as $\sin^{-1}$, $\cos^{-1}$ and $\tan^{-1}$.

For the sine function, the domain is restricted to $-90° \leq x \leq 90°$ or $-\frac{1}{2}\pi \leq x \leq \frac{1}{2}\pi$.

For the cosine function, the domain is restricted to $0° \leq x \leq 180°$ or $0 \leq x \leq \pi$.

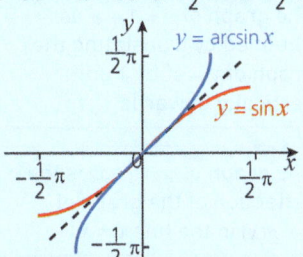
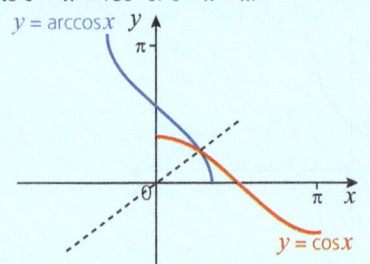

> Trigonometrical functions are more often given in radians than degrees. See Chapter 6 for revision of radians.

> These graphs are drawn in radians so that the same scale can be used for both axes. Remember the line $y = x$ is the line of symmetry.

For the tangent function, the domain is restricted to $-90° < x < 90°$ or $-\frac{1}{2}\pi < x < \frac{1}{2}\pi$.

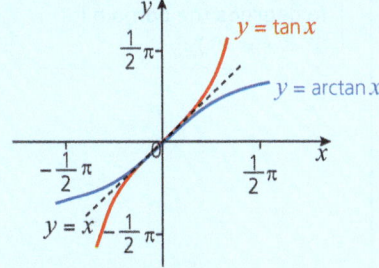

Worked examples

1 Finding an inverse function

The function f is defined as $f(x) = 3x^2 - 2 \quad x \geq 0$.

Find the inverse of f.

Hint: The domain is restricted to only positive numbers (and zero) because if you included negative numbers then the function would be many–one and so would not have an inverse.

Solution

Write the function in the form $y = f(x)$: $\quad y = 3x^2 - 2$

Interchange y and x, to give $x = f(y)$: $\quad x = 3y^2 - 2$

Edexcel A Level Mathematics (Pure)

Rearrange to make y the subject: $x + 2 = 3y^2$

$$\frac{x+2}{3} = y^2$$

$$\sqrt{\frac{x+2}{3}} = y$$

The range of the function f is $f(x) \geq -2$, so the domain of f^{-1} is $x \geq -2$.

The inverse of f is $f^{-1}(x) = \sqrt{\frac{x+2}{3}}$, $x \geq -2$.

Common mistake: The symbol $\sqrt{\ }$ means the positive square root only.

$y = \pm\sqrt{\frac{x+2}{3}}$ is not a function because for each input there would be two outputs, so it is a one–many mapping.

The graph of $y = f(x)$ and $y = f^{-1}(x)$ are shown in Key fact 3.

2 Sketching the graph of an inverse function

The function g is defined as $g(x) = 1 + e^x$.
  i Find an expression for $g^{-1}(x)$, stating the domain of this function.
  ii Sketch the graphs of $y = g(x)$ and $y = g^{-1}(x)$ on the same axes.

Solution

  i Write the function in the form $y = g(x)$: $y = 1 + e^x$
     Interchange y and x, to give $x = g(y)$: $x = 1 + e^y$
     Rearrange to make y the subject: $x - 1 = e^y$
                                                    $\ln(x - 1) = y$

The range of the function g is $g(x) > 1$ so the domain of g^{-1} is $x > 1$.
The inverse of g is $g^{-1}(x) = \ln(x - 1)$ $x > 1$.

Remember that $e^x > 0$ for all x. Since $g(x) = 1 + e^x$, $g(x) > 1$.

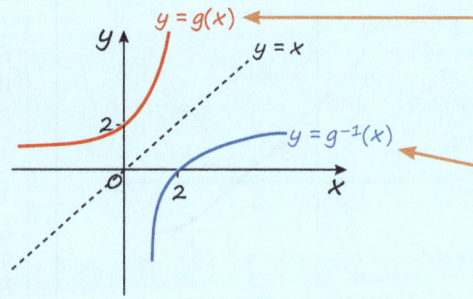

The graph of $y = 1 + e^x$ is obtained by translating the graph of $y = e^x$ by 1 unit vertically upwards.

The graph of $y = g^{-1}(x)$ is the reflection of the graph of $y = g(x)$ in the line $y = x$.

3 Finding inverse trigonometric functions

The function $f(x)$ is defined by $f(x) = 1 + 2\cos x$ for $0 \leq x \leq \pi$.

Find an expression for $f^{-1}(x)$, and state the domain of this function.

In degrees the domain is $0° \leq x \leq 180°$.

Solution

$y = 1 + 2\cos x$

Interchanging x and y: $x = 1 + 2\cos y$

$x - 1 = 2\cos y$

$\frac{x-1}{2} = \cos y$

$\arccos\left(\frac{x-1}{2}\right) = y$

The range of the function f is $-1 \leq f(x) \leq 3$, so the domain of f^{-1} is $-1 \leq x \leq 3$.

$f^{-1}(x) = \arccos\left(\frac{x-1}{2}\right)$ for $-1 \leq x \leq 3$.

$-1 \leq \cos x \leq 1$, hence $-2 \leq 2\cos x \leq 2$ and $-1 \leq 1 + 2\cos x \leq 3$.

Test yourself

1. The function f is defined by $f(x) = 2x^3 - 1$. Find the inverse function f^{-1}.

   A $f^{-1}(x) = \dfrac{\sqrt[3]{x+1}}{2}$ B $f^{-1}(x) = \dfrac{1}{2x^3 - 1}$ C $f^{-1}(x) = \sqrt[3]{\dfrac{x+1}{2}}$ D $f^{-1}(x) = \dfrac{\sqrt[3]{x}+1}{2}$

2. The diagram shows the graph of $y = g(x)$.
   Which of the diagrams below shows the graph of $y = g^{-1}(x)$?

   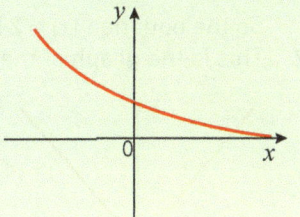

   A 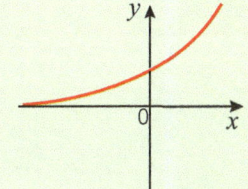   B 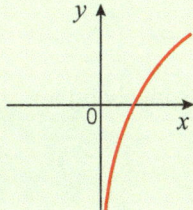   C 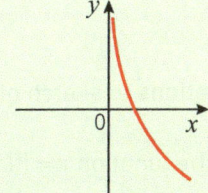   D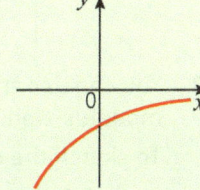

3. The function f is defined by $f(x) = \ln(2x - 1)$ $x > \dfrac{1}{2}$.
   The inverse function f^{-1} is defined by

   A $f^{-1}(x) = e^{2x-1}$ $x > \dfrac{1}{2}$ B $f^{-1}(x) = e^{2x-1}$ $x \in \mathbb{R}$

   C $f^{-1}(x) = \dfrac{1}{2}(e^x + 1)$ $x > \dfrac{1}{2}$ D $f^{-1}(x) = \dfrac{1}{2}(e^x + 1)$ $x \in \mathbb{R}$

4. The function f(x) is defined by $f(x) = 1 + \sin 2x$.
   The domain of x is restricted so that the function has an inverse, $f^{-1}(x)$.
   Three of the following statements are false and one is true. Which one is true?

   A $f^{-1}(x) = \arcsin\left(\dfrac{x-1}{2}\right)$.   B The domain of x could be $-360° \leqslant x \leqslant 360°$.

   C The domain of $f^{-1}(x)$ is $-1 \leqslant x \leqslant 3$.   D If the range of f(x) is $a \leqslant x \leqslant b$, then $b - a \leqslant 2$.

5. The function g is defined by $g(x) = 2\tan\left(x - \dfrac{\pi}{2}\right)$ $0 < x < \pi$.
   The inverse function g^{-1} is defined by

   A $g^{-1}(x) = \arctan\left(\dfrac{1}{2}x\right) + \dfrac{\pi}{2}$ $x \in \mathbb{R}$ B $g^{-1}(x) = \arctan\left(\dfrac{1}{2}x\right) + \dfrac{\pi}{2}$ $0 < x < \pi$

   C $g^{-1}(x) = \dfrac{\pi}{2} + \dfrac{1}{2}x \arctan x$ $x \in \mathbb{R}$ D $g^{-1}(x) = \dfrac{\pi}{2} + \dfrac{1}{2}x \arctan x$ $0 < x < \pi$.

Full worked solutions online   CHECKED ANSWERS

Exam-style question

The function f is defined by $f(x) = 1 + \cos 2x$ for $0 \leqslant x \leqslant \dfrac{1}{2}\pi$.

> In degrees, the domain is $0° \leqslant x \leqslant 90°$, but it is important to work in radians for this question.

i Sketch the graph of $y = f(x)$.
ii Find an expression for $f^{-1}(x)$, and state its domain and range.
iii Add a sketch of $y = f^{-1}(x)$ to your sketch from i.

Short answers on page 215

Full worked solutions online   CHECKED ANSWERS

Edexcel A Level Mathematics (Pure)   69

The modulus function

REVISED

Key facts

1. The **modulus** of x means the positive (or absolute) value of x, and is written as $|x|$.

   So the equation $|x| = 2$ has two roots: $x = 2$ and $x = -2$.

2. This is the **graph** of the modulus function.

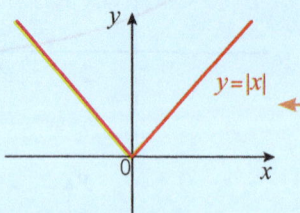

   Notice that the graph has two branches. The branch on the right, for $x > 0$, is part of the line $y = x$, and the branch on the left, for $x < 0$, is part of the line $y = -x$.

3. You can use **transformations** to sketch other graphs involving a modulus sign.

   To sketch the graph of the function $y = |f(x)|$
   - sketch the graph of $y = f(x)$
   - then reflect the negative part of the graph in the x-axis.

   See Worked example 2.

4. **Inequalities involving a modulus sign**.

   Look at the diagram, which shows the graphs of $y = |x|$ and $y = 2$.
   The solution to the inequality $|x| < 2$ is $-2 < x < 2$.
   The solution to the inequality $|x| > 2$ is $x < -2$ or $x > 2$.

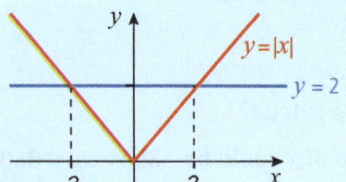

   You want the set of values of x for which the red line lies below the blue line.

   There are two regions where the red graph is above the blue line, so you need two inequalities.

Worked examples

1 Sketching the graph of a modulus function

Sketch the graphs of

   i $y = 1 - |x|$ ii $y = |x - 2|$ iii $y = |x^2 - 2|$.

Solution

i   Start with the graph $y = |x|$.   Reflect $y = |x|$ in the x-axis to obtain $y = -|x|$.

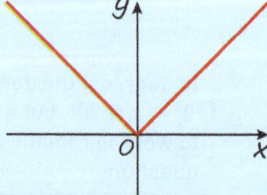

Hint: Remember that when you sketch a graph, you should show the coordinates of the points where the graph cuts the coordinate axes.

Translate $y = -|x|$ by $\begin{pmatrix} 0 \\ 1 \end{pmatrix}$ to obtain $y = 1 - |x|$.

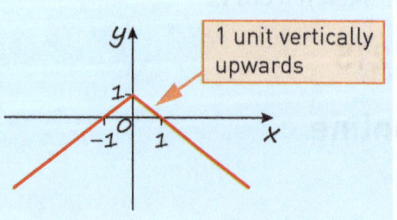

1 unit vertically upwards

ii Start with the graph of $y = x - 2$.

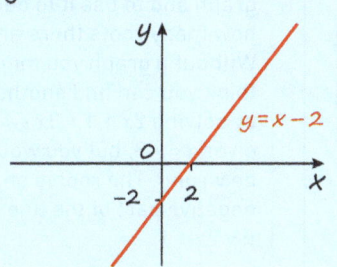

Now reflect the negative part of the line in the x-axis

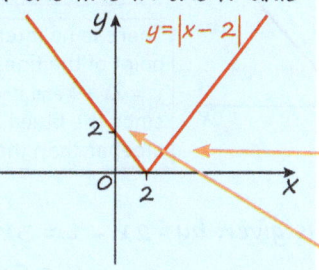

This graph passes through the points (0, 2) and (2, 0).

This branch is part of the line $y = -(x - 2)$.

iii Start with the graph of $y = x^2 - 2$

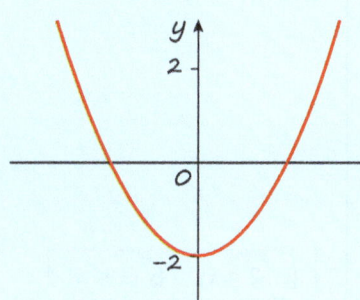

Now reflect the negative part of the curve in the x-axis.

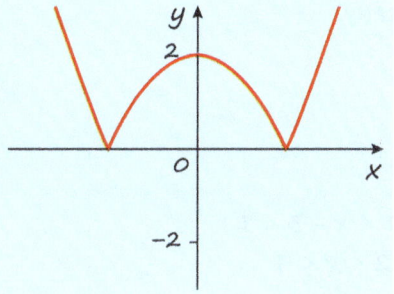

2 Solving equations involving a modulus sign

i Solve the equation $|x - 2| = 3$.

ii Sketch a graph to illustrate your answer.

Solution

i $|x - 2| = 3$
  $x - 2 = 3$ or $x - 2 = -3$
  $x = 5$ $x = -1$
  The roots are $x = 5$ and $x = -1$.

ii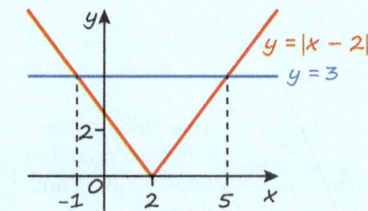

3 Using a graph to solve an equation

i Sketch the graph of $y = |2x + 1|$.

ii Add the graph of $y = 3x + 4$ to your sketch, and hence solve the equation $|2x + 1| = 3x + 4$.

Solution

i Start by sketching the graph of $y = 2x + 1$. This graph passes through the points $(0, 1)$ and $(-\frac{1}{2}, 0)$.

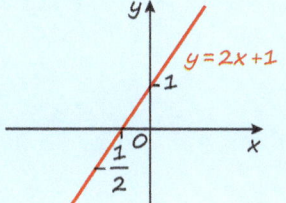

Now reflect the negative part of the graph in the x-axis.

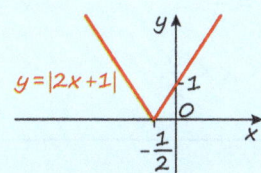

Hint: If the right-hand side of the equation is anything other than a simple number, you should always start by drawing a graph.

ii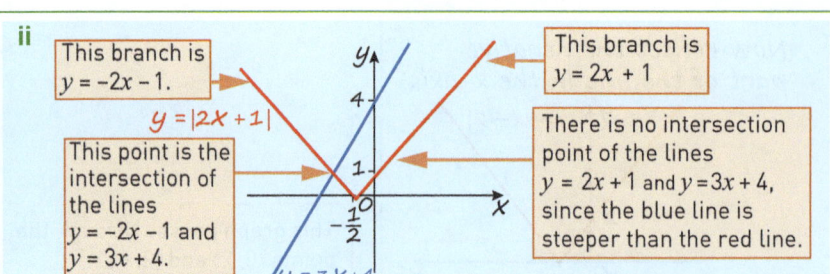

There is only one root. It is given by $-2x - 1 = 3x + 4$
$$-5 = 5x$$
$$x = -1.$$

Common mistake: This shows how important it is to draw a graph and to use it to decide how many roots there are. Without a graph you might think you can find another root by solving $2x + 1 = 3x + 4$, which gives $x = -3$, but you would be wrong. The root is on the negative part of the line $y = 2x + 1$.

4 Solving inequalities algebraically

Solve the inequalities
i $|x - 3| < 1$
ii $|2x + 1| \geq 3$.

Solution

i $|x - 3| < 1 \Rightarrow -1 < x - 3 < 1$
$$2 < x < 4$$

$\{x : 2 < x\} \cap \{x : x < 4\}$

ii $|2x + 1| \geq 3 \Rightarrow 2x + 1 \leq -3$ or $2x + 1 \geq 3$
$$2x \leq -4 \qquad 2x \geq 2$$
$$x \leq -2 \qquad x \geq 1$$

$\{x : x \leq -2\} \cup \{x : x \geq 1\}$

5 Using a graph to solve an inequality

Solve the inequality $|3x - 2| < x + 1$.

Hint: If the right-hand side of the inequality is not just a simple number, you must always start by drawing a sketch graph.

Solution

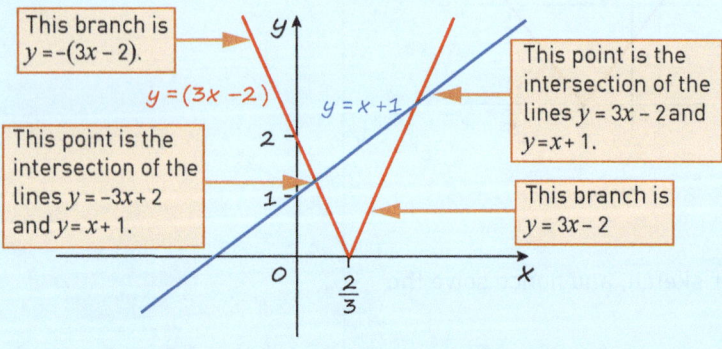

The solution of the inequality is the set of values of x for which the red graph lies below the blue graph.
The intersection points are $-3x + 2 = x + 1$ and $3x - 2 = x + 1$
$$1 = 4x \qquad 2x = 3$$
$$x = \frac{1}{4} \qquad x = \frac{3}{2}$$

The solution of the inequality is $\frac{1}{4} < x < \frac{3}{2}$.

6 Using upper and lower bounds

Write the inequality $-1 \leq x \leq 7$ in the form $|x - a| \leq b$.

Solution

$|x - a| \leq b \Rightarrow -b \leq x - a \leq b$
$\Rightarrow a - b \leq x \leq a + b$

Solving $a - b = -1$ and $a + b = 7$ simultaneously gives $a = 3$ and $b = 4$.

So $-1 \leq x \leq 7$ can be written as $|x - 3| \leq 4$.

Test yourself

1 Which of the following could be the equation of the graph below?

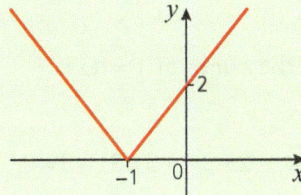

- A $y = |2x + 1|$
- B $y = 2|x| + 1$
- C $y = |x| + 1$
- D $y = 2|x + 1|$
- E $y = |x| + 2$

2 Solve the inequality $|3x - 2| \geq 5$.
- A $x \geq \frac{7}{3}$
- B $x \geq \frac{7}{3}$ or $x \geq -1$
- C $x \leq -1$ or $x \geq \frac{7}{3}$
- D $-1 \leq x \leq \frac{7}{3}$

3 Solve the equation $3|x + 1| = 2x + 5$.
- A $x = -\frac{8}{5}$ or $x = 2$
- B $x = 2$
- C $x = 4$
- D $x = -\frac{4}{5}$ or $x = 4$
- E $x = -\frac{2}{5}$ or $x = 2$

4 Write the inequality $-5 < x < 4$ in the form $|x - a| < b$.
- A $|x - 0.5| < 4.5$
- B $|x - 4.5| < -0.5$
- C $|x + 4.5| < -0.5$
- D $|x + 0.5| < 4.5$

5 Solve the inequality $|x - 2| > 2x - 1$
- A $x > 1$
- B $x < 1$
- C $x < -1$
- D $-1 < x < 1$

Full worked solutions online   CHECKED ANSWERS

Exam-style question

i Sketch the graph of $y = |2x - 1|$.
ii Solve
  a $|2x - 1| \leq 3x + 2$
  b $|2x - 1| = |3x + 2|$

Short answers on page 215—216

Full worked solutions online   CHECKED ANSWERS

Review questions (Chapters 1–5)

1. **a** Prove or disprove the following conjecture:
   Given that p is prime, then $2p + 1$ is also prime.
   b i Write down the first four prime numbers: p_1, p_2, p_3 and p_4.
   Show that

   A $p_1 \times p_2 + 1$ **B** $p_1 \times p_2 \times p_3 + 1$ **C** $p_1 \times p_2 \times p_3 \times p_4 + 1$

   are prime numbers.
   ii Show that 59 is a prime factor of $p_1 \times p_2 \times p_3 \times p_4 \times p_5 \times p_6 + 1$.
   iii Use proof by contradiction to prove that there are an infinite number of primes.

2. Given that $\dfrac{5}{\sqrt{2}-1} - \dfrac{x}{\sqrt{2}+1} = 8 + y\sqrt{2}$, find the values of x and y.

3. **a** Solve the following equations.
   i $2^{x+1} = 3^x$ **ii** $\ln x + \ln 4x = 3\ln 4$
   b Find the value of $\log_a a^3 - 2\log_a \dfrac{1}{a}$.

4. **i** Express $f(x) = -2x^2 + 5x + 3$ in the form $f(x) = a(x+b)^2 + c$. Hence sketch the curve of $y = f(x)$.
   ii Solve

   a $-2x^4 + 5x^2 + 3 = 0$ **b** $-2x + 5\sqrt{x} + 3 = 0$ **c** $-2 \times 3^{2x} + 5 \times 3^x + 3 = 0$

5. Simplify $\dfrac{2}{x} + \dfrac{1}{4x} - \dfrac{1}{2x-1}$. Hence solve $\dfrac{2}{x} + \dfrac{1}{4x} - \dfrac{1}{2x-1} = 1$.

6. You are given that $f(x) = 2x^3 - 7x^2 - 17x + 10$.
   i Show that $(x - 5)$ is a factor of $f(x)$. Hence factorise $f(x)$ fully.

   The curve $y = f(x)$ is translated by the vector $\begin{pmatrix} 3 \\ 0 \end{pmatrix}$ to give the curve $y = g(x)$.

   ii Solve $g(x) = 0$.

7. Three points A, B and C have coordinates $(1, -2)$, $(5, 0)$ and $(4, -8)$ respectively.
   i Find the distance AB and BC.
   ii Hence show that triangle ABC is right-angled and find the area of triangle ABC.

8. A circle has equation $(x + 3)^2 + (y - 5)^2 = 36$.
   i Find the equations of the two tangents to the circle which are parallel to the y-axis.
   ii Show that the line $y = 2x - 5$ does not intersect the circle.

9. The functions $f(x)$ and $g(x)$ are defined for all real numbers x by
   $f(x) = \cos x$, $g(x) = 1 - x$ where x is measured in radians.
   i Find the composite functions $fg(x)$ and $gf(x)$.
   ii State a sequence of two transformations that would map the curve $y = f(x)$ onto the curve $y = gf(x)$.
   Sketch the curves $y = f(x)$ and $y = gf(x)$ on the same axes, for $-2\pi \leq x \leq 2\pi$, indicating clearly which is which.
   iii State a sequence of two transformations that would map the curve $y = f(x)$ onto the curve $y = fg(x)$.
   Sketch the curves $y = f(x)$ and $y = fg(x)$ on the same axes, for $-2\pi \leq x \leq 2\pi$, indicating clearly which is which.

Short answers on page 216

Full worked solutions online

CHECKED ANSWERS

SECTION 2

Target your revision (Chapters 6–8)

1. **Use exact values of sin θ, cos θ and tan θ in degrees and radians**
   Without using your calculator find the exact values of
   i $\cos \dfrac{5\pi}{3}$ ii $\sin\left(\dfrac{3\pi}{4}\right)$ iii $\tan 315°$.
   (see page 78)

2. **Solve trigonometric equations in degrees**
   Solve the following equations for $0° \leq x \leq 360°$.
   i $\sin^2 x = \dfrac{3}{4}$ ii $\sqrt{3} - 3\tan 2x = 0$
   iii $2\cos(x - 45°) = 1$
   (see page 78)

3. **Solve trigonometric equations in radians**
   Solve the following equations for $0 \leq x \leq 2\pi$.
   i $\sin x = 0.8$
   ii $\cos 2x = \dfrac{\sqrt{3}}{2}$
   iii $\sin\left(x - \dfrac{\pi}{2}\right) = \sqrt{3}\cos\left(x - \dfrac{\pi}{2}\right)$
   (see page 87)

4. **Use trigonometric identities**
   Prove that $\dfrac{\tan^2 \theta - 1}{\tan^2 \theta + 1} \equiv 2\sin^2 \theta - 1$.
   (see page 78)

5. **Solve problems involving triangles without right-angles**
   In triangle PQR, PQ = 5.6 cm, PR = 7.4 cm and angle PRQ = 35°.
   i   Find the possible values for angle PQR.
   ii Given PQR is acute, find the area of triangle PQR.
   (see page 83)

6. **Solve problems involving sectors of circles**

   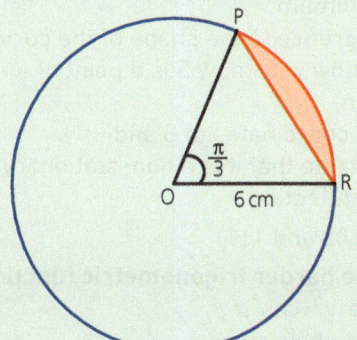

   Find the exact area and perimeter of the shaded region.
   (see page 87)

7. **Use the reciprocal trigonometric functions**
   i Sketch the graph of $y = \operatorname{cosec} x$ for $0 \leq x \leq 2\pi$.
   ii Solve $\operatorname{cosec} x = -2$ for $0 \leq x \leq 2\pi$.
   (see page 91)

8. **Prove identities involving the reciprocal trigonometric functions**
   Prove that $\operatorname{cosec}^2 \theta + \sec^2 \theta \equiv \operatorname{cosec}^2 \theta \sec^2 \theta$.
   (see page 91)

9. **Use the small angle approximations**
   Find an approximation for the following expressions when θ is small and in radians.
   i $\dfrac{1 - \cos \theta}{\theta}$ ii $2\cot \theta - \sin \theta$
   (see page 91)

10. **Use compound angle formulae**
    Use the fact that $15° = 45° - 30°$ to find the exact value of $\cos 15°$.
    (see page 95)

11. **Use compound angle formulae to solve equations**
    i Expand $\cos\left(x + \dfrac{\pi}{6}\right)$ and $\sin\left(\dfrac{\pi}{6} - x\right)$.
    ii Hence solve $\cos\left(x + \dfrac{\pi}{6}\right) = \sin\left(\dfrac{\pi}{6} - x\right)$ for $0 \leq x \leq 2\pi$.
    (see page 95)

12. **Use double angle formulae**
    Use the expansion of $\tan(A + B)$ to prove that $\tan 2A = \dfrac{2\tan A}{1 - \tan^2 A}$.
    (see page 96)

13. **Use the forms $r\cos(\theta \pm \alpha)$ and $r\sin(\theta \pm \alpha)$**
    Express $16\sin \theta - 30\cos \theta$ in the form $r\sin(\theta - \alpha)$, where $r > 0$ and $0 < \alpha < 90°$. Hence state the maximum value of $16\sin \theta - 30\cos \theta$.
    (see page 99)

14. **Differentiate functions involving powers of x**
    Differentiate
    i $y = 5x^2 - 3x + \dfrac{4}{x} + 1$ ii $y = \dfrac{1}{x^3} + \sqrt{x} - x$.
    (see page 103)

15. **Find the gradient of a curve at a point**
    i Find the gradient of the curve $y = 6x - 3\sqrt{x}$ at the point where $x = 9$.
    ii Given $f(x) = \dfrac{x^2}{\sqrt{x}}$, find the coordinates of the point with a gradient of 9.
    (see page 103)

Edexcel A Level Mathematics (Pure) 75

16 Find the equation of a tangent and a normal to a curve
  i Find the equation of the tangent to the curve $y = 2x^2 - x$ at the point (3, 15).
  ii Find the equation of the normal to the curve $y = \dfrac{2}{x}$ at the point where $x = 2$.

(see page 106)

17 Find the second derivative
Given $f(x) = x^2 + \dfrac{1}{x} - \sqrt{x}$, find $f''(x)$.
(see page 107)

18 Identify where a function is increasing or decreasing
Find the values of x for which $f(x) = 2x^3 + 3x^2 - 36x$ is a decreasing function.
(see page 107)

19 Sketch the graph of the gradient function
The diagram shows the graph of $y = f(x)$. Sketch the gradient function, $y = f'(x)$.

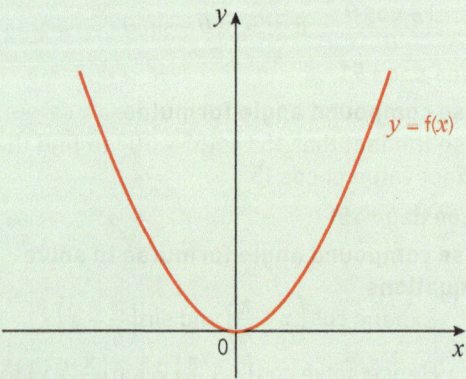

(see page 109)

20 Differentiate from first principles
  i Expand $(x + h)^3$.
  ii Given $f(x) = 2x^3$, find an expression for $f(x + h) - f(x)$.
  iii Differentiate $f(x) = 2x^3$ from first principles.
  You must show all your working.

(see page 114)

21 Use the chain rule
Differentiate $y = (4x - 5)^5$.
(see page 119)

22 Use the product rule
Differentiate $y = 3x \sin x$.
(see page 124)

23 Use the quotient rule
A curve has equation $y = \dfrac{x^3 + 1}{3x - 1}$.
  i Find $\dfrac{dy}{dx}$.
  ii Find the gradient of the curve at $x = 2$.
(see page 124)

24 Differentiate e^x and identify stationary points
Differentiate the following:
  i a $y = e^{2x}$ b $y = xe^{2x}$ c $y = \dfrac{e^{2x}}{x}$
  ii For each of the following curves, state the coordinates of the stationary point. Identify the nature of each stationary point.
   a $y = xe^{2x}$
   b $y = \dfrac{e^{2x}}{x}$
(see pages 107, 117 and 122)

25 Differentiate $\ln x$
Differentiate the following.
  i $y = \ln 2x$ ii $y = x \ln 2x$ iii $y = \dfrac{\ln 2x}{x}$
(see pages 117 and 125)

26 Differentiate trigonometric functions
Differentiate the following.
  i $y = \sin(3x + \pi)$ ii $y = \cos(4x^2)$
  iii $y = \tan 2x$
(see pages 114 and 121)

27 Classify stationary points and points of inflection
The diagram shows part of the curve of $y = x + \cos 2x$ for $0 \leq x \leq 2$.

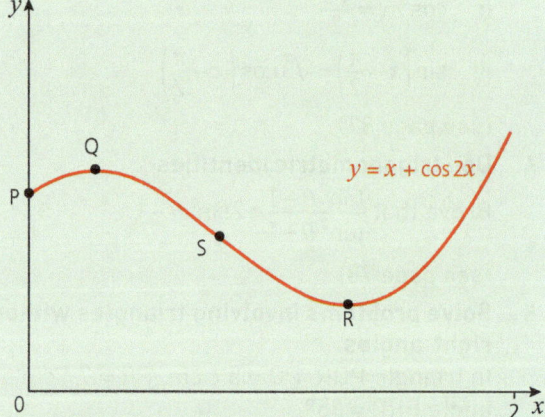

  i Find the coordinates of P.
  ii Find $\dfrac{dy}{dx}$ and $\dfrac{d^2y}{dx^2}$.
  ii Hence find the coordinates of Q and R and prove that Q is a local maximum and R is a local minimum.
  iii With reference to the shape of the curve, explain how you know S is a point of inflection.
  iv Find the coordinates of S and demonstrate that it is a non-stationary point of inflection.
(see pages 108 and 111)

28 Differentiate harder trigonometric functions
Differentiate $y = \cos^2 x$
  i using the chain rule
  ii using the product rule.
(see pages 121 and 125)

29 **Differentiate a^x using $\frac{dy}{dx} = \frac{1}{\frac{dx}{dy}}$**

Differentiate i $y = 3^x$ ii $y = 3^{2x}$.

(see page 119)

30 **Use implicit differentiation**
Given that $x^2 = \cos 2y + xy$, find $\frac{dy}{dx}$.

(see page 127)

31 **Find indefinite integrals involving powers of x**
Find

i $\int (2x-3)(x+5) dx$ ii $\int \left(\sqrt{x} + \frac{1}{x^2}\right) dx$.

(see page 131)

32 **Integrate using e^x and $\ln x$**
Find

i $\int \left(\frac{1+2x}{x^2}\right) dx$ ii $\int \left(10e^{5x} + \frac{1}{e^x}\right) dx$.

(see page 132)

33 **Find the equation of a curve given its gradient function, $\frac{dy}{dx}$**

The gradient of a curve is given by $\frac{dy}{dx} = \frac{2}{x} + e^2$. The curve passes through $(1, e^2)$. Find the equation of the curve.

(see page 133)

34 **Evaluate definite integrals**
Find the exact value of

i $\int_1^4 \left(\frac{8}{x^2} - 2\sqrt{x}\right) dx$ ii $\int_{\frac{1}{2}}^1 \left(e^{2x} - \frac{1}{x}\right) dx$.

(see page 132)

35 **Find the area under a curve**
The diagram shows the graph of $y = x^3 - 5x^2 + 2x + 8$.

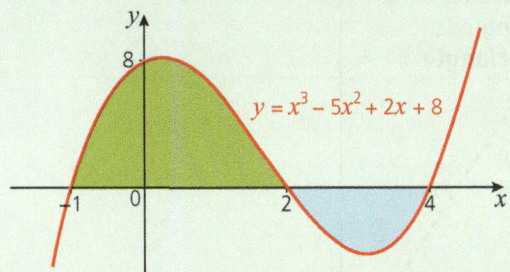

i Find the area of the green shaded region.
ii Find the area of the blue shaded region.
iii Find the total area of the shaded regions.
iv Evaluate $\int_{-1}^4 (x^3 - 5x^2 + 2x + 8) dx$.

(see page 134)

36 **Find the area between two curves**
The diagram shows the curves $y = 5 + 4x - x^2$ and $y = x^2 - 4x + 5$.

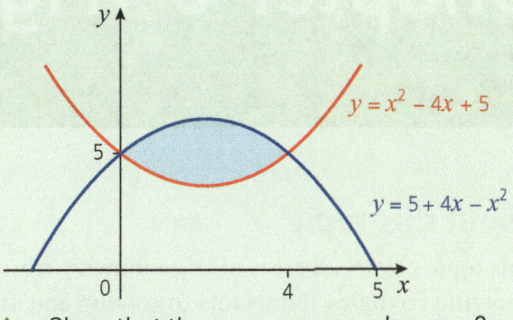

i Show that the curves cross when $x = 0$ and $x = 4$.
ii Find the area between the x-axis, the lines $x = 0$ and $x = 4$ and
 a $y = 5 + 4x - x^2$ b $y = x^2 - 4x + 5$.
iii Hence find the area between the two curves.

(see page 134)

37 **Find the area between a curve and the y-axis**
The diagram shows the curve $y = x^2 + 2$ and the lines $y = 11$ and $y = 3$.
Find the exact area of the shaded region.

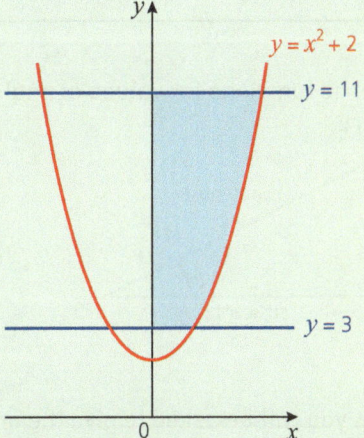

(see page 138)

38 **Use integration by substitution**
Find $\int x e^{x^2} dx$.

(see page 140)

39 **Integrate trigonometric functions**
Find $\int_0^{\frac{\pi}{2}} (2\cos x - \sin x) dx$.

(see page 144)

40 **Use trigonometric identities in integration**

i Show that $\int_{\frac{\pi}{4}}^{\frac{\pi}{2}} \cot \theta \, d\theta = \frac{1}{2} \ln 2$.

ii Show that $\sin^3 \theta \equiv \sin \theta - \sin \theta \cos^2 \theta$.
Hence find $\int \sin^3 \theta \, d\theta$.

(see page 146)

41 **Use integration by parts**
Find $\int x e^{2x} dx$.

(see page 148)

Short answers on pages 216–217
Full worked solutions online

CHECKED ANSWERS

Edexcel A Level Mathematics (Pure)

Chapter 6 Trigonometry

About this topic

This topic covers all areas of trigonometry from solving problems involving triangles and sectors to solving equations and proving identities. For much of this work, you will need to use radians as your measure of angle, rather than degrees. You will also use trigonometric identities to simplify expressions in order to integrate them in Chapter 8.

Before you start, remember

- the trigonometric functions: sine, cosine and tangent
- surds
- solving quadratic equations.

Working with trigonometric functions

REVISED

Key facts

1. Trigonometric functions for values of angle θ between 0° and 90° inclusive are:

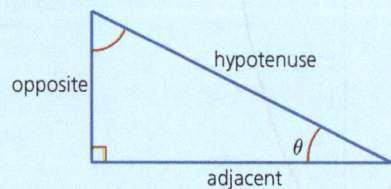

$$\sin\theta = \frac{\text{opposite}}{\text{hypotenuse}} = \frac{O}{H}$$

$$\cos\theta = \frac{\text{adjacent}}{\text{hypotenuse}} = \frac{A}{H}$$

$$\tan\theta = \frac{\text{opposite}}{\text{adjacent}} = \frac{O}{A}$$

2. Sometimes you will be asked to give the answers in surd form, so you need to know the trigonometric ratios of special angles:

   i **equilateral triangle**   ii **equilateral triangle**

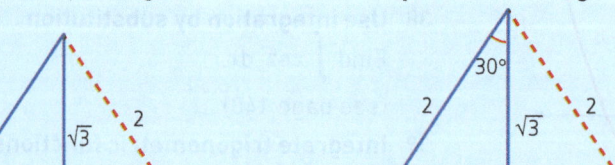

$\sin 60° = \dfrac{O}{H} = \dfrac{\sqrt{3}}{2}$ $\sin 30° = \dfrac{O}{H} = \dfrac{1}{2}$

$\cos 60° = \dfrac{A}{H} = \dfrac{1}{2}$ $\cos 30° = \dfrac{A}{H} = \dfrac{\sqrt{3}}{2}$

$\tan 60° = \dfrac{O}{A} = \dfrac{\sqrt{3}}{1} = \sqrt{3}$ $\tan 30° = \dfrac{O}{A} = \dfrac{1}{\sqrt{3}} = \dfrac{\sqrt{3}}{3}$

iii **isosceles triangle**

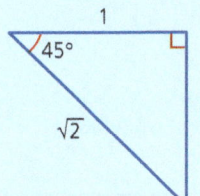

$\sin 45° = \dfrac{O}{H} = \dfrac{1}{\sqrt{2}} = \dfrac{\sqrt{2}}{2}$

$\cos 45° = \dfrac{A}{H} = \dfrac{1}{\sqrt{2}} = \dfrac{\sqrt{2}}{2}$

$\tan 45° = \dfrac{O}{A} = \dfrac{1}{1} = 1$

3 The unit circle is a circle with radius 1 unit. The following are true for any point P(x, y) on the unit circle and acute angle θ between OP and the x-axis.

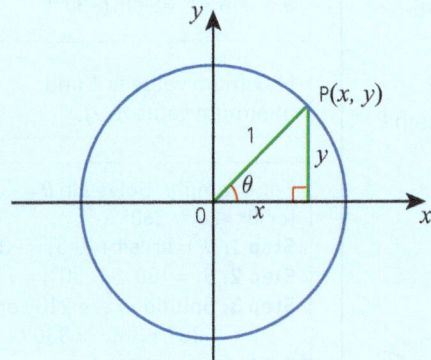

$\sin \theta = \dfrac{y}{1} = y \qquad \sin 0° = 0$

$\cos \theta = \dfrac{x}{1} = x \qquad \cos 0° = 1$

$\tan \theta = \dfrac{y}{x}$

$y^2 + x^2 = 1$ ◄──── Using Pythagoras' theorem.

Two important identities are

$\sin^2 \theta + \cos^2 \theta \equiv 1$

and

$\tan \theta \equiv \dfrac{\sin \theta}{\cos \theta}, \quad \cos \theta \neq 0.$

4 The angles in the anticlockwise direction from the x-axis are positive and in the clockwise direction are negative.

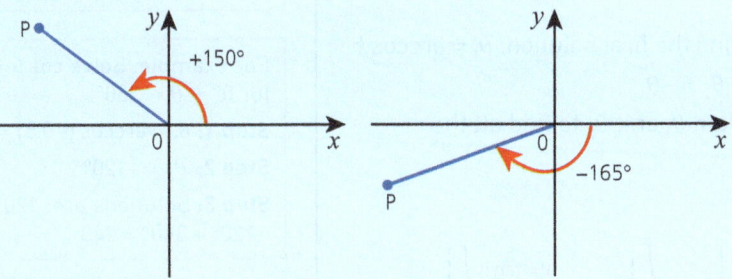

5 You can use a **CAST** diagram to decide whether sin, cos or tan of an angle is positive or negative.

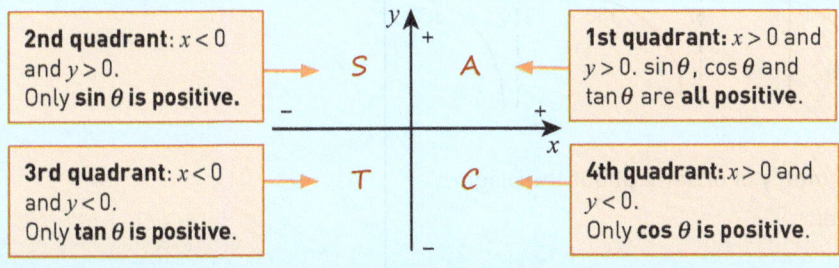

2nd quadrant: $x < 0$ and $y > 0$.
Only $\sin \theta$ is positive.

1st quadrant: $x > 0$ and $y > 0$. $\sin \theta$, $\cos \theta$ and $\tan \theta$ are **all positive**.

3rd quadrant: $x < 0$ and $y < 0$.
Only $\tan \theta$ is positive.

4th quadrant: $x > 0$ and $y < 0$.
Only $\cos \theta$ is positive.

6 You can use the properties of the graphs of $y = \sin \theta$, $y = \cos \theta$ and $y = \tan \theta$ to help you solve equations.

Edexcel A Level Mathematics (Pure)

7 **Graph of $y = \sin\theta$**

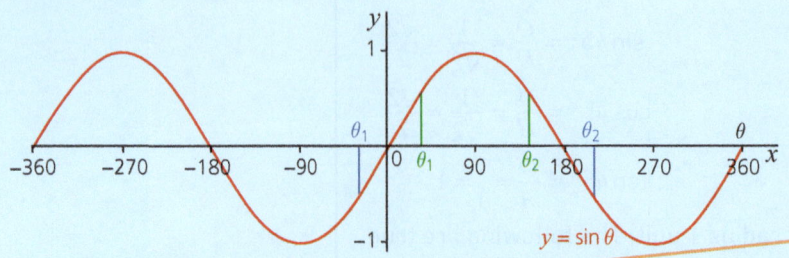

- Period of $y = \sin\theta$ is 360°.
- $y = \sin\theta$ has rotational symmetry of order 2 about the origin.
- $-1 \leq \sin\theta \leq 1$
- To solve $\sin\theta = k$
  Step 1: Use your calculator to find the first solution: $\theta_1 = \arcsin k$.
  Step 2: The second solution is $\theta_2 = 180° - \theta_1$.
  Step 3: Add or subtract 360° from θ_1 and θ_2 to find all the solutions.

It repeats every 360°.

$\sin\theta = -\sin(-\theta)$
e.g. $\sin 30° = -\sin(-30°)$.

Maximum value is 1 and minimum value is –1.

For example: Solve $\sin\theta = -0.5$ for $0° \leq \theta \leq 360°$
Step 1: $\theta_1 = \arcsin(-0.5) = -30°$
Step 2: $\theta_2 = 180° - (-30°) = 210°$
Step 3: Solutions are 210° and $-30° + 360° = 330°$.

8 **Graph of $y = \cos\theta$**

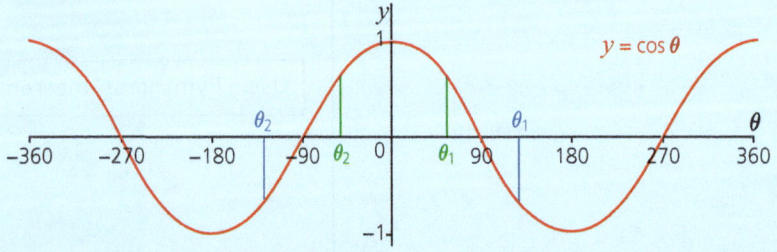

- Period of $y = \cos\theta$ is 360°.
- $y = \cos\theta$ is symmetrical about the y-axis.
- $-1 \leq \cos\theta \leq 1$
- To solve $\cos\theta = k$
  Step 1: Use your calculator to find the first solution: $\theta_1 = \arccos k$.
  Step 2: The second solution is $\theta_2 = -\theta_1$.
  Step 3: Add or subtract 360° from θ_1 and θ_2 to find all the solutions.

$\cos\theta = \cos(-\theta)$
e.g. $\cos 30° = \cos(-30°)$

Maximum value is 1 and minimum value is –1.

For example: Solve $\cos\theta = -0.5$ for $0° \leq \theta \leq 360°$
Step 1: $\theta_1 = \arccos(-0.5) = 120°$
Step 2: $\theta_2 = -120°$
Step 3: Solutions are: 120° and $-120° + 360° = 240°$.

9 **Graph of $y = \tan\theta$**

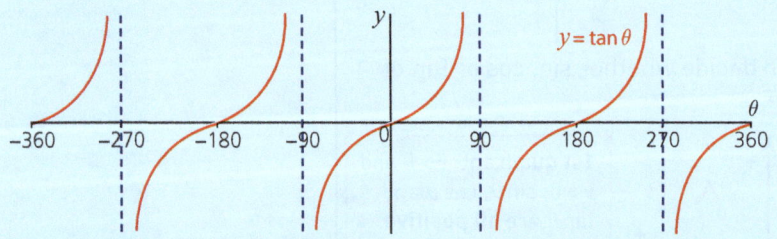

- Period of $y = \tan\theta$ is 180°.
- $y = \tan\theta$ has rotational symmetry of order 2 about the origin.
- $\tan\theta$ can be any real number.
  As $\theta \to 90°$, $\tan\theta \to +\infty$.
  As $\theta \to -90°$, $\tan\theta \to -\infty$.
- There are asymptotes at $x = \pm 90°$, $x = \pm 270°\ldots$

- To solve $\tan\theta = k$

  Step 1: Use your calculator to find the first solution: $\theta_1 = \arctan k$.

  Step 2: Add or subtract $180°$ from θ_1 to find all the solutions.

> For example: Solve $\tan\theta = -0.5$ for $0° \leq \theta \leq 360°$
>
> **Step 1**: $\theta_1 = \arctan(-0.5) = -26.6°$
>
> **Step 2**: Solutions are:
> $-26.6° + 180° = 153.4°$ and
> $153.4° + 180° = 333.4°$

Worked examples

1 Using the CAST diagram

Given that $\cos\theta = -\dfrac{5}{13}$ and θ is obtuse, find the exact value of

  i $\sin\theta$ and   **ii** $\tan\theta$

> An **obtuse** angle is greater than $90°$ but less than $180°$.

Solution

  i Using the identity $\sin^2\theta + \cos^2\theta \equiv 1$

$\Rightarrow \sin^2\theta \equiv 1 - \cos^2\theta$

$\Rightarrow \sin^2\theta = 1 - \left(-\dfrac{5}{13}\right)^2 = 1 - \dfrac{25}{169} = \dfrac{144}{169}$

$\Rightarrow \sin\theta = \pm\dfrac{12}{13}$.

Using the CAST diagram, when $\cos\theta$ is negative and θ is obtuse then θ is in the 2nd quadrant where only $\sin\theta$ is positive.
So $\sin\theta = \dfrac{12}{13}$.

> You need to work out if $\sin\theta$ is positive or negative.

  ii Using the identity $\tan\theta \equiv \dfrac{\sin\theta}{\cos\theta}$

$\Rightarrow \tan\theta = \dfrac{\frac{12}{13}}{-\frac{5}{13}} = -\dfrac{12}{5}$.

> To divide by a fraction, you multiply by its reciprocal.
> $\dfrac{12}{13} \times \left(-\dfrac{13}{5}\right) = -\dfrac{12}{5}$.

2 Proving identities

Prove that $\dfrac{\sin^2\theta - \cos^2\theta}{1 - \sin^2\theta} \equiv \tan^2\theta - 1$.

Solution

Using the identity $\sin^2\theta + \cos^2\theta \equiv 1 \Rightarrow 1 - \sin^2\theta \equiv \cos^2\theta$

$\dfrac{\sin^2\theta - \cos^2\theta}{1 - \sin^2\theta} \equiv \dfrac{\sin^2\theta - \cos^2\theta}{\cos^2\theta}$

$\equiv \dfrac{\sin^2\theta}{\cos^2\theta} - \dfrac{\cos^2\theta}{\cos^2\theta}$

$\equiv \tan^2\theta - 1$ as required.

> $\dfrac{\sin\theta}{\cos\theta} \equiv \tan\theta \Rightarrow \dfrac{\sin^2\theta}{\cos^2\theta} \equiv \tan^2\theta$

3 Solving trigonometric equations (1)

Solve the equation $2\sin\theta\cos\theta - \sin\theta = 0$, for $0° \leq \theta \leq 360°$.

Solution

$$2\sin\theta\cos\theta - \sin\theta = 0$$
$$\Rightarrow \sin\theta(2\cos\theta - 1) = 0$$

Either $\sin\theta = 0$ or $2\cos\theta - 1 = 0$
$\Rightarrow \theta = 0°$ or $180°$ $\Rightarrow \cos\theta = \frac{1}{2}$
$\Rightarrow \theta = 60°$ or $300°$

So $\theta = 0°, 60°, 180°$ or $300°$.

> **Common mistake:** Don't divide by $\sin\theta$, otherwise you lose the roots of $\sin\theta = 0$.

> $-60° + 360° = 300°$

4 Solving trigonometric equations (2)

Solve
  i $\cos(x + 45°) = \frac{\sqrt{3}}{2}$, for $0° \leq x \leq 360°$
  ii $\tan^2 2x = 3$, for $0° \leq x \leq 180°$.

Solution

i Let $\theta = x + 45° \Rightarrow \cos\theta = \frac{\sqrt{3}}{2}$.

$\theta = \arccos\left(\frac{\sqrt{3}}{2}\right) = 30°$ or $330°$ or $390°$

So $x + 45° = 330°$ or $x + 45° = 390°$.
   $x = 285°$, $x = 345°$

So $x = 285°$ or $345°$

ii Let $\theta = 2x \Rightarrow \tan^2\theta = 3$
$\Rightarrow \tan\theta = \pm\sqrt{3}$
$\tan\theta = \sqrt{3} \Rightarrow \theta = 60°, 240°$
$\tan\theta = -\sqrt{3} \Rightarrow \theta = -60°, 120°, 300°$
$2x = 60°, 120°, 240°, 300°$
$\Rightarrow x = 30°, 60°, 120°, 150°$

> $-30° + 360° = 330°$

> Look for all solutions for θ in the range $45° \leq \theta \leq 405°$ so that when you subtract 45° you get the values of x in the range $0° \leq x \leq 360°$.

> **Common mistake:** Don't forget the negative square root!

> Look for all solutions for θ in the range $0° \leq \theta \leq 360°$ so that when you divide by 2 you get the values of x in the range $0° \leq x \leq 180°$.

> Once you have the first value, just keep adding 180° to find all the other values.

Test yourself

TESTED

1 Four of these diagrams below are correct. Which one of the diagrams is incorrect?

A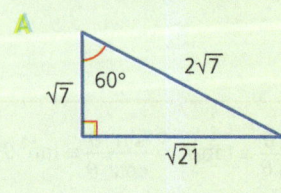
B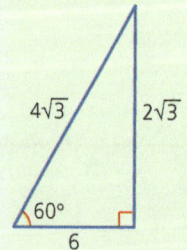
C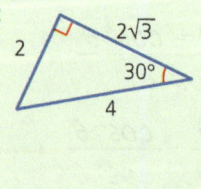
D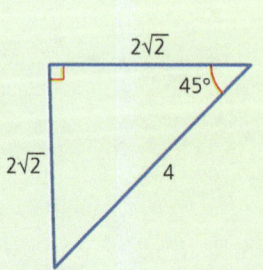
E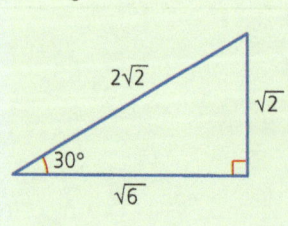

2 Which one of the following has the same value as tan 315°?
   A tan(−315°)   B −tan 45°   C tan 225°   D tan 45°

3 Solve the equation cos 2x = −0.5 to the nearest degree for 0° ⩽ x ⩽ 360°.
   A 60°   B 60°, 120°, 240° and 300°   C 120° and 240°
   D 120°   E 60°, 120° and 240°

4 Solve the equation $\sin^2 \theta = \sin \theta$ to the nearest degree for 0° ⩽ θ ⩽ 360°.
   A 0° and 90°   B 0°, 90°, 180° and 360°   C 0°, 180° and 360°   D 90°

5 Which one of the following is the exact value of $\dfrac{1 - \sin 240°}{1 + \sin 240°}$?
   A $7 - 4\sqrt{3}$ B 1   C $7 + 4\sqrt{3}$ D 3   E 7

Full worked solutions online   CHECKED ANSWERS

Exam-style question

i   Express $2\sin^2 x - \cos x$ as a quadratic function of cos x.
ii Hence solve the equation $2\sin^2 2x - \cos 2x = 1$ for 0° ⩽ x ⩽ 180°.

Short answers on page 218

Full worked solutions online   CHECKED ANSWERS

Triangles without right angles   REVISED

Key facts

1 Usually the vertices of any triangle are labelled with capital letters, and the opposite sides with corresponding small letters.

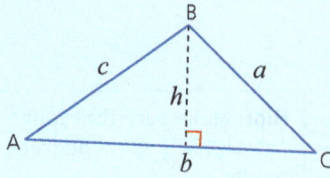

2 The **area of triangle** ABC is $\frac{1}{2}ab\sin C$

   You can find the area of any triangle if you know 2 sides and the angle between them.

3 The **sine rule** for triangle ABC is
   $$\frac{\sin A}{a} = \frac{\sin B}{b} = \frac{\sin C}{c}$$

   Use this form to find a missing angle...

   $$\frac{a}{\sin A} = \frac{b}{\sin B} = \frac{c}{\sin C}$$

   ...and this form to find a missing side.

   When you use the sine rule to find a missing angle, θ, always check whether 180° − θ is also a solution.

   *This is called the **ambiguous case**.*

4 The **cosine rule** for the triangle ABC is
   $a^2 = b^2 + c^2 - 2bc\cos A$
   or $\cos A = \dfrac{b^2 + c^2 - a^2}{2bc}$

   Use the cosine rule when you know:
   - 2 sides and the angle between them and you need the 3rd side
   - all 3 sides and you need to find any angle.

Edexcel A Level Mathematics (Pure) 83

Worked examples

1 Find the area of a triangle

Find the area of the triangle ABC shown below.

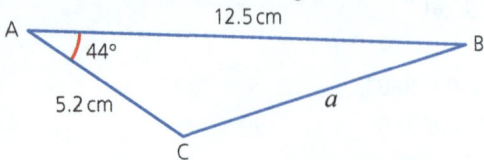

Solution

Area = $\frac{1}{2} ab \sin C$ or $\frac{1}{2} bc \sin A$ ← You know two sides and the angle between them.

Area = $\frac{1}{2} \times 5.2 \times 12.5 \times \sin 44°$

= 22.57...

Area = 22.6 cm² (to 3 s.f.) ← Be careful with the units.

2 Using the sine rule to find a missing side

← You can only use the sine rule when you know one angle and the side opposite it. Here you know angle Z and side XY.

Find the side x in this triangle.

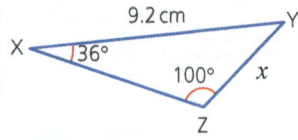

Solution

∠X = 36° and ∠Z = 100°.

XY = z = 9.2 cm.

Use the sine rule in the form: $\dfrac{x}{\sin X} = \dfrac{z}{\sin Z}$

$\dfrac{x}{\sin 36°} = \dfrac{9.2}{\sin 100°}$

$x = \sin 36° \times \dfrac{9.2}{\sin 100°}$

= 5.491...

x = 5.49 cm (to 3 s.f.) ←

Hint: Make sure that your calculator is set in degrees mode.

Round off only in the final answer.

3 Using the sine rule to find a missing angle

In triangle PQR, PR = 3.8 cm, QR = 5.1 cm and ∠Q = 42°.

 i Draw the triangle.
 ii Find the possible size of angle P.

Solution

 i From the diagram there are two possible angles and two possible positions for point P. These are marked P_1 and P_2.

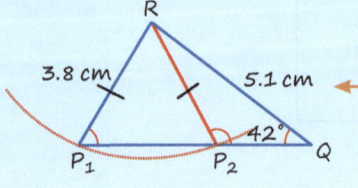

← You can see from the diagram that P lies on the circle with centre R and radius of 3.8 cm.

ii $\dfrac{\sin P}{5.1} = \dfrac{\sin 42°}{3.8}$

$\sin P = 5.1 \times \dfrac{\sin 42°}{3.8}$

$\sin P = 0.8980...$

$P = 63.90...°$ or $180° - 63.90...° = 116.09°$

So $\angle P_1 = 63.9°$ or $\angle P_2 = 116.1°$ (both to 1 d.p.)

Hint: Always check for the ambiguous case. There are two possible values of P between $0°$ and $180°$.

4 Using the cosine rule to find a missing angle

Find the angle ABC in the triangle shown below:

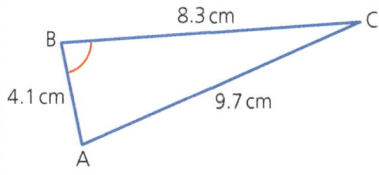

Solution

Using cosine rule:

$\cos B = \dfrac{c^2 + a^2 - b^2}{2ca}$

$= \dfrac{4.1^2 + 8.3^2 - 9.7^2}{2 \times 4.1 \times 8.3} = -0.1232...$

$B = 97.1°$ (to 1 d.p.)

Hint: You know three sides and you need to find a missing angle so you should use the cosine rule.

It is usual to give angles correct to 1 decimal place.

5 Using the cosine rule to find a missing side

Find the length of the side p in the triangle given below.

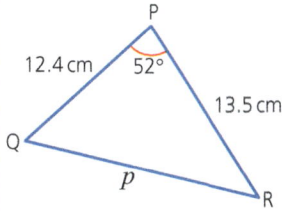

Solution

Use the cosine rule in the form
$p^2 = q^2 + r^2 - 2qr \cos P$

$p^2 = 13.5^2 + 12.4^2 - 2 \times 13.5 \times 12.4 \times \cos 52°$

$\Rightarrow p^2 = 129.88...$

$\Rightarrow p = 11.39...$

$= 11.4 \text{ cm}$ (to 3 s.f.).

Hint: You know two sides and the angle between them so use the cosine rule to find the 3rd side.

Common mistake: Don't round until you get to your final answer.

Keep all other values stored in your calculator.

Test yourself

TESTED

1. In the triangle ABC, ∠CAB = 37°, ∠ABC = 56° and CB = 4 cm. Find the length of AC.

   A 3.24 cm   B 8.02 cm   C 5.51 cm   D 0.18 cm   E 2.90 cm

2. In the triangle XYZ, XY = 3.8 cm, YZ = 4.5 cm and ∠YZX = 40°. Three of the following statements are false and one is true. Which one is true?

   A A possible value for the area of the triangle is exactly 8.55 cm^2.
   B The only possible value of ∠XYZ is 90° to the nearest degree.
   C You can find the remaining side and angles of the triangle using only the cosine rule.
   D The possible values of ∠YXZ are 50° and 130° (to the nearest degree).

3. In the triangle MNP, MN = 5.4 cm, NP = 6 cm and MP = 7 cm. Find angle MNP correct to 3 s.f.

   A 48.3°   B 75.6°   C 56.1°   D 1.32°

4. For the triangle given below three of the statements are true and one is false. Which one is false?

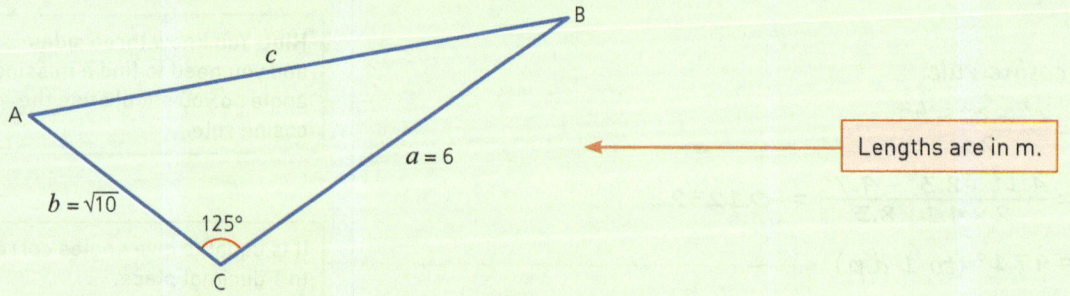

Lengths are in m.

   A The area of the triangle is 7.77 m^2 (to 3 s.f.).
   B AB is 8.23 m (to 3 s.f.).
   C Using only the sine rule you can find the value of c.
   D ∠B = 18.34° (to 2 d.p.).

5. At 12 noon a ship is at a point M which is on a bearing of 148° from a lighthouse, L. The ship travels due East at 20 km per hour and at 1230 hours it is at point N, on a bearing of 127° from the lighthouse. Three of the following statements are false and one is true. Which one is true?

   A At 1230 hours the ship is 17 km from the lighthouse.
   B The area LMN is 100 km^2, to the nearest whole number.
   C At noon the ship is 16.8 km from L to the nearest km.
   D If the ship continues at the same speed and on the same course, it will be on a bearing of 106° from the lighthouse at 1300 hours.

Full worked solutions online   **CHECKED ANSWERS**

Exam-style question

In a quadrilateral ABCD, AD = 14.7 cm, DC = 12.2 cm, ∠ADC = 45°, ∠ABC = 125° and ∠BAC = 27°.
i Find the length of AC.
ii Find the angle CAD.
iii Find the length of AB.
iv Find the area of the quadrilateral.

Short answers on page 218

Full worked solutions online   **CHECKED ANSWERS**

Radians and circular measure

REVISED

Key facts

1. An angle of 1 **radian** at the centre of a circle subtends an arc equal in length to the radius of the circle.

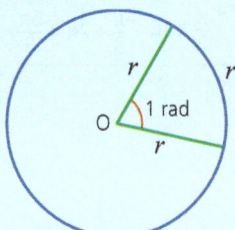

Since the circumference of a circle is $2\pi r$, a complete turn is 2π radians.

2. You need to know these **common values**.

θ	0	$30° = \frac{\pi}{6}$	$45° = \frac{\pi}{4}$	$60° = \frac{\pi}{3}$	$90° = \frac{\pi}{2}$	$180° = \pi$	$360° = 2\pi$
$\sin\theta$	0	$\frac{1}{2}$	$\frac{\sqrt{2}}{2}$	$\frac{\sqrt{3}}{2}$	1	0	0
$\cos\theta$	1	$\frac{\sqrt{3}}{2}$	$\frac{\sqrt{2}}{2}$	$\frac{1}{2}$	0	-1	1
$\tan\theta$	0	$\frac{1}{\sqrt{3}}$	1	$\sqrt{3}$	Undefined	0	0

3. To **convert** degrees into radians multiply by $\frac{\pi}{180°}$.

   To convert radians into degrees multiply by $\frac{180°}{\pi}$.

4. **Arcs:**

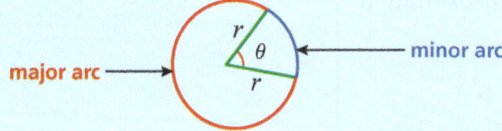

Arc length, s, is given by $s = r\theta$ where θ is in radians.

> The fraction of the whole circle is $\frac{\theta}{2\pi}$.
> So arc length $= \frac{\theta}{2\pi} \times 2\pi r = r\theta$

5. **Sectors:**

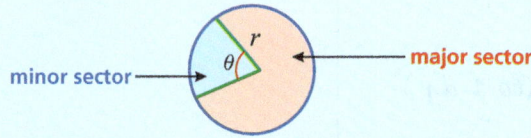

Area of the sector $= \frac{1}{2}r^2\theta$ where θ is in radians.

> Area $= \frac{\theta}{2\pi} \times \pi r^2 = \frac{1}{2}r^2\theta$

6. You can **solve trigonometric equations** using radians in a similar way to those in degrees.
   - To solve $\sin\theta = k$
     - **Step 1:** Use your calculator in radians mode to find the first solution: $\theta_1 = \arcsin k$.
     - **Step 2:** The second solution is $\theta_2 = \pi - \theta_1$.
     - **Step 3:** Add or subtract 2π from θ_1 and θ_2 to find all the solutions.

- To solve $\cos\theta = k$
  Step 1: Use your calculator in radians mode to find the first solution: $\theta_1 = \arccos k$.
  Step 2: The second solution is $\theta_2 = -\theta_1$.
  Step 3: Add or subtract 2π from θ_1 and θ_2 to find all the solutions.
- To solve $\tan\theta = k$
  Step 1: Use your calculator in radians mode to find the first solution: $\theta_1 = \arctan k$.
  Step 2: Add or subtract π from θ_1 to find all the solutions.

For example, solve $\cos\theta = 0.5$ for $0 \leqslant \theta \leqslant 2\pi$
Step 1: $\theta_1 = \arccos(0.5) = \dfrac{\pi}{3}$.
Step 2: $\theta_2 = -\dfrac{\pi}{3}$.
Step 3: Solutions are:
$\dfrac{\pi}{3}$ and $-\dfrac{\pi}{3} + 2\pi = \dfrac{5\pi}{3}$.

Worked examples

1 Converting degrees to radians

Express the following angles in radians, leaving your answers in terms of π where appropriate.

   i 23°   ii 120°   iii 150°   iv 207°

Solution

To convert degrees into radians multiply by $\dfrac{\pi}{180}$.

i $23° \times \dfrac{\pi}{180°} = 0.4014\ldots \Rightarrow 23° = 0.401$ radians (to 3 s.f.)

ii $120° \times \dfrac{\pi}{180°} = 2 \times \dfrac{\pi}{3} \Rightarrow 120° = \dfrac{2\pi}{3}$ radians

iii $150° \times \dfrac{\pi}{180°} = 5 \times \dfrac{\pi}{6} \Rightarrow 150° = \dfrac{5\pi}{6}$

iv $207° \times \dfrac{\pi}{180°} = 3.612\ldots \Rightarrow 207° = 3.61$ radians (to 3.s.f).

$1° = \dfrac{\pi}{180}$ radians.

You cannot cancel down 23 and 180, so use your calculator.

2 Converting radians to degrees

Express the following angles in degrees, using suitable approximation where necessary.

   i $\dfrac{\pi}{9}$ ii 3 rad   iii 5π

Solution

To convert radians into degrees multiply by $\dfrac{180}{\pi}$.

So

i $\dfrac{\pi}{9} \times \dfrac{180°}{\pi} = 20° \Rightarrow \dfrac{\pi}{9} = 20°$

ii $3 \times \dfrac{180°}{\pi} = 171.887\ldots° \Rightarrow 3\text{ rad} = 171.9°$ (to 1 d.p.)

iii $5\pi \times \dfrac{180°}{\pi} = 900° \Rightarrow 5\pi = 900°$.

1 radian $= \dfrac{180°}{\pi}$.

3 Using exact values in radians

Without using the calculator evaluate the following.

   i $\sin\dfrac{\pi}{4}$ ii $\tan\dfrac{\pi}{4}$ iii $\cos\dfrac{\pi}{3}$ iv $\sin\dfrac{2\pi}{3}$

Solution

i $\sin\dfrac{\pi}{4} = \dfrac{\sqrt{2}}{2}$ ii $\tan\dfrac{\pi}{4} = 1$

iii $\cos\dfrac{\pi}{3} = 0.5$ iv $\sin\dfrac{2\pi}{3} = \dfrac{\sqrt{3}}{2}$

Hint: You may find it helps to 'think' in degrees:
$\dfrac{\pi}{4} = 45°, \dfrac{\pi}{3} = 60°, \dfrac{2\pi}{3} = 120°$.

4 Solving equations in radians

Solve the following equations for $-\pi < \theta < \pi$.

i $\cos\theta = 0.2$ **ii** $\tan^2\theta = 3$

Hint: Notice the use of π here. It tells you to work in radians.

Solution

i $\cos\theta = 0.2$
$\theta = \arccos(0.2)$
$\theta = 1.369\ldots$
$\theta = 1.37$ or $\theta = -1.37$ (to 3 s.f.)

Be sure that your calculator is in radians mode.

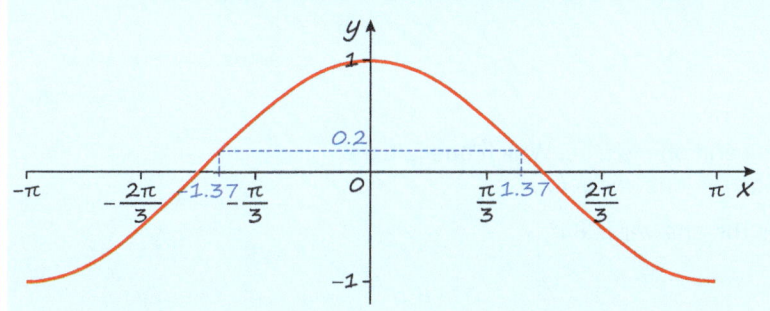

ii $\tan^2\theta = 3$
$\tan\theta = \pm\sqrt{3}$
$\tan\theta = \sqrt{3}$ or $\tan\theta = -\sqrt{3}$
$\theta = \arctan(\sqrt{3})$ or $\theta = \arctan(-\sqrt{3})$

Don't forget the negative square root.

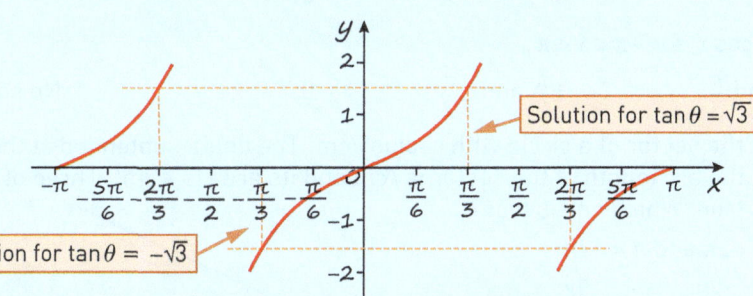

So $\theta = \dfrac{\pi}{3}$ or $\theta = -\dfrac{\pi}{3}$

or $\theta = \dfrac{\pi}{3} - \pi = -\dfrac{2\pi}{3}$ or $\theta = -\dfrac{\pi}{3} + \pi = \dfrac{2\pi}{3}$.

You should recognize these exact values!

Add and subtract π to find all other solutions.

5 Find area and perimeter of a sector

For the given sector, calculate

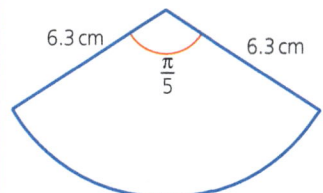

i the arc length
ii the perimeter of the sector
iii the area of the sector.

Solution

i Arc length $= r\theta$
$= 6.3 \times \dfrac{\pi}{5}$
$= 3.958\ldots\text{cm}$
$= 3.96\,\text{cm}$ (to 3 s.f.).

Keep the result without rounding it off. You will need it to calculate the perimeter.

ii Perimeter = 3.958... + 6.3 + 6.3
= 16.558...
= 16.6 cm (to 3 s.f.).

iii Area = $\frac{1}{2}r^2\theta$
= $\frac{1}{2} \times (6.3)^2 \times \frac{\pi}{5}$
= 12.46898124
= 12.5 cm² (to 3 s.f.).

Test yourself

TESTED

1. Four of the statements below are false and one is true. Which one is true?
   A When you convert 540° into radians the answer is 6π.
   B When you convert $\frac{7\pi}{15}$ into degrees the answer is 84°.
   C $\left(\tan\frac{\pi}{4} - \cos\frac{\pi}{2}\right) = \left(\cos\frac{\pi}{2} - \tan\frac{\pi}{4}\right)$.
   D $\cos\frac{5\pi}{6}$ is the same as $\cos 150°$ and the result is a positive number.
   E $\sin(\pi + \theta) = \sin\theta$ is true for all values of θ.

2. How many solutions are there for the equation $\sin\theta = \frac{1}{\sqrt{2}}$ for $-2\pi \leq \theta \leq 2\pi$?
   A 2   B 3   C 6   D 4

3. Solve the equation $2 - 2\sin^2 x = \cos x$, for $-\pi < x < \pi$.
   A $-\frac{\pi}{2}$ and $\frac{\pi}{2}$ B $-\frac{\pi}{3}$ and $\frac{\pi}{3}$ C $\pm\frac{\pi}{3}$ and $\pm\frac{\pi}{2}$ D 0   E No solution.

4. Here are four statements about the sector of a circle with radius r cm. The angle subtended at the centre of the circle is θ radians, the arc length of the sector is l cm and its area is A cm². Three of the statements are false and one is true. Which one is true?
   A When $r = 6$ and $\theta = \frac{\pi}{3}$ then $l = 2\pi$ and $A = 12\pi$.
   B When $r = 9$ and $l = \frac{\pi}{3}$ then $\theta = \frac{20}{3}$ and $A = \frac{3\pi}{2}$.
   C When $\theta = 0.6$ and $l = 2.4$ then $r = 4$ and $A = 4.8$.
   D When $r = 2\sqrt{2}$ and $l = 2$ then the area $A = 2\sqrt{2}$ and $\theta = \sqrt{2}$.

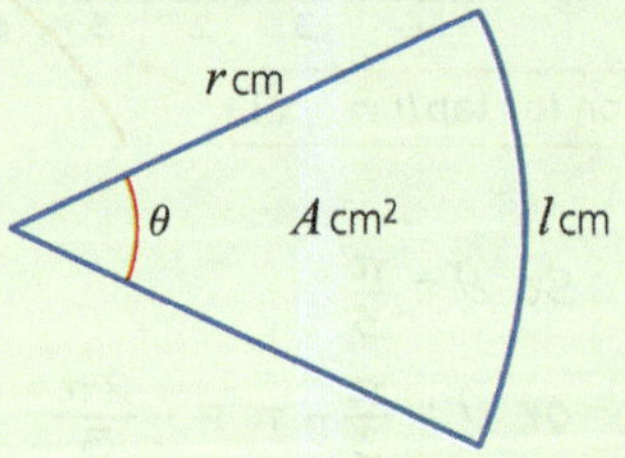

5. What is the period of the function $f(x) = \tan 3x$?
   A $\frac{1}{3}$ B 3π C 3   D $\frac{2}{3}\pi$ E $\frac{1}{3}\pi$

Full worked solutions online   CHECKED ANSWERS

Exam-style question

The figure on the right shows a circle with centre O and radius 12.6 cm. ST and RT are tangents to the circle and the angle SOR is 1.82 radians.
i Show that ST = 16.2 cm to 3 significant figures.
ii Find the area and perimeter of the shaded shape, SRT.

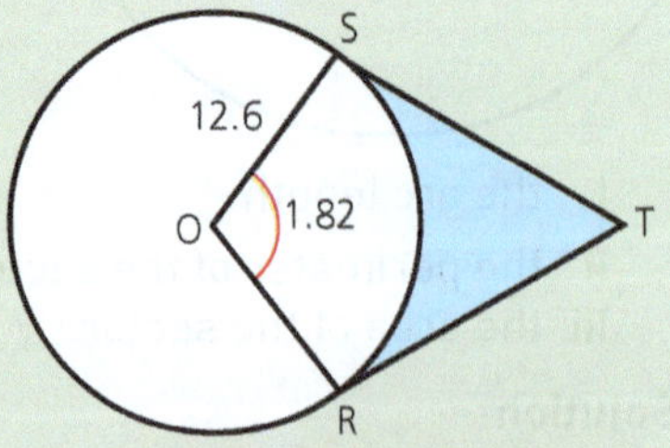

Short answers on page 218
Full worked solutions online   CHECKED ANSWERS

Reciprocal trig functions and small angle approximations

REVISED

Key facts

1 Definitions

$$\text{cosec}\,\theta = \frac{\text{hypotenuse}}{\text{opposite}}$$

$$\sec\theta = \frac{\text{hypotenuse}}{\text{adjacent}}$$

$$\cot\theta = \frac{\text{adjacent}}{\text{opposite}}$$

- $\sec\theta = \dfrac{1}{\cos\theta}$, $\theta \neq (180n + 90)°$ (in radians $\theta \neq \left(n + \tfrac{1}{2}\right)\pi$)
- $\text{cosec}\,\theta = \dfrac{1}{\sin\theta}$, $\theta \neq 180n°$ (in radians $\theta \neq n\pi$)
- $\cot\theta = \dfrac{1}{\tan\theta}\left(=\dfrac{\cos\theta}{\sin\theta}\right)$, $\theta \neq 180n°$ (in radians $\theta \neq n\pi$)

2 You need to know the graphs of the **reciprocal trigonometric functions**.

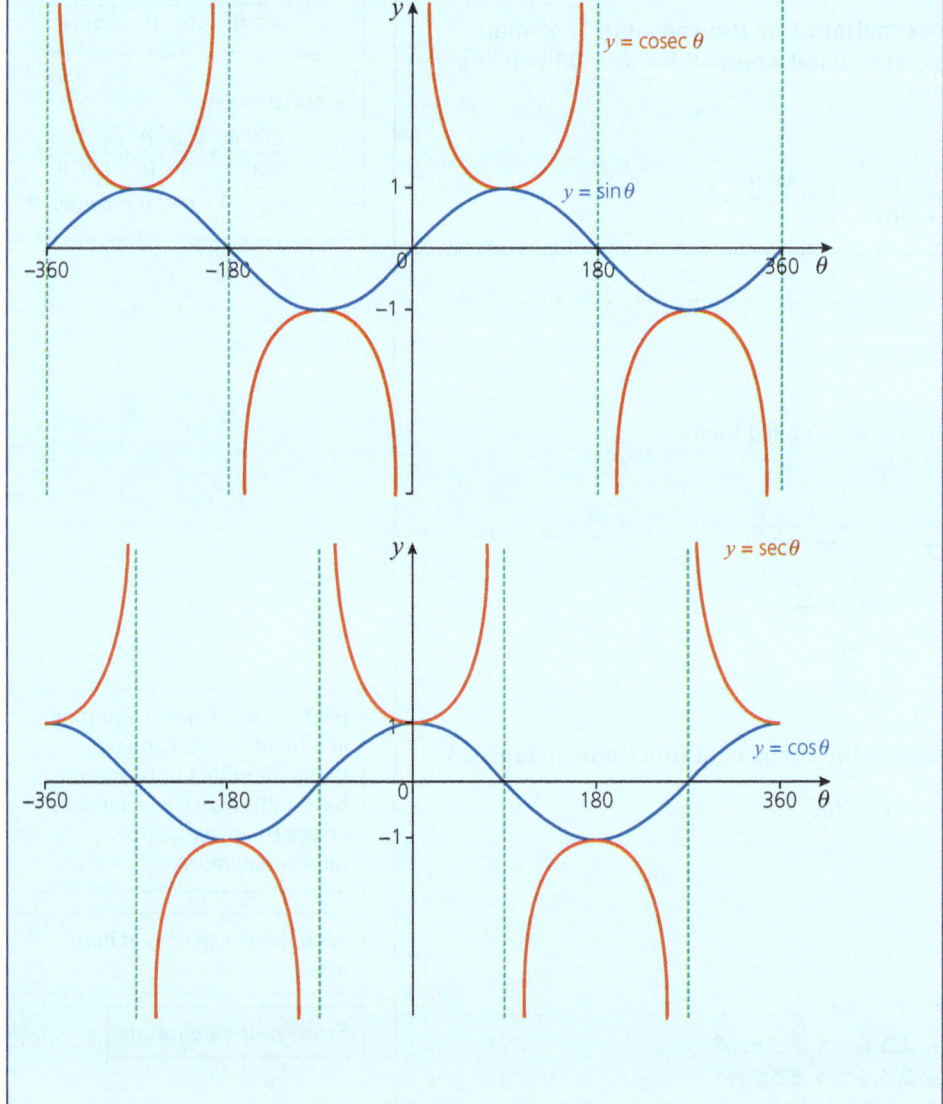

Hint: You can use the 3rd letter of se**c**, co**s**ec and co**t** to remind you which trigonometric function it is the reciprocal of.

Chapter 6 Trigonometry

Edexcel A Level Mathematics (Pure)

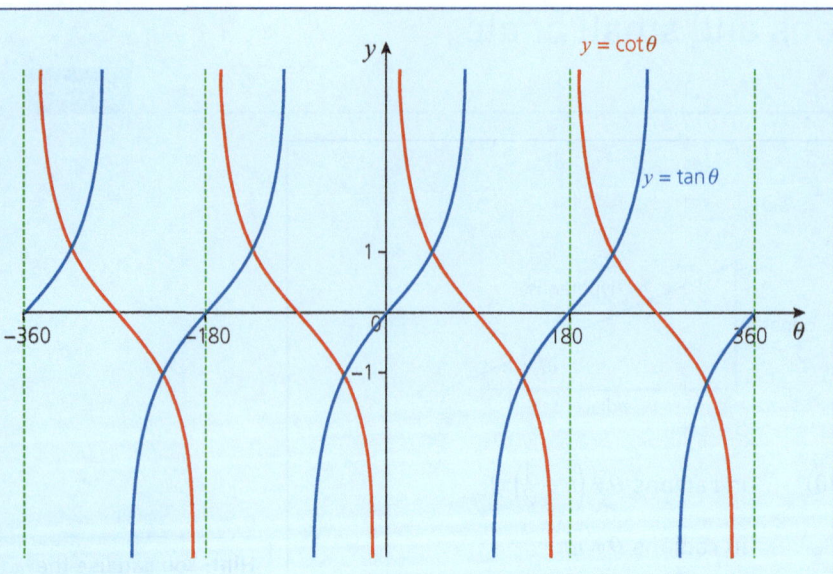

3. **Identities**
   - $\sin^2\theta + \cos^2\theta \equiv 1$
   - $\tan^2\theta + 1 \equiv \sec^2\theta$
   - $1 + \cot^2\theta \equiv \csc^2\theta$

4. The **small angle approximations** (for θ in radians). The small angle approximations can be used when $-0.1 \leq \theta \leq 0.1$ radians.
   - $\sin\theta \approx \theta$
   - $\tan\theta \approx \theta$
   - $\cos\theta \approx 1 - \dfrac{\theta^2}{2}$

   So you can say that $\lim\limits_{\theta \to 0}\dfrac{\theta}{\sin\theta} = \lim\limits_{\theta \to 0}\dfrac{\sin\theta}{\theta} = 1$.

Hint: These are straightforward to derive if you can't remember them.
Divide $\sin^2\theta + \cos^2\theta \equiv 1$ through by
- $\cos^2\theta$ to get:
$$\dfrac{\sin^2\theta}{\cos^2\theta} + \dfrac{\cos^2\theta}{\cos^2\theta} \equiv \dfrac{1}{\cos^2\theta}$$
$$\Rightarrow \tan^2\theta + 1 \equiv \sec^2\theta$$

- $\sin^2\theta$ to get:
$$\dfrac{\sin^2\theta}{\sin^2\theta} + \dfrac{\cos^2\theta}{\sin^2\theta} \equiv \dfrac{1}{\sin^2\theta}$$
$$\Rightarrow 1 + \cot^2\theta \equiv \csc^2\theta$$

Worked example

1 Using exact values

Find $\sec 210°$ leaving your answer in surd form.

Solution

$$\sec 210° = \dfrac{1}{\cos 210°} = \dfrac{1}{-\cos 30°}$$

$$= 1 \div \dfrac{-\sqrt{3}}{2}$$

$$= \dfrac{-2}{\sqrt{3}}$$

2 Solving equations involving the reciprocal functions in degrees

Solve $\csc x = -2.5$ for $0° \leq x \leq 360°$

Solution

i $\csc x = -2.5$
$\Rightarrow \dfrac{1}{\sin x} = -2.5$
$\Rightarrow \sin x = -0.4$
$\Rightarrow x = -23.6°$
$\Rightarrow x = 180° + 23.6° = 203.6°$
or $x = 360° - 23.6° = 336.4°$
So $x = 203.6°$ or $336.4°$

Hint: To solve a trig equation, you need to rearrange it using identities until you have either $\sin\theta = $ a number or $\cos\theta = $ a number or $\tan\theta = $ a number.

Take the reciprocal of both sides.

From your calculator.

3 Solving equations involving the reciprocal functions in radians

Solve $\csc^2 x = \cot x + 3$ for $0 \leq x \leq 2\pi$.

Solution

$$\csc^2 x = \cot x + 3$$
$$\Rightarrow \quad 1 + \cot^2 x = \cot x + 3$$ — Use the trig identity $1 + \cot^2 x \equiv \csc^2 x$.
$$\Rightarrow \quad \cot^2 x - \cot x - 2 = 0$$
$$\Rightarrow \quad (\cot x - 2)(\cot x + 1) = 0$$ — Factorising.
$$\Rightarrow \quad \cot x = 2 \text{ or } \cot x = -1$$ — Remember that $\cot x = \dfrac{1}{\tan x}$.
$$\Rightarrow \quad \tan x = \tfrac{1}{2} \text{ or } \tan x = -1$$

$\tan x = \tfrac{1}{2} \Rightarrow x = 0.4636... = 0.464$ (to 3 s.f.) — Don't forget to have your calculator in radians mode.
or $x = 0.4636... + \pi = 3.605... = 3.61$ (to 3 s.f.)

$\tan x = -1 \Rightarrow x = -\tfrac{\pi}{4}$ (not in range) — This is one of the exact values, so give the answers in terms of π.
or $\quad x = -\tfrac{\pi}{4} + \pi = \tfrac{3\pi}{4}$
or $\quad x = -\tfrac{\pi}{4} + 2\pi = \tfrac{7\pi}{4}$

so $\quad x = 0.464, \tfrac{3\pi}{4}, 3.61 \text{ or } \tfrac{7\pi}{4}$

4 Proving an identity

Prove that $\dfrac{1}{1-\cos\theta} + \dfrac{1}{1+\cos\theta} \equiv 2\csc^2\theta$.

Hint: Work with the more complicated side of the identity and show it simplifies to give the other side.

Solution

$$\dfrac{1}{1-\cos\theta} + \dfrac{1}{1+\cos\theta} = \dfrac{(1+\cos\theta)}{(1-\cos\theta)(1+\cos\theta)} + \dfrac{(1-\cos\theta)}{(1+\cos\theta)(1-\cos\theta)}$$

The common denominator is $(1-\cos\theta)(1+\cos\theta)$.

$$= \dfrac{(1+\cos\theta)+(1-\cos\theta)}{(1-\cos\theta)(1+\cos\theta)}$$

Remember the difference of two squares: $(a+b)(a-b) = a^2 - b^2$.

$$= \dfrac{2}{1-\cos^2\theta}$$

Using the trig identity $\sin^2\theta + \cos^2\theta \equiv 1 \Rightarrow \sin^2\theta \equiv 1 - \cos^2\theta$.

$$= \dfrac{2}{\sin^2\theta}$$

$$= 2\csc^2\theta$$

Remember $\dfrac{1}{\sin\theta} = \csc\theta$.

5 Using the small angle approximations

If θ is a small angle given in radians:
  i find an approximate expression for $\dfrac{2 - 2\cos\theta}{\sin\theta + \tan\theta}$
  ii when $\theta = 0.1$, calculate the percentage error in using the approximate expression.

Solution

i Using the small angle approximations $\sin\theta \approx \theta$, $\tan\theta \approx \theta$ and $\cos\theta \approx 1 - \dfrac{\theta^2}{2}$

$$\dfrac{2 - 2\cos\theta}{\sin\theta + \tan\theta} \approx \dfrac{2 - 2 + \theta^2}{\theta + \theta}$$

$$\approx \dfrac{\theta^2}{2\theta}$$

So $\dfrac{2 - 2\cos\theta}{\sin\theta + \tan\theta} \approx \dfrac{\theta}{2}$.

Edexcel A Level Mathematics (Pure)

ii When $\theta = 0.1$,
$\frac{2 - 2\cos\theta}{\sin\theta + \tan\theta} = 0.049916394...$ and $\frac{\theta}{2} = 0.05$
so the error is $0.05 - 0.049916394... = 0.0000836...$
$\Rightarrow$ the percentage error is $\frac{0.0000836...}{0.049916...} \times 100\%$
$= 0.167\%$ (to 3 s.f.)

Test yourself

TESTED

1. Find, without using a calculator, the exact value of cosec 240°.

   A -2
   B $-\frac{2}{\sqrt{3}}$
   C $\frac{2}{\sqrt{3}}$
   D $-\frac{\sqrt{3}}{2}$

2. What are the solutions of the equation $\tan^2\theta = \sec\theta + 5$ in the range $0 \leqslant \theta \leqslant 2\pi$?

   A 1.23, 2.09
   B 70.5°, 120°, 240°, 289.5°
   C 1.17, 2.27, 4.02, 5.11
   D 1.23, 2.09, 4.19, 5.05

3. Which of the following statements are true?
   Statement P: $\sec\theta$, $\csc\theta$ and $\cot\theta$ all have the same sign for $0 < \theta < \pi$.
   Statement Q: As $1 + \cot^2\theta \equiv \csc^2\theta$ then $\csc\theta > \cot\theta$.
   Statement R: $f(x) = \sec x$ is symmetrical about the y-axis.
   Statement S: $\lim_{x \to 0} \frac{\tan x}{x} = 1$.

   A Statements Q, R and S
   B Statements P, Q and R
   C Statements P, Q and S
   D Statements R and S

4. Find the value of $\lim_{x \to 0} \frac{\cos(2x) - 1}{\sin^2 x + \tan^2 x}$.

   A -1
   B $-\frac{1}{2}$
   C 0
   D ∞

Full worked solutions online CHECKED ANSWERS

Exam-style question

The function f is defined by $f(\theta) = \frac{4\csc\theta - 2}{3\csc\theta + 2}$ for $0 < \theta < 2\pi$, $\theta \neq \pi$.

i a Solve the equation $f(\theta) = 1$.
   b Show $f(\theta) = -1$ has no solution.

ii Prove that, where it is defined, $f(\theta)$ can be written as $\frac{4 - 2\sin\theta}{3 + 2\sin\theta}$.

Hence verify that $f(\theta)$ can also be written as $-1 + \frac{7}{3 + 2\sin\theta}$.

The graph shows the curve $y = -1 + \frac{7}{3 + 2\sin\theta}$ for $0 < \theta < 2\pi$.

P is the minimum point and Q is the maximum point.

iii Find the coordinates of P and Q and hence write down the range of $f(\theta)$.

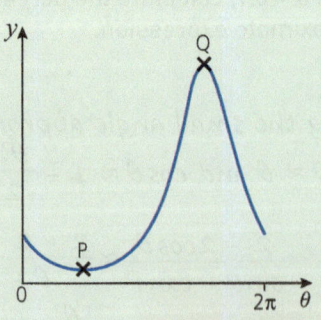

Short answers on page 218

Full worked solutions online CHECKED ANSWERS

Compound angle formulae

REVISED

Key facts

1 Compound angle formulae
- $\sin(A+B) = \sin A \cos B + \cos A \sin B$
- $\sin(A-B) = \sin A \cos B - \cos A \sin B$
- $\cos(A+B) = \cos A \cos B - \sin A \sin B$
- $\cos(A-B) = \cos A \cos B + \sin A \sin B$
- $\tan(A+B) = \dfrac{\tan A + \tan B}{1 - \tan A \tan B}$ $A+B \neq (k+\tfrac{1}{2}\pi)\ldots$ or $A+B \neq 90°, 270°, \ldots$
- $\tan(A-B) = \dfrac{\tan A - \tan B}{1 + \tan A \tan B}$ $A-B \neq (k+\tfrac{1}{2}\pi)\ldots$ or $A-B \neq 90°, 270°, \ldots$

Hint: These formulae are often given using θ and ϕ instead of A and B. You will see A and B on the formula sheet.

2 Double angle formulae
- $\sin 2A = 2 \sin A \cos A$
- $\cos 2A = \cos^2 A - \sin^2 A$
  $= 1 - 2\sin^2 A$
  $= 2\cos^2 A - 1$
- $\tan 2A = \dfrac{2\tan A}{1 - \tan^2 A}$ $A \neq 45°, 135°, \ldots$

Hint: You can find these by replacing B with A in the expansions above:
$\sin(A+A) = \sin A \cos A + \cos A \sin A$
$\Rightarrow \sin(2A) = 2\sin A \cos A$
$\tan(A+A) = \dfrac{\tan A + \tan A}{1 - \tan A \tan A}$
$\Rightarrow \tan 2A = \dfrac{2\tan A}{1 - \tan^2 A}$

Worked examples

1 Using the compound angle formulae to find exact values (1)

Use the compound angle formulae to calculate the **exact** value of $\cos 105°$.

Notice the word '**exact**' – leave your answer as a surd or a fraction.

Solution

$\cos 105° = \cos(60° + 45°)$

Using the compound angle formula
$\cos(A+B) = \cos A \cos B - \sin A \sin B$ with $A = 60°$ and $B = 45°$

$\cos 105° = \cos 60° \cos 45° - \sin 60° \sin 45°$

$\cos 105° = \dfrac{1}{2} \times \dfrac{\sqrt{2}}{2} - \dfrac{\sqrt{3}}{2} \times \dfrac{\sqrt{2}}{2}$

$= \dfrac{\sqrt{2} - \sqrt{6}}{4}$

Remember the triangles

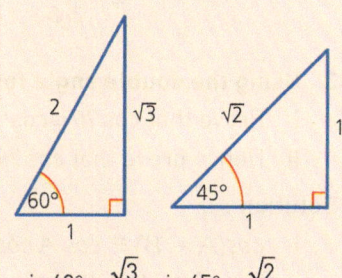

$\sin 60° = \dfrac{\sqrt{3}}{2}$ $\sin 45° = \dfrac{\sqrt{2}}{2}$
$\cos 60° = \dfrac{1}{2}$ $\cos 45° = \dfrac{\sqrt{2}}{2}$

2 Using the compound angle formulae to find exact values (2)

You are given that $\sin x = \dfrac{15}{17}$ and $\cos y = \dfrac{3}{5}$, where x and y are both acute angles. Find the exact values of **i** $\sin(x-y)$ and **ii** $\tan(x+y)$.

Solution

Using $\sin^2 x + \cos^2 x = 1$, $\left(\dfrac{15}{17}\right)^2 + \cos^2 x = 1$

$\Rightarrow \cos^2 x = 1 - \dfrac{225}{289} = \dfrac{64}{289}$

$\Rightarrow \cos x = \pm\dfrac{8}{17}$

As x and y are acute, x and y must be in the first quadrant. So $\sin x$, $\cos x$, $\tan x$, $\sin y$, $\cos y$ and $\tan y$ must all be positive.

Start by finding the values of $\cos x$ and $\tan x$.

$\cos \theta$, $\tan \theta$ and $\sin \theta$ are ALL positive when θ is in the 1st quadrant (acute).

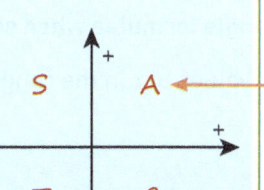

Edexcel A Level Mathematics (Pure)

So $\cos x = +\frac{8}{17}$

and $\tan x = \frac{\sin x}{\cos x} = \frac{\frac{15}{17}}{\frac{8}{17}} = \frac{15}{8}$

Similarly, using $\sin^2 y + \cos^2 y = 1$, $\sin^2 y + \left(\frac{3}{5}\right)^2 = 1$,

$\Rightarrow \sin^2 y = \frac{16}{25}$

$\Rightarrow \sin y = \pm \frac{4}{5}$

As y is an acute angle, $\sin y = +\frac{4}{5}$

and $\tan y = \frac{\sin y}{\cos y} = \frac{\frac{4}{5}}{\frac{3}{5}} = \frac{4}{3}$

i $\sin(x - y) = \sin x \cos y - \cos x \sin y$

$= \frac{15}{17} \times \frac{3}{5} - \frac{8}{17} \times \frac{4}{5}$

$= \frac{45 - 32}{85}$

$= \frac{13}{85}$

ii $\tan(x + y) = \frac{\tan x + \tan y}{1 - \tan x \tan y}$

$= \frac{\frac{15}{8} + \frac{4}{3}}{1 - \frac{15}{8} \times \frac{4}{3}}$

$= \frac{\frac{77}{24}}{-\frac{36}{24}}$

$= -\frac{77}{36}$

Hint: You can also find $\tan x$ and $\tan y$ by using right-angled trigonometry:

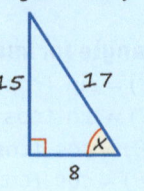
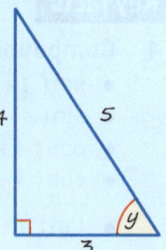

Take care when dividing by a fraction! Turn it 'upside down' and multiply.

$\frac{\frac{77}{24}}{-\frac{36}{24}} = \frac{77}{24} \times \left(-\frac{24}{36}\right) = -\frac{77}{36}$.

3 Using the double angle formulae

i Prove that $\cos 2\theta \equiv \cos^2 \theta - \sin^2 \theta$.

ii Hence prove that $\cos 2\theta \equiv 2\cos^2 \theta - 1$ and $\cos 2\theta \equiv 1 - 2\sin^2 \theta$.

Solution

i $\cos(A + B) \equiv \cos A \cos B - \sin A \sin B$

Writing $A = B = \theta$ gives

$\cos(\theta + \theta) \equiv \cos \theta \cos \theta - \sin \theta \sin \theta$

and so $\cos 2\theta \equiv \cos^2 \theta - \sin^2 \theta$

ii Since $\sin^2 \theta + \cos^2 \theta \equiv 1$, $\sin^2 \theta \equiv 1 - \cos^2 \theta$

and $\cos^2 \theta \equiv 1 - \sin^2 \theta$

So $\cos 2\theta \equiv \cos^2 \theta - \sin^2 \theta$ $\cos 2\theta \equiv \cos^2 \theta - \sin^2 \theta$

$\equiv \cos^2 \theta - (1 - \cos^2 \theta)$ $\equiv 1 - \sin^2 \theta - \sin^2 \theta$

$\equiv 2\cos^2 \theta - 1$ $\equiv 1 - 2\sin^2 \theta$

Hint: Notice that there are three common ways of writing $\cos 2\theta$ and you often need to choose the most appropriate form.

$\cos 2\theta \equiv \cos^2 \theta - \sin^2 \theta$,
$\cos 2\theta \equiv 1 - 2\sin^2 \theta$ or
$\cos 2\theta \equiv 2\cos^2 \theta - 1$

4 Using compound and double angle formulae when solving equations

Solve the following equations for values of x in the range $0° \leq x \leq 360°$.

i $\sin(x - 30°) = \cos(x + 45°)$

ii $3 \sin 2x = \cos x$

iii $\tan(3x) \tan x = 1$.

Solution

i $\sin(x - 30°) = \cos(x + 45°)$

$\sin x \cos 30° - \cos x \sin 30° = \cos x \cos 45° - \sin x \sin 45°$

$\sin x \times \dfrac{\sqrt{3}}{2} - \cos x \times \dfrac{1}{2} = \cos x \times \dfrac{\sqrt{2}}{2} - \sin x \times \dfrac{\sqrt{2}}{2}$

$\Rightarrow \sqrt{3} \sin x + \sqrt{2} \sin x = \cos x + \sqrt{2} \cos x$

$\Rightarrow (\sqrt{3} + \sqrt{2}) \sin x = (1 + \sqrt{2}) \cos x$

$\Rightarrow \tan x = \dfrac{1 + \sqrt{2}}{\sqrt{3} + \sqrt{2}} = 0.76732...$

$\Rightarrow x = 37.5°$ or $217.5°$

> $\sin(A - B) = \sin A \cos B - \cos A \sin B$
>
> **Hint:** Remember to use exact values.
>
> Multiplying through by 2.
>
> Using $\dfrac{\sin x}{\cos x} = \tan x$.
>
> Use your calculator to find the first solution and then add 180° to find the 2nd solution.

ii $3 \sin 2x = \cos x$

$\Rightarrow 6 \sin x \cos x = \cos x$

$\Rightarrow 6 \sin x \cos x - \cos x = 0$

$\Rightarrow \cos x (6 \sin x - 1) = 0$

$\Rightarrow \cos x = 0$ or $\sin x = \dfrac{1}{6}$

$\cos x = 0 \Rightarrow x = 90°$ or $270°$

$\sin x = \dfrac{1}{6} \Rightarrow x = 9.6°$ or $170.4°$

$\Rightarrow x = 9.6°, 90°, 170.4°$ or $270°$

> Using $\sin 2x = 2 \sin x \cos x$.
>
> Be careful! Don't cancel $\cos x$ as you will lose solutions.

iii $\tan 3x \tan x = 1$

$\Rightarrow \dfrac{\sin 3x}{\cos 3x} \times \dfrac{\sin x}{\cos x} = 1$

$\Rightarrow \sin 3x \sin x = \cos 3x \cos x$

$\Rightarrow 0 = \cos 3x \cos x - \sin 3x \sin x$

$\Rightarrow 0 = \cos(3x + x)$

$\Rightarrow 0 = \cos 4x$

As $0° \leq x \leq 360°$, $0° \leq 4x \leq 1440°$

$4x = 90°, 270°, 450°, 630°, 810°, 990°, 1170°, 1350°$

$\Rightarrow x = 22.5°, 67.5°, 112.5°, 157.5°, 202.5°, 247.5°, 292.5°, 337.5°$

> Notice that this is the form $\cos(A + B) = \cos A \cos B - \sin A \sin B$ with $A = 3x$ and $B = x$
>
> Find all the values of $4x$ first, and then divide each value by 4 to find the values of x.

5 Using half-angles

i Prove that $\cos^2\left(\dfrac{\theta}{2}\right) \equiv \dfrac{1}{2}(1 + \cos\theta)$.

ii By letting $\theta = \dfrac{\pi}{4}$, find the exact value of $\cos\dfrac{\pi}{8}$.

Solution

i $\cos 2A \equiv 2 \cos^2 A - 1$

Let $A = \dfrac{\theta}{2}$, so $\cos\left(2 \times \dfrac{\theta}{2}\right) \equiv 2 \cos^2\left(\dfrac{\theta}{2}\right) - 1$

$\Rightarrow \cos\theta \equiv 2 \cos^2\left(\dfrac{\theta}{2}\right) - 1$

$\Rightarrow 2 \cos^2\left(\dfrac{\theta}{2}\right) \equiv 1 + \cos\theta$

$\Rightarrow \cos^2\left(\dfrac{\theta}{2}\right) \equiv \dfrac{1}{2}(1 + \cos\theta)$

ii Replace θ with $\frac{\pi}{4}$: $\cos^2 \frac{\pi}{8} = \frac{1}{2}\left(1 + \cos\frac{\pi}{4}\right)$

$$\cos^2 \frac{\pi}{8} = \frac{1}{2}\left(1 + \frac{\sqrt{2}}{2}\right)$$

$$\cos^2 \frac{\pi}{8} = \frac{1}{2} + \frac{\sqrt{2}}{4}$$

$$= \frac{2 + \sqrt{2}}{4}$$

So $\cos \frac{\pi}{8} = \sqrt{\frac{2 + \sqrt{2}}{4}} = \frac{\sqrt{2 + \sqrt{2}}}{2}$

Note: you want the positive square root as $\cos \theta$ is positive when θ is less than 90°.

Test yourself

TESTED

1 You are given that $\tan\theta = \frac{24}{7}$ and θ is an acute angle.
Find the values of $\sin 2\theta$ and $\cos 2\theta$.

A $\sin 2\theta = \frac{48}{25}$, $\cos 2\theta = \frac{14}{25}$

B $\sin 2\theta = \frac{336}{625}$, $\cos 2\theta = -\frac{527}{625}$

C $\sin 2\theta = -\frac{527}{625}$, $\cos 2\theta = \frac{336}{625}$

D $\sin 2\theta = \frac{336}{625}$, $\cos 2\theta = \frac{289}{625}$

2 Solve the equation $\cos 2\theta = \cos \theta - 1$ for $0 \leq \theta \leq 2\pi$.

A 60°, 90°, 270°, 300°

B $\frac{\pi}{3}, \frac{\pi}{2}$

C $\frac{\pi}{3}, \frac{5\pi}{3}$

D $\frac{\pi}{3}, \frac{\pi}{2}, \frac{3\pi}{2}, \frac{5\pi}{3}$

3 Three of the following statements are false and one is true. Which one is true?

A $\frac{1 - \cos 2\theta}{\sin 2\theta} = \tan \theta$

B $\cos(A+B) - \cos(A-B) = 2\sin A \sin B$

C $\sin 3x = 3\sin x \cos x$

D $\tan\left(\theta + \frac{\pi}{4}\right) = \tan\theta + 1$

4 You are given that $\sin P = \frac{3}{5}$, where P is an obtuse angle and $\cos Q = \frac{5}{13}$, where Q is an acute angle.
Find the exact value of $\cos(P - Q)$.

A $-\frac{56}{65}$

B $-\frac{77}{65}$

C $\frac{16}{65}$

D $\frac{56}{65}$

Full worked solutions online

CHECKED ANSWERS

Exam-style question

In this question you must show detailed reasoning.

In triangle PQR, $PQ = a$, $PR = b$, $PS = h$ and the angle $QPR = A + B$.

i a Find, in terms of a, b, A and B, the area of triangle
   A PQR
   B PQS
   C PRS.

  b Hence prove that $\sin(A + B) = \sin A \cos B + \cos A \sin B$.

ii Solve $\sin(\theta + 60°) = \cos\theta$ for $0° \leq \theta \leq 360°$.

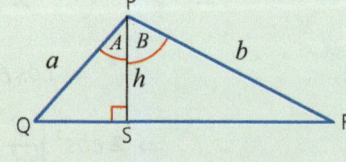

Short answers on page 218

Full worked solutions online

CHECKED ANSWERS

The forms $r\cos(\theta \pm \alpha)$ and $r\sin(\theta \pm \alpha)$

REVISED

Key facts

1 **The r, α formulae**
   - $a\sin\theta + b\cos\theta = r\sin(\theta + \alpha)$
   - $a\sin\theta - b\cos\theta = r\sin(\theta - \alpha)$
   - $a\cos\theta + b\sin\theta = r\cos(\theta - \alpha)$
   - $a\cos\theta - b\sin\theta = r\cos(\theta + \alpha)$

   where $r = \sqrt{a^2 + b^2}$, $\cos\alpha = \frac{a}{r}$ and $\sin\alpha = \frac{b}{r}$

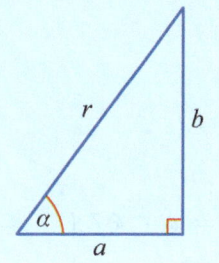

Hint: Notice that r is positive.

Worked examples

1 Choosing an appropriate form

Write $12\cos\theta + 5\sin\theta$ as a single term.

Solution

As the first term has a $\cos\theta$, the second term has a $\sin\theta$ and there is a 'plus' sign between them, choose the expression $r\cos(\theta - \alpha)$.

So $12\cos\theta + 5\sin\theta = r\cos(\theta - \alpha)$
$\qquad = r(\cos\theta\cos\alpha + \sin\theta\sin\alpha)$

Comparing the coefficients of $\cos\theta$: $12 = r\cos\alpha$

Comparing the coefficients of $\sin\theta$: $5 = r\sin\alpha$

Using the right-angled triangle, you can calculate the values of r and α.

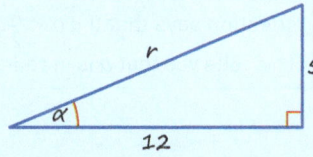

$r = \sqrt{5^2 + 12^2} = 13$

Hint: Remember that $\cos(A - B) = \cos A\cos B + \sin A\sin B$.

Using Pythagoras' theorem. Note that r is positive.

and $\sin\alpha = \frac{5}{13}$, $\cos\alpha = \frac{12}{13}$ $\Rightarrow$ $\alpha = 22.6°$

So a single term for $12\cos\theta + 5\sin\theta$ is $13\cos(\theta - 22.6°)$.

2 Using the form $r\sin(\theta + \alpha)$

Express $12\cos\theta + 5\sin\theta$ in the form $r\sin(\theta + \alpha)$, where $r > 0$ and $0 < \alpha < 90°$.

Solution

$12\cos\theta + 5\sin\theta = r\sin(\theta + \alpha)$
$\qquad = r(\sin\theta\cos\alpha + \cos\theta\sin\alpha)$

Comparing the coefficients of $\sin\theta$: $5 = r\cos\alpha$

Comparing the coefficients of $\cos\theta$: $12 = r\sin\alpha$

Remember that $\sin(A + B) = \sin A\cos B + \cos A\sin B$.

Edexcel A Level Mathematics (Pure)

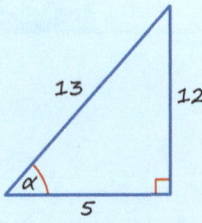

$r = \sqrt{5^2 + 12^2} = 13$
and $\sin\alpha = \frac{12}{13}$, $\cos\alpha = \frac{5}{13} \Rightarrow \alpha = 67.4°$
So $12\cos\theta + 5\sin\theta = 13\sin(\theta + 67.4°)$

Hint: Worked examples 1 and 2 both start with the expression $12\cos\theta + 5\sin\theta$. They show that an expression of the form $a\cos\theta \pm b\sin\theta$ can be written in the form $r\cos(\theta \pm \alpha)$ or $r\sin(\theta \pm \alpha)$. There is no single correct answer.

Using Pythagoras' theorem.

3 Using the form $r\sin(\theta - \alpha)$

i Express $\sin\theta - \cos\theta$ in the form $r\sin(\theta - \alpha)$, where $r > 0$ and $0 < \alpha < \frac{\pi}{2}$.

ii State the maximum and minimum values of $\sin\theta - \cos\theta$.

iii Sketch the graph of $y = \sin\theta - \cos\theta$ for $0 \leqslant \theta \leqslant 2\pi$.

iv Solve the equation $\sin\theta - \cos\theta = \frac{1}{\sqrt{2}}$ for $0 \leqslant \theta \leqslant 2\pi$.

Hint: You can think of $\sin\theta - \cos\theta$ as $1\sin\theta - 1\cos\theta$

Solution

i $\sin\theta - \cos\theta = r\sin(\theta - \alpha)$
$= r(\sin\theta\cos\alpha - \cos\theta\sin\alpha)$
Comparing the coefficients of $\sin\theta$: $1 = r\cos\alpha$
Comparing the coefficients of $\cos\theta$: $1 = r\sin\alpha$

Remember that $\sin(A - B) = \sin A \cos B - \cos A \sin B$

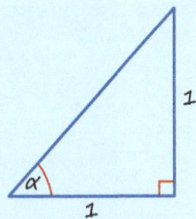

Hint: The triangle shows you that $r = \sqrt{1^2 + 1^2} = \sqrt{2}$ and the angle $\alpha = 45°$. However, the question says that $0 < \alpha < \frac{\pi}{2}$. This tells you that α is in radians.

So $r = \sqrt{1^2 + 1^2} = \sqrt{2}$ and $\alpha = \frac{\pi}{4}$, giving
$\sin\theta - \cos\theta = \sqrt{2}\sin\left(\theta - \frac{\pi}{4}\right)$

ii $\sin\theta - \cos\theta = \sqrt{2}\sin\left(\theta - \frac{\pi}{4}\right)$
As the sine function oscillates between 1 and -1, $\sqrt{2}\sin\left(\theta - \frac{\pi}{4}\right)$ will oscillate between $\sqrt{2}$ and $-\sqrt{2}$.
So the maximum value of $\sin\theta - \cos\theta$ is $\sqrt{2}$ and the minimum value of $\sin\theta - \cos\theta$ is $-\sqrt{2}$.

iii The graph of $y = \sqrt{2}\sin\left(\theta - \frac{\pi}{4}\right)$ in the range $0 \leqslant \theta \leqslant 2\pi$ can be obtained from the graph of $y = \sin\theta$ by a translation of $\begin{pmatrix}\frac{\pi}{4}\\0\end{pmatrix}$ and a one-way stretch, scale factor $\sqrt{2}$, parallel to the y-axis.

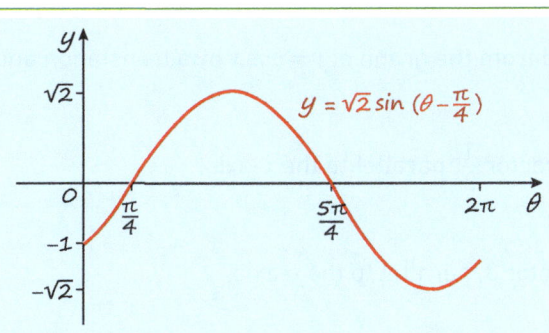

iv $\sin\theta - \cos\theta = \dfrac{1}{\sqrt{2}}$

$\Rightarrow \sqrt{2}\sin\left(\theta - \dfrac{\pi}{4}\right) = \dfrac{1}{\sqrt{2}}$

$\Rightarrow \sin\left(\theta - \dfrac{\pi}{4}\right) = \dfrac{1}{2}$

$\Rightarrow \left(\theta - \dfrac{\pi}{4}\right) = \dfrac{\pi}{6}$ or $\dfrac{5\pi}{6}$

$\Rightarrow \theta = \dfrac{5\pi}{12}$ or $\dfrac{13\pi}{12}$

So the roots of the equation $\sin\theta - \cos\theta = \dfrac{1}{\sqrt{2}}$ in the range $0 \leq \theta \leq 2\pi$ are $\dfrac{5\pi}{12}$ and $\dfrac{13\pi}{12}$.

Hint: You might find it easier to think in degrees:

$\dfrac{\pi}{6}$ radians = 30°, sin 30° = $\dfrac{1}{2}$,

$\dfrac{5\pi}{6}$ radians = 150°, sin 150° = $\dfrac{1}{2}$.

Test yourself

TESTED

1. What are the values of r and α when $4\cos\theta + 3\sin\theta$ is expressed in the form $r\cos(\theta - \alpha)$, where $r > 0$ and $0 < \alpha < \dfrac{\pi}{2}$?

   A $r = 5, \alpha = 53.1°$ B $r = 5, \alpha = 0.644$ C $r = 25, \alpha = 0.644$ D $r = 5, \alpha = 36.9°$

2. Solve the equation $3\sin 2\theta + 5\cos 2\theta = 4$ in the range $0° \leq \theta \leq 180°$. Give your answers correct to 1 decimal place.

   A 38.8°, 172.1°   B 77.6°   C 38.8°   D 6.2°, 52.9°

3. Find the maximum and minimum points of the function $f(\theta) = \sqrt{3}\cos\theta + \sin\theta$ in the range $0 < \theta < 2\pi$.

   A min $\left(\dfrac{\pi}{6}, 2\right)$, max $\left(\dfrac{7\pi}{6}, -2\right)$

   B min (210°, −2), max (30°, 2)

   C min $\left(\dfrac{7\pi}{6}, -2\right)$, max $\left(\dfrac{\pi}{6}, 2\right)$

   D min $\left(\dfrac{4\pi}{3}, -2\right)$, max $\left(\dfrac{\pi}{3}, 2\right)$

Edexcel A Level Mathematics (Pure)

4 The graph of $y = \frac{3\sqrt{3}}{2}\cos x - \frac{3}{2}\sin x$ can be obtained from the graph of $y = \cos x$ by a translation and a stretch. Describe the translation and the stretch.

A translation $\begin{pmatrix} -\frac{\pi}{6} \\ 0 \end{pmatrix}$ and one-way stretch, scale factor $\frac{1}{3}$, parallel to the x-axis.

B translation $\begin{pmatrix} \frac{\pi}{6} \\ 0 \end{pmatrix}$ and one-way stretch, scale factor 3, parallel to the y-axis.

C translation $\begin{pmatrix} \frac{\pi}{6} \\ 0 \end{pmatrix}$ and one-way stretch, scale factor $\frac{1}{3}$, parallel to the x-axis.

D translation $\begin{pmatrix} -\frac{\pi}{6} \\ 0 \end{pmatrix}$ and one-way stretch, scale factor 3, parallel to the y-axis.

Full worked solutions online   CHECKED ANSWERS

Exam-style question

The two parallel lines AB and CD are d cm apart.
PQRS is a rectangle with P on AB and R on CD.
PQ = 12 cm, QR = 5 cm and ∠BPQ = θ.
BQD is a straight line at right angles to the two parallel lines.

Diagram is not to scale

i Show that $d = 12\sin\theta + 5\cos\theta$.
ii Express $12\sin\theta + 5\cos\theta$ in the form $r\sin(\theta + \alpha)$, where $r > 0$ and $0° < \theta < 90°$.
iii Sketch the curve with equation $y = 12\sin\theta + 5\cos\theta$ for $0 \leq \theta \leq 360°$.
   Mark on your sketch the coordinates of
   a the points where the curve meets the axes
   b the maximum and minimum points.
iv a Find the distance between the parallel lines when the angle θ is 30°.
   b Find the value of the acute angle θ when the lines are 10 cm apart.

Short answers on page 218
Full worked solutions online   CHECKED ANSWERS

Chapter 7 Differentiation

About this topic

You can use differentiation to find the gradient of the tangent to a curve at any point. You can use this to find equations of tangents and normals to a curve. Differentiation is also used to give important information about the shape of a curve: you can find the coordinates of stationary points, and find where a function is increasing or decreasing.

In this topic, you will also revise how to differentiate more complicated functions such as exponentials and trigonometric functions using the product, quotient and chain rules.

Before you start, remember

- the laws of indices
- how to find the equation of a straight line from its gradient and a point on the line
- the rule that if two lines with gradients m_1 and m_2 are perpendicular, then $m_1 m_2 = -1$
- how to solve equations.

Tangents and normals

REVISED

Key facts

1. The **gradient function**, $\frac{dy}{dx}$ or $f'(x)$, gives the gradient of the curve $y = f(x)$, and measures the rate of change of y with respect to x.

   The gradient function is sometimes called the **derivative**.

   As well as telling you the gradient of a curve, the derivative $\frac{dy}{dx}$ tells you the rate of change of y with respect to x.

   The process of finding a derivative is called **differentiation**.

   Hint: For example, for the curve $y = x^2$, the gradient function is given by $\frac{dy}{dx} = 2x$. This means that at the point where $x = 1$, the gradient of the curve is 2; at the point where $x = 2$, the gradient of the curve is 4, and so on.

2. The gradient function of $y = kx^n$ is given by $\frac{dy}{dx} = knx^{n-1}$

   This is true for all values of n, including negative and fractional values.

   A constant is any number like 2, -0.2, $\frac{1}{2}$ or π.

3. When you differentiate a **constant term**, k, the result is 0.

   $y = k \Rightarrow \frac{dy}{dx} = 0$

   So when $y = \pi$ then $\frac{dy}{dx} = 0$.

4. You can find the gradient m_1, of a tangent to a curve at a given point by differentiating and then substituting the x coordinate into the derivative. The **equation of the tangent** to a curve at the point (x_1, y_1) is given by $y - y_1 = m_1(x - x_1)$.

   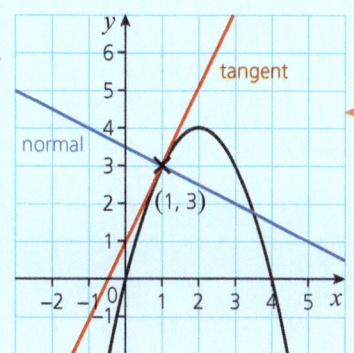

   The diagram shows the curve $y = 4x - x^2$ and the tangent and normal at the point $(1, 3)$.
   $\frac{dy}{dx} = 4 - 2x$ and when $x = 1$, the gradient m_1 of the tangent is 2
   So the equation of the tangent is
   $y - y_1 = m_1(x - x_1)$
   $y - 3 = 2(x - 1)$
   $y = 2x + 1$

Edexcel A Level Mathematics (Pure) 103

5. The **normal** to a curve at a given point is the straight line which is perpendicular to the tangent at that point.
   - The gradient, m_2, of a **normal** to a curve at a given point can be found by first finding the **gradient m_1** of the **tangent**, and then using the relationship $m_2 = -\dfrac{1}{m_1}$.
   - The **equation of the normal** to a curve at the point (x_1, y_1) is given by $y - y_1 = m_2(x - x_1)$.

Worked examples

1 Finding the derivative

Differentiate the function $y = 2x^3 + \sqrt[3]{x} - \dfrac{3}{x^4} - 5$.

Solution

$$y = 2x^3 + \sqrt[3]{x} - \dfrac{3}{x^4} - 5$$

$$= 2x^3 + x^{\frac{1}{3}} - 3x^{-4} - 5$$

$$\dfrac{dy}{dx} = 6x^2 + \dfrac{1}{3}x^{-\frac{2}{3}} - 3 \times (-4)x^{-5}$$

$$= 6x^2 + \dfrac{1}{3}x^{-\frac{2}{3}} + 12x^{-5}$$

$$= 6x^2 + \dfrac{1}{3x^{\frac{2}{3}}} + \dfrac{12}{x^5}$$

Hint: When differentiating a fractional or negative power of x, you use the standard result in the same way that you do for an integer power of x:
$y = kx^n \Rightarrow \dfrac{dy}{dx} = knx^{n-1}$
You can think of this as 'multiply by the power and reduce the power by 1'.

Rewrite the expression using fractional and negative indices.

Differentiate each term. Be careful with signs. Remember the derivative of a constant is 0.

You could write this expression using roots, but it is fine to leave it as it is.

2 Differentiation with other variables

i Differentiate $V = \dfrac{4}{3}\pi r^3$.

ii Given $f(t) = \dfrac{2t^4 - 3t^2}{t^3}$, find $f'(-2)$.

Solution

i $V = \dfrac{4}{3}\pi r^3 \Rightarrow \dfrac{dV}{dr} = \dfrac{4}{3}\pi \times 3r^2$

$= 4\pi r^2$

ii $f(t) = \dfrac{2t^4 - 3t^2}{t^3} = 2t - 3t^{-1}$

$\Rightarrow f'(t) = 2 - 3 \times (-1)t^{-2}$

$= 2 + \dfrac{3}{t^2}$

$f'(-2) = 2 + \dfrac{3}{(-2)^2} = 2.75$

Hint: When $y = f(x)$ you can use the notation $f'(x)$ instead of $\dfrac{dy}{dx}$. In this case you have $f(t)$ so you should write $f'(t)$.

You need the rate of change of V with respect to r.

Common mistake: You need to simplify before you can differentiate: $\dfrac{2t^4 - 3t^2}{t^3} = \dfrac{2t^4}{t^3} - \dfrac{3t^2}{t^3}$.

Substitute $t = -2$ into the expression for $f'(t)$.

3 Finding the gradient at a point

Find the gradient of $y = 3x^2 - \dfrac{16}{x^3} - 7\sqrt{x} - 3x + \dfrac{1}{4}$ at the point (4, 22).

Solution

$$y = 3x^2 - \dfrac{16}{x^3} - 7\sqrt{x} - 3x + \dfrac{1}{4}$$

$$= 3x^2 - 16x^{-3} - 7x^{\frac{1}{2}} - 3x + \dfrac{1}{4}$$ ← Rewrite the expression using fractional indices.

$$\dfrac{dy}{dx} = 6x + 48x^{-4} - \dfrac{7}{2}x^{-\frac{1}{2}} - 3$$

$$= 6x + \dfrac{48}{x^4} - \dfrac{7}{2\sqrt{x}} - 3$$ ← It is easier to substitute values of x into this form.

When $x = 4$, $\dfrac{dy}{dx} = 6 \times 4 + \dfrac{48}{4^4} - \dfrac{7}{2\sqrt{4}} - 3$

$$= 24 + \dfrac{48}{256} - \dfrac{7}{4} - 3$$

$$= \dfrac{311}{16}$$

4 Finding the coordinates of a point with a given gradient

Find the coordinates of the point(s) on the curve $y = 2x^3 + 5x^2 - 3$ at which the gradient of the curve is 4.

Solution

$$y = 2x^3 + 5x^2 - 3 \Rightarrow \dfrac{dy}{dx} = 6x^2 + 10x$$

At the point where the gradient is 4:

$$6x^2 + 10x = 4$$
$$6x^2 + 10x - 4 = 0$$
$$3x^2 + 5x - 2 = 0$$
$$(3x - 1)(x + 2) = 0$$
$$x = \dfrac{1}{3} \text{ or } x = -2$$

When $x = \dfrac{1}{3}$, $y = 2 \times \left(\dfrac{1}{3}\right)^3 + 5 \times \left(\dfrac{1}{3}\right)^2 - 3 = -\dfrac{64}{27}$. ← Substitute the x coordinates into the equation of the curve to find the y coordinates.

When $x = -2$, $y = 2 \times (-2)^3 + 5 \times (-2)^2 - 3 = 1$.

The points at which the gradient of the curve is 4 are $\left(\dfrac{1}{3}, -\dfrac{64}{27}\right)$ and $(-2, 1)$.

Common mistakes: Don't mix up the equation of the curve and the gradient function. If you want to find the y coordinate of the point on the curve, make sure you use the equation of the curve and not the gradient function!

5 Finding the equation of a tangent and a normal to a curve

Find the equations of the tangent and normal to the curve $y = \dfrac{16}{x^2} + 3\sqrt{x}$ at the point where $x = 4$.

Solution

When $x = 4$, $y = \dfrac{16}{4^2} + 3\sqrt{4} = \dfrac{16}{16} + 3 \times 2 = 7.$ ← **Step 1**: Find the y coordinate of the point where $x = 4$.

So the point is $(4, 7)$.

$y = \dfrac{16}{x^2} + 3\sqrt{x} \Rightarrow y = 16x^{-2} + 3x^{\frac{1}{2}}$ ← **Step 2**: Differentiate to find the gradient function. Make sure you rewrite the function using fractional and negative powers first.

$\Rightarrow \dfrac{dy}{dx} = -32x^{-3} + \dfrac{3}{2}x^{-\frac{1}{2}}$

$= -\dfrac{32}{x^3} + \dfrac{3}{2\sqrt{x}}$ ← **Hint**: It is easier to substitute in for x when you write the answer in this form.

When $x = 4$, the gradient m_1 of the tangent is

$m_1 = -\dfrac{32}{4^3} + \dfrac{3}{2\sqrt{4}} = -\dfrac{1}{2} + \dfrac{3}{4} = \dfrac{1}{4}$ ← **Step 3**: Substitute $x = 4$ to find the gradient of the tangent.

So the tangent has gradient $\dfrac{1}{4}$ and passes through the point $(4, 7)$.

The equation of the tangent is $y - y_1 = m_1(x - x_1)$ ← **Step 4**: Find the equation of the tangent.

$y - 7 = \dfrac{1}{4}(x - 4)$

$4y - 28 = x - 4$ ← **Hint**: It is easier to deal with the fraction by multiplying through by 4.

$4y - x = 24$

The gradient m_2 of the normal is $m_2 = -\dfrac{1}{m_1} = -4$ ← **Step 5**: Find the gradient of the normal.

The normal has gradient -4 and passes through the point $(4, 7)$

so the equation of the normal is $y - y_1 = m_2(x - x_1)$ ← **Step 6**: Find the equation of the normal.

$y - 7 = -4(x - 4)$

$y - 7 = -4x + 16$

$y = 23 - 4x$

Test yourself TESTED

1. Find the coordinates of the point on the curve $y = 4 - 3x + x^2$ at which the tangent to the curve has gradient -1.
   - A $(1, 2)$
   - B $(1, -1)$
   - C $(-2, 14)$
   - D $(-2, -1)$

2. Find the gradient of the curve $y = x^5(2x+1)$ at the point at which $x = -1$.
   - A 10
   - B -3
   - C -7
   - D -17
   - E 7

3. Find the gradient of the function $y = 5\sqrt{x} - \dfrac{4}{\sqrt{x}}$ at the point where $x = 4$.
   - A $\dfrac{19}{12}$
   - B 1
   - C $\dfrac{3}{2}$
   - D $-\dfrac{59}{4}$
   - E $\dfrac{9}{4}$

4. Find the coordinates of the point(s) at which the graph $y = x - \dfrac{1}{x}$ has gradient 5.
   - A $\left(\dfrac{1}{2}, -\dfrac{3}{2}\right)$
   - B $\left(2, \dfrac{3}{2}\right)$ and $\left(-2, -\dfrac{3}{2}\right)$
   - C $\left(2, \dfrac{3}{2}\right)$
   - D $\left(-\dfrac{1}{2}, \dfrac{3}{2}\right)$
   - E $\left(\dfrac{1}{2}, -\dfrac{3}{2}\right)$ and $\left(-\dfrac{1}{2}, \dfrac{3}{2}\right)$

5. Find the equation of the tangent to the curve $y = x^3 - 3x^2 + x + 4$ at the point where $x = 1$.
   - A $y + 2x = 5$
   - B $y + 2x = 1$
   - C $y = -2x$
   - D $y + 2x = 7$
   - E $y + 2x = 3$

6. Find the equation of the normal to the curve $y = x^2 + 7x + 6$ at the point where $x = -2$.
   - A $y = 3x + 2$
   - B $3y + 10 = x$
   - C $3y + x + 38 = 0$
   - D $y + 6 = 3x$
   - E $3y + x + 14 = 0$

7 Find the equation of the tangent to the curve $y = \sqrt{x}$ at the point where $x = 4$.
   A $4y = x - 8$ B $y + x = 6$ C $4y = x + 8$ D $4y = x - 2$ E $4y = x + 4$

8 Find the equation of the normal to the curve $y = \dfrac{1}{x}$ at the point where $x = 2$.
   A $y + x = 4$ B $2y = 8x - 15$ C $2y + 8x = 17$ D $4y = x$ E $2y = 8x - 17$

Full worked solutions online   CHECKED ANSWERS

Exam-style question

The diagram shows the cubic curve with equation $y = 2x^3 + 3x^2 - 3$.

i Show that the tangent to the curve at the point P(1, 2) has gradient 12.
ii Find the coordinates of the other point, Q, on the curve at which the tangent has gradient 12.
iii Find the equation of the normal to the curve at Q.

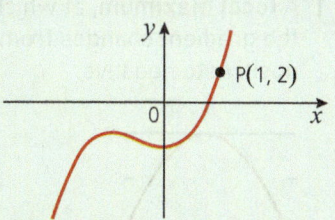

Short answers on page 218

Full worked solutions online   CHECKED ANSWERS

Curve sketching and stationary points   REVISED

Key facts

1 The **second derivative** is found by differentiating the gradient function $\dfrac{dy}{dx}$ or $f'(x)$.

   It is written as $\dfrac{d^2y}{dx^2}$ or $f''(x)$.

   The second derivative tells you the rate of change of the gradient function.

2 A function is **increasing** if the gradient function is positive. ◀ So y increases as x increases.

   A function is **decreasing** if the gradient function is negative. ◀ So y decreases as x increases.
   Many functions are increasing functions for some values of x, and decreasing functions for other values of x.

This is an **increasing** function, $\dfrac{dy}{dx} > 0$

This is a **decreasing** function, $\dfrac{dy}{dx} < 0$

This function is decreasing for negative values of x, and increasing for positive values of x.

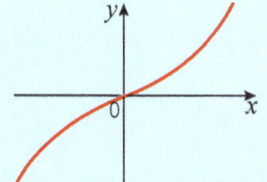
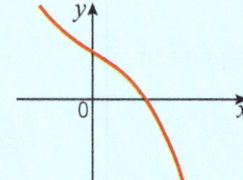
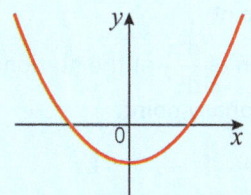

3 A section of curve which has an **increasing gradient** is said to be **convex** (or **concave upwards**).
   The rate of change of the gradient function is positive, so $\dfrac{d^2y}{dx^2} > 0$

If you join two points **on the curve** with **a straight line**, the **line** lies **above** the **curve**.

Edexcel A Level Mathematics (Pure)   107

4 A section of curve which has a **decreasing gradient** is said to be **concave** (**downwards**).
   The rate of change of the gradient function is negative, so $\dfrac{d^2y}{dx^2} < 0$

   *If you join two points **on the curve** with **a straight line**, the **curve** lies **above** the **line**.*

5 **Stationary points** on a curve are points at which the gradient of the curve is zero.
   This means that the tangent to the curve is horizontal and $\dfrac{dy}{dx} = 0$ at that point.
   There are three different types of stationary point:

   1 A **local maximum**, at which the gradient changes from positive to negative.

   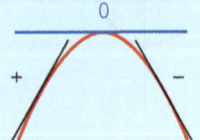

   2 A **local minimum**, at which the gradient changes from negative to positive.

   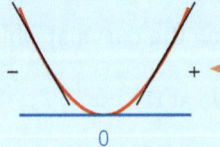

   *At a maximum or a minimum, the curve turns, so these are also called **turning points**.*

   At a maximum, the curve is **concave**, so $\dfrac{d^2y}{dx^2} < 0$.

   At a minimum, the curve is **convex**, so $\dfrac{d^2y}{dx^2} > 0$.

   *At any **point of inflection**, the curve changes from convex to concave (or vice versa) and so the second derivative is zero.*

   3 A **stationary point of inflection** is where the gradient on both sides of a stationary point has the same sign.

   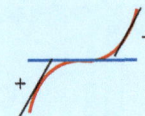 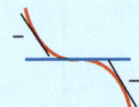

6 You can **find the coordinates of a stationary point** by
   - finding $\dfrac{dy}{dx}$
   - then finding the value(s) of x for which $\dfrac{dy}{dx} = 0$
   - and then substituting the value(s) of x into the equation of the curve to find the y coordinates.

   $y = ...$

7 The **nature of a stationary point** can be found by considering the sign of $\dfrac{dy}{dx}$ on either side of the point.

   The gradient function.

   At a local **minimum point** ⌣, $\dfrac{dy}{dx}$ is negative to the left of the point and positive to the right of the point.

   At a local **maximum point** ⌢, $\dfrac{dy}{dx}$ is **positive** to the left of the point and **negative** to the right of the point.

8 Alternatively, you can use the sign of $\dfrac{d^2y}{dx^2}$ at the stationary point to determine the nature of the stationary point.
   - If $\dfrac{d^2y}{dx^2} > 0$, it is a local minimum.
   - If $\dfrac{d^2y}{dx^2} < 0$, it is a local maximum.
   - If $\dfrac{d^2y}{dx^2} = 0$, you need to look at the gradient on either side to find out the nature of the stationary point.

   The point could be a point of inflection but it could also be a turning point.

9 You can use graph of the $y = f(x)$ to sketch the graph of the gradient function.
   - Look for any **stationary points**, where $\frac{dy}{dx} = 0$.
   - Look for regions where the function is **increasing**, where $\frac{dy}{dx} > 0$.
     The graph of $\frac{dy}{dx}$ will **lie above** the x-axis at the points where $y = f(x)$ is **increasing**.
   - Look for regions where the function is **decreasing**, where $\frac{dy}{dx} < 0$.
     The graph of $\frac{dy}{dx}$ will **lie below** the x-axis at the points where $y = f(x)$ is **decreasing**.

> The graph of $\frac{dy}{dx}$ will **cross** the x-axis at the points where $y = f(x)$ has a **stationary point**.

10 Not all points of inflection are stationary.

   At a **point of inflection**, the curve changes from concave to convex (or vice versa) and so the second derivative is zero.

   To find a non-stationary point of inflection.
   - Find a point P where the 2nd derivative is 0 by solving $\frac{d^2y}{dx^2} = 0$.
   - Show the sign of $\frac{d^2y}{dx^2}$ just before P and just after P changes from negative to positive (or vice versa).

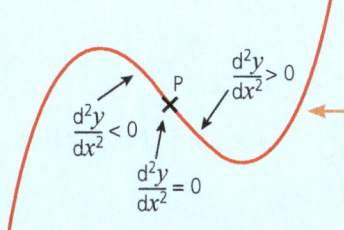

> This curve is concave just before P and convex just after P, hence P is a point of inflection.

Worked examples

1 Sketching the graph of the gradient function

The diagram shows the graph of $y = f(x)$.
Sketch the graph of $y = f'(x)$.

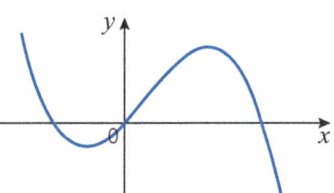

Solution

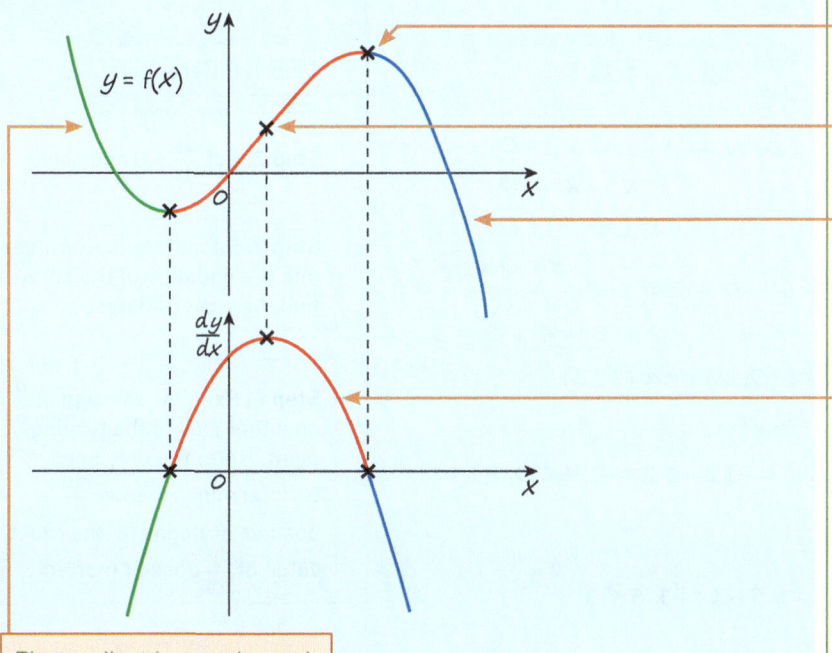

> The gradient is 0 at the turning points.

> At this point the gradient is at its steepest.

> The gradient is negative and increasing in magnitude.

> The gradient of y is positive and increases to a maximum value and then decreases again.

> The gradient is negative and decreasing in magnitude.

Edexcel A Level Mathematics (Pure)

2 Finding where a function is decreasing

Find the range of values of x for which the function $y = 6x - \dfrac{x^2}{2} - \dfrac{x^3}{3}$ is a decreasing function of x.

Solution

$y = 6x - \dfrac{x^2}{2} - \dfrac{x^3}{3} \Rightarrow \dfrac{dy}{dx} = 6 - x - x^2$

The function is decreasing if $\dfrac{dy}{dx} < 0$

First find $\dfrac{dy}{dx} = 0$,

$6 - x - x^2 = 0$

$x^2 + x - 6 = 0$ ← Multiply through by –1 and reverse the inequality sign.

$(x + 3)(x - 2) = 0$

$\Rightarrow$ critical values are $x = -3$ or $x = 2$

A sketch of the graph of $\dfrac{dy}{dx} = 6 - x - x^2$ shows that $\dfrac{dy}{dx}$ is negative when $x < -3$ or $x > 2$.

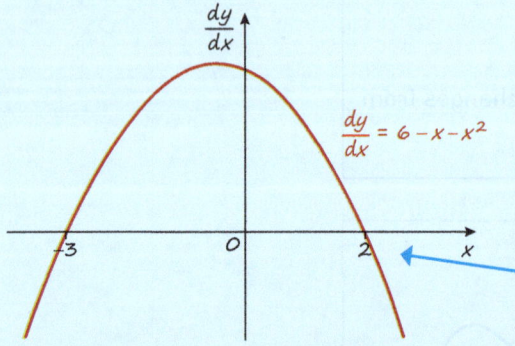

Common mistake: There are two regions where the gradient function is negative so you need two inequalities for your answer. Do not try and combine them into one inequality.

Hint: A function is neither increasing nor decreasing if the gradient function is zero.

So the function is decreasing for $x < -3$ or $x > 2$.

3 Finding stationary points

i Find the coordinates of the stationary points on the curve $y = 3x^4 - 4x^3 + 2$ and determine their nature.

ii Sketch the curve.

Solution

i $y = 3x^4 - 4x^3 + 2 \Rightarrow \dfrac{dy}{dx} = 12x^3 - 12x^2$ ← **Step 1:** Differentiate.

At stationary points, $\dfrac{dy}{dx} = 0$, so $12x^3 - 12x^2 = 0$ ← **Step 2:** Set $\dfrac{dy}{dx} = 0$ and solve.

$x^3 - x^2 = 0$

$x^2(x - 1) = 0$

$x = 0$ or $x = 1$

When $x = 0$, $y = 3 \times 0 - 4 \times 0 + 2 = 2$ ← **Step 3:** Substitute the x-values into the equation of the curve to find the y coordinates.

When $x = 1$, $y = 3 \times 1^4 - 4 \times 1^3 + 2 = 3 - 4 + 2 = 1$

The stationary points are $(0, 2)$ and $(1, 1)$.

At the point where $x = -1$:

$\dfrac{dy}{dx} = 12(-1)^3 - 12(-1)^2 = -12 - 12 = -24 < 0$ ← **Step 4:** Examine the sign of $\dfrac{dy}{dx}$ on either side of the turning point. Note: You only need to determine whether $\dfrac{dy}{dx}$ is positive or negative, the exact value of $\dfrac{dy}{dx}$ doesn't matter.

At the point where $x = \dfrac{1}{2}$:

$\dfrac{dy}{dx} = 12\left(\dfrac{1}{2}\right)^3 - 12\left(\dfrac{1}{2}\right)^2 = 1.5 - 3 = -1.5 < 0$

At the point where $x = 2$:

$\dfrac{dy}{dx} = 12 \times 2^3 - 12 \times 2^2 = 96 - 48 = 48 > 0$

	$x<0$	$x=0$	$0<x<1$	$x=1$	$x>1$
Sign of $\frac{dy}{dx}$	−ve \	0 —	−ve \	0 —	+ve /
Stationary point		Point of inflection		Local minimum	

(0, 2) is a stationary point of inflection, and (1, 1) is a local minimum point.

ii

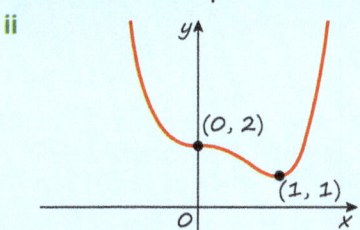

Hint: It is a good idea to set your work out in a table.

Common mistake: Be careful if there is more than one stationary point close together. For example, if there are stationary points at $x = \frac{1}{2}$ and $x = 1$, then you can't use $x = 0$ to look at the gradient on the left of the point $x = 1$. You need to use a point between $x = \frac{1}{2}$ and $x = 1$, such as $x = \frac{3}{4}$.

4 Using the second derivative to find the nature of stationary points

The curve $y = 2x^5 + 5x^4 - 1$ has stationary points at (−2, 15) and (0, −1).

Determine the nature of these stationary points.

Solution

$y = 2x^5 + 5x^4 - 1 \Rightarrow \frac{dy}{dx} = 10x^4 + 20x^3 \Rightarrow \frac{d^2y}{dx^2} = 40x^3 + 60x^2$

When $x = -2$, $\frac{d^2y}{dx^2} = 40(-2)^3 + 60(-2)^2 = -320 + 240 = -80$

Since the second derivative is negative, (−2, 15) is a local maximum point.

When $x = 0$, $\frac{d^2y}{dx^2} = 40 \times 0^3 + 60 \times 0^2 = 0$

Since the second derivative is zero, test the gradient function on either side.

When $x = -1$, $\frac{dy}{dx} = 10(-1)^4 + 20(-1)^3 = 10 - 20 = -10$

When $x = 1$, $\frac{dy}{dx} = 10 \times 1^4 + 20 \times 1^3 = 10 + 20 = 30$

The gradient function is going from negative to positive, so (0, −1) is a local minimum point.

Common mistake: When the second derivative is zero, it is crucial that you test the gradient function on either side of it. The point could be either a point of inflection or a maximum or minimum!

5 Finding non-stationary points of inflection

Given that $y = x^3 - x^2$, find the coordinates of the point, P, where $\frac{d^2y}{dx^2} = 0$ and demonstrate that P is a non-stationary point of inflection.

Solution

$y = x^3 - x^2 \Rightarrow \frac{dy}{dx} = 3x^2 - 2x$

$\Rightarrow \frac{d^2y}{dx^2} = 6x - 2$

$\frac{d^2y}{dx^2} = 0 \Rightarrow 6x - 2 = 0 \Rightarrow x = \frac{1}{3}$

When $x = \frac{1}{3}$, $y = \left(\frac{1}{3}\right)^3 - \left(\frac{1}{3}\right)^2 = -\frac{2}{27}$

So P is the point $\left(\frac{1}{3}, -\frac{2}{27}\right)$

Edexcel A Level Mathematics (Pure)

Test the sign of $\frac{d^2y}{dx^2}$ on either side of $x = \frac{1}{3}$:

At $x = 0$, $\frac{d^2y}{dx^2} = 6 \times 0 - 2 = -2 < 0$ ← So the curve is concave just before P...

At $x = 1$, $\frac{d^2y}{dx^2} = 6 \times 1 - 2 = 4 > 0$ ← ...and convex just after P.

The sign of $\frac{d^2y}{dx^2}$ changes ⇒ P is a point of inflection.

Now check that P is non-stationary:

At $x = \frac{1}{3}$, $\frac{dy}{dx} = 3 \times \left(\frac{1}{3}\right)^2 - 2 \times \frac{1}{3} = -\frac{1}{3}$

So at P, $\frac{dy}{dx} \neq 0$ and $\frac{d^2y}{dx^2} = 0$ and the sign of $\frac{d^2y}{dx^2}$ changes on either side of P ⇒ P is a non-stationary point of inflection.

Test yourself

TESTED

1 Find the range of values of x for which the function $y = x^3 + 2x^2 + x + 2$ is a decreasing function of x.

A $\quad x < -1$ and $x > -\frac{1}{3}$

B $\quad x \leqslant -1$ and $x \geqslant -\frac{1}{3}$

C $\quad x < -2$

D $\quad -1 \leqslant x \leqslant -\frac{1}{3}$

E $\quad -1 < x < -\frac{1}{3}$

2 The diagram shows the graph of $y = f(x)$.

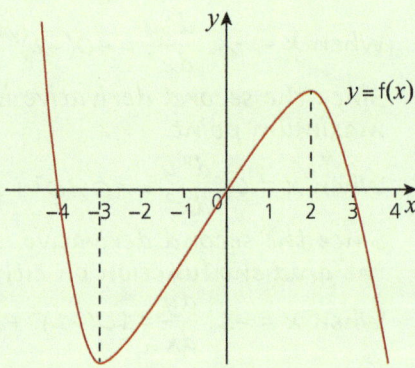

Which of the following shows the graph of $y = f'(x)$?

A
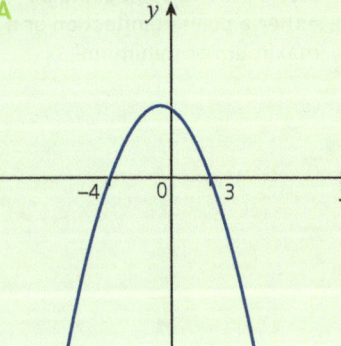

B
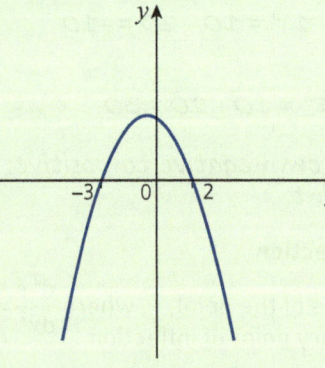

C
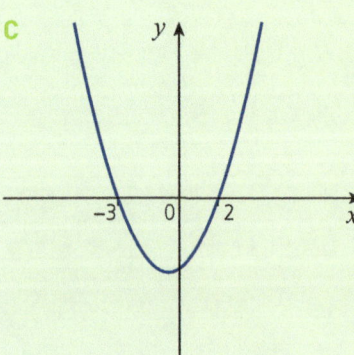

D
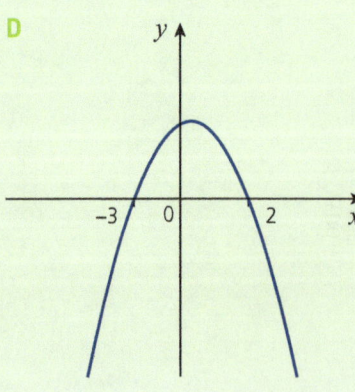

E
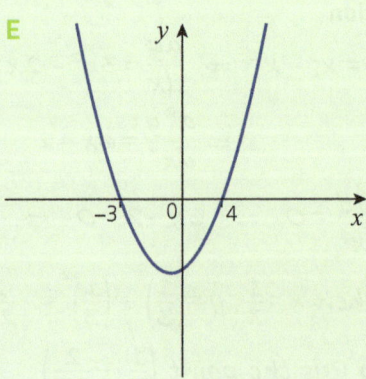

3 Find the x coordinates of the turning point(s) on the curve $y = x + \frac{1}{x^3}, x \neq 0$ and identify the nature of each turning point.
   A There is a local maximum at $x = -\sqrt[4]{3}$ and a local minimum at $x = \sqrt[4]{3}$.
   B There is a local minimum at $x = \sqrt[4]{3}$ only.
   C There is a local minimum at $x = -\sqrt[4]{3}$ and a local maximum at $x = \sqrt[4]{3}$.
   D There is a local maximum at $x = -\sqrt{3}$ and a local minimum at $x = \sqrt{3}$.
   E There is a local minimum at $x = \sqrt{3}$ only.

4 Find the y coordinate of the point on the curve $y = x^3 + 6x^2 + 5x - 3$ at which the second derivative is zero.
   A -2 B -45 C 3 D -7 E 9

5 Find the x coordinate of the stationary point of the curve $y = x - \sqrt{x} + 2$.
   Use the second derivative to identify its nature.
   A $x = \frac{1}{4}$; minimum
   B $x = \frac{1}{4}$; maximum
   C $x = \sqrt{2}$; minimum
   D $x = \frac{1}{\sqrt{2}}$; minimum
   E $x = \frac{1}{\sqrt{8}}$; maximum

6 Find the coordinates of the stationary point(s) on the curve $y = 3x^4 + 2x^3 + 1$.
   A $\left(-\frac{1}{2}, \frac{15}{16}\right)$ only
   B $(0, 1)$ and $\left(-\frac{1}{2}, \frac{9}{16}\right)$
   C $\left(-\frac{1}{2}, \frac{9}{16}\right)$ only
   D $(0, 1)$ and $\left(-\frac{1}{2}, \frac{15}{16}\right)$
   E $(0, 1)$ and $\left(\frac{1}{2}, \frac{23}{16}\right)$

7 The curve $y = x^3 - 2x^2 + x$ has stationary points at $\left(\frac{1}{3}, \frac{4}{27}\right)$ and $(1, 0)$.
   What is the nature of these stationary points?
   A $\left(\frac{1}{3}, \frac{4}{27}\right)$ is a stationary point of inflection, and $(1, 0)$ is a stationary point of inflection.
   B $\left(\frac{1}{3}, \frac{4}{27}\right)$ is a local maximum point, and $(1, 0)$ is a local minimum point.
   C $\left(\frac{1}{3}, \frac{4}{27}\right)$ is a local minimum point, and $(1, 0)$ is a local maximum point.
   D $\left(\frac{1}{3}, \frac{4}{27}\right)$ is a local maximum point, and $(1, 0)$ is a stationary point of inflection.
   E $\left(\frac{1}{3}, \frac{4}{27}\right)$ is a stationary point of inflection, and $(1, 0)$ is a local minimum point.

Questions 8 and 9 are to do with the following situation.

A rectangular sheet of sides 24 cm and 15 cm has four equal squares of sides x cm cut from the corners. The sides are then turned up to make an open rectangular box.

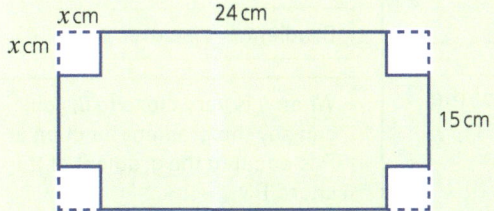

8 Find an expression in terms of x for the volume of the rectangular box.
   A $4x^2 - 78x + 360$ B $360x$ C $4x^3 - 78x^2 + 360x$ D $2x^3 - 39x^2 + 180x$

9 Find the value of x so that the volume of the box is a maximum.
   A 3   B 10   C 9.75   D 2.55

Full worked solutions online CHECKED ANSWERS

Exam-style question

A curve has equation $y = 3x^4 - 8x^3 + 5$.

i   Find the coordinates of the points P and Q at which $\dfrac{d^2y}{dx^2} = 0$.

    Demonstrate that one of these points is a non-stationary point of inflection and that the other is a stationary point of inflection.

ii Find the coordinates of the turning point, R, of this curve. Determine also the nature of this turning point.

iii Sketch the curve.

Short answers on page 218

Full worked solutions online

CHECKED ANSWERS

First principles and differentiating $\sin x$, $\cos x$ and $\tan x$

REVISED

Key facts

1 You need to be able to differentiate using **first principles**.

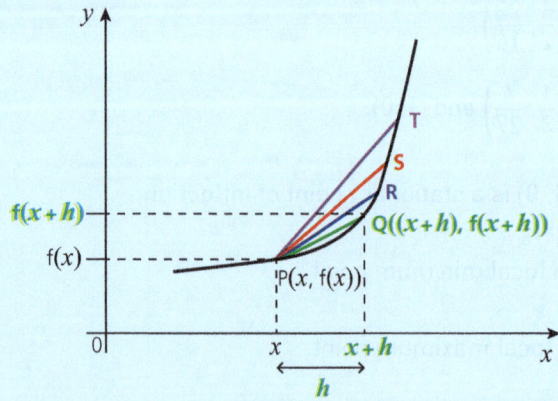

This diagram shows that the gradient of a series of chords from P converges to the gradient of the tangent at P.

You can see the gradient of the chord PS is a better approximation for the gradient at P than the chord PT and the gradient of the chord PR is better still.

The closer a point is to P the closer the gradient of the chord is to the gradient of the tangent at P. So, the gradient of the chord PQ where point Q, is only a short distance, h, away from P is very close to the gradient of the tangent at P.

The gradient of the chord PQ is $\dfrac{f(x+h) - f(x)}{h}$.

> Gradient is 'rise over run'.

As $h \to 0$ then the gradient of the chord PQ $\to$ the gradient of the tangent at P.
So you can write $f'(x) = \lim\limits_{h \to 0} \dfrac{f(x+h) - f(x)}{h}$.

> When h is very close to 0, you can say the gradient function at P is equal to the gradient of the chord PQ.

2 You need to know the standard results for **differentiating sine, cosine and tangent**.

- $y = \sin x \Rightarrow \dfrac{dy}{dx} = \cos x$
- $y = \cos x \Rightarrow \dfrac{dy}{dx} = -\sin x$
- $y = \tan x \Rightarrow \dfrac{dy}{dx} = \sec^2 x$

> Further differentiation of trigonometric functions can be found in the section on the chain rule.

> **Common mistake:** When you differentiate or integrate a trigonometric function you must use radians.

> Proof of this is shown on page 125.

Worked examples

1 Differentiating a small power of x from first principles

Differentiate $f(x) = 4x^2$ from first principles.

Solution

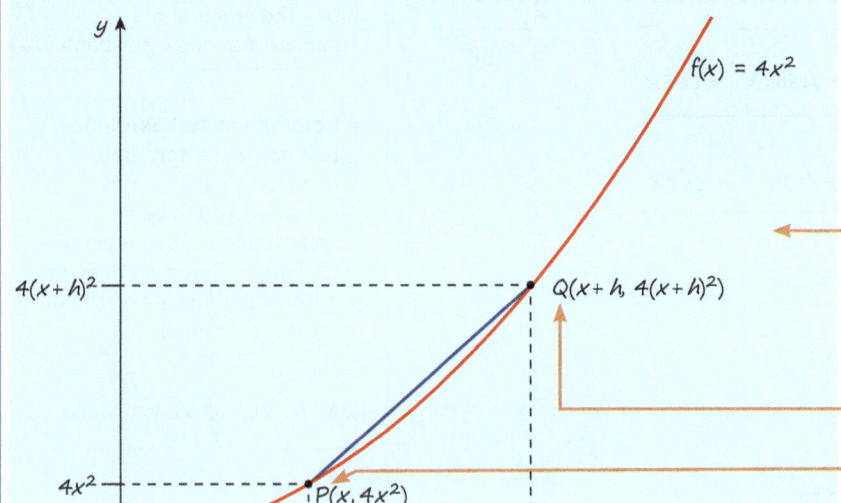

Start by drawing a sketch showing the point P on the curve $y = f(x)$ and the point Q a distance h away from P.

The curve is $y = 4x^2$ so the y coordinate of P is $4x^2$, and the y coordinate of Q is $4(x+h)^2$.

$$\text{The gradient of the chord } PQ = \frac{4(x+h)^2 - 4x^2}{h}$$

Gradient is 'rise over run'.

$$= \frac{4(x^2 + 2xh + h^2) - 4x^2}{h}$$

$$= \frac{4x^2 + 8xh + 4h^2 - 4x^2}{h}$$

$$= \frac{8xh + 4h^2}{h}$$

The $4x^2$ terms cancel.

$$= 8x + 4h$$

Cancel h.

When $h \to 0$, the gradient of the chord PQ $\to$ the gradient of the tangent at P.

This is the derivative of $f(x) = 4x^2$.

When $h \to 0$, $8x + 4h \to 8x$

When h is very close to 0, $4h$ is also very close to 0.

So when $f(x) = 4x^2$, $f'(x) = 8x$.

2 Differentiating $\cos x$ from first principles

Differentiate $f(x) = \cos x$ from first principles.

Solution

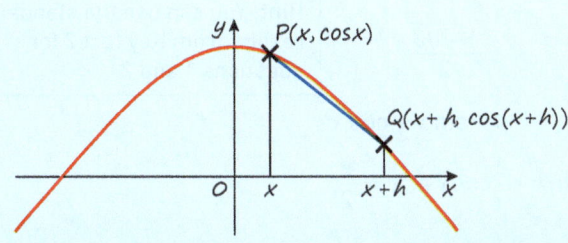

$$\text{The gradient of the chord } PQ = \frac{\cos(x+h) - \cos x}{h}$$

Gradient is 'rise over run'.

$$= \frac{(\cos x \cos h - \sin x \sin h) - \cos x}{h}$$

Using the compound angle formula $\cos(A+B) = \cos A \cos B - \sin A \sin B$.

When $h \to 0$, the gradient of the chord PQ $\to$ the gradient of the tangent at P

So $y = \cos x \Rightarrow \dfrac{dy}{dx} = \lim\limits_{h \to 0} \dfrac{(\cos x \cos h - \sin x \sin h) - \cos x}{h}$

When h is small and in radians then $\cos h \approx 1 - \frac{1}{2}h^2$ and $\sin h \approx h$

> Use the small angle approximations – see page 92

$\Rightarrow \dfrac{dy}{dx} = \lim\limits_{h \to 0} \dfrac{\left(\left(1 - \frac{1}{2}h^2\right)\cos x - h \sin x\right) - \cos x}{h}$

$= \lim\limits_{h \to 0} \dfrac{\cos x - \frac{1}{2}h^2 \cos x - h \sin x - \cos x}{h}$

$= \lim\limits_{h \to 0} \dfrac{-\frac{1}{2}h^2 \cos x - h \sin x}{h}$

$= \lim\limits_{h \to 0} \left(-\frac{1}{2}h \cos x - \sin x\right)$

$= -\sin x$

> **Common mistakes:** Notice that you were only able to differentiate $\cos x$ by considering h to be small and in radians. You must use radians for any questions on differentiating sine or cosine or related functions.

> As $h \to 0$, $-\frac{1}{2}h \cos x \to 0$.

3 Differentiating sin x, cos x and tan x

Differentiate
 i $y = 10 + 8 \sin x$
 ii $y = x^5 + 3\cos x$
 iii $y = 3\sin x - \tan x$.

Solution

i $y = 10 + 8 \sin x$

$\dfrac{dy}{dx} = 8 \cos x$

ii $y = x^5 + 3 \cos x$

$\dfrac{dy}{dx} = 5x^4 - 3 \sin x$

iii $y = 3 \sin x - \tan x$

$\dfrac{dy}{dx} = 3 \cos x - \sec^2 x$

Test yourself

TESTED

1 Given $f(x) = \sin x - \cos x$, find $f''\left(\dfrac{\pi}{3}\right)$.

 A $\dfrac{\sqrt{3} - 1}{2}$
 B $\dfrac{1 - \sqrt{3}}{2}$
 C $\dfrac{\sqrt{3} + 1}{2}$
 D $-\sin x + \cos x$

> **Hint:** You can use the standard results from Key fact 2 for questions 1 and 2.

2 Find the equation of the normal to the curve $f(x) = \sin x + 2\cos x$ at $x = \dfrac{\pi}{2}$.

 A $y + 2x = 1 + \pi$
 B $2y + 2x = 4 + \pi$
 C $4y + 2x = 4 + \pi$
 D $4y - 2x = 4 - \pi$

3 The graph shows part of the curve $y = \sin x + \cos x$
Three of the following statements are false and one is true.
Which one is true?

A At the point P, $x = \dfrac{\pi}{2}$.

B The gradient at Q is 2.

C At R the value of y is -1.

D The gradient of the curve always lies between a minimum value of $-\sqrt{2}$ and a maximum value of $\sqrt{2}$.

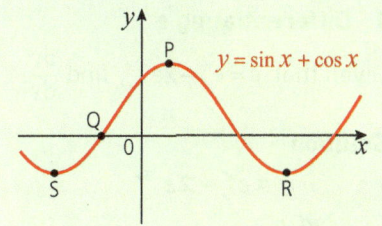

Questions 4–6 are to do with the curve on the right.

P and Q are two points on the curve $y = x^3$.

P is the point (x, x^3) and the x coordinate of Q is $(x + h)$.

4 Find the expansion of $(x + h)^3$
  A $x^3 + h^3$
  B $x^3 + 6x^2h + h^3$
  C $x^3 + 3x^2h + 3hx^2 + h^3$
  D $x^3 + 3xh + 3hx + h^3$

5 Find an expression for the gradient of the chord PQ
  A $\dfrac{x^3 + 3x^2h + 3hx^2 + h^3}{h}$
  B $\dfrac{3x^2h + 3h^2x + h^3}{x + h}$
  C $3x^2$
  D $3x^2 + 3hx + h^2$

6 Write down the limit of the gradient of the chord as $h \to 0$
  A 0   B $3x^2 + 3x$ C x^3 D $3x^2$

Full worked solutions online   CHECKED ANSWERS

Exam-style question

Given $f(x) = \sin x$, use differentiation from first principles to find $f'\left(\dfrac{\pi}{3}\right)$.

Short answers on page 218

Full worked solutions online   CHECKED ANSWERS

Differentiating $\ln x$ and e^x REVISED

Key facts

1 $y = e^{kx} \Rightarrow \dfrac{dy}{dx} = ke^{kx}$

2 $y = \ln x \Rightarrow \dfrac{dy}{dx} = \dfrac{1}{x}$

Hint: See page 14 for the rules of logs.

See page 121 for a proof of this.

Edexcel A Level Mathematics (Pure)   117

Worked examples

1 Differentiating e^x

Given that $y = e^x - 2e^{-3x}$, find $\dfrac{dy}{dx}$.

Solution

$$y = e^x - 2e^{-3x}$$
$$\Rightarrow \dfrac{dy}{dx} = e^x - 2 \times (-3)e^{-3x} = e^x + 6e^{-3x}$$

2 Differentiating $\ln x$

Differentiate:
  i $y = \ln 5x$
  ii $y = \ln x^7$

Solution

i $y = \ln 5x$
   $= \ln 5 + \ln x$ ← Using the rule $\ln(ab) = \ln a + \ln b$.
   $\Rightarrow \dfrac{dy}{dx} = 0 + \dfrac{1}{x} = \dfrac{1}{x}$

   Common mistake: $\ln 5$ is a constant so differentiating it gives zero.

ii $y = \ln x^7 = 7\ln x$ ← Using the rule $\ln(x^n) = n\log x$.
   $\Rightarrow \dfrac{dy}{dx} = 7 \times \dfrac{1}{x} = \dfrac{7}{x}$

Test yourself

1 Differentiate $y = \dfrac{6}{e^{2x}}$.

A $\dfrac{dy}{dx} = -3e^{-2x}$ B $\dfrac{dy}{dx} = -12e^x$ C $\dfrac{dy}{dx} = -\dfrac{1}{12e^{2x}}$ D $\dfrac{dy}{dx} = -\dfrac{12}{e^{2x}}$ E $\dfrac{dy}{dx} = 12e^{2x}$

2 Differentiate $y = \ln 2x$.

A $\dfrac{dy}{dx} = \ln 2 + \dfrac{1}{x}$ B $\dfrac{dy}{dx} = \dfrac{2}{x}$ C $\dfrac{dy}{dx} = \dfrac{1}{2x}$ D $\dfrac{dy}{dx} = \dfrac{1}{x}$

3 Differentiate $y = 3\ln\left(\dfrac{1}{x^5}\right)$.

A $\dfrac{dy}{dx} = -\dfrac{15}{x}$ B $\dfrac{dy}{dx} = -\dfrac{5}{x}$ C $\dfrac{dy}{dx} = \dfrac{15}{x}$ D $\dfrac{dy}{dx} = -15\ln\left(\dfrac{1}{x^6}\right)$

4 Given that $y = 6e^{4x}$, find $\dfrac{dy}{dx}$.

A $4e^{4x}$ B $\dfrac{3}{2}e^{4x}$ C $24e^{4x}$ D $24e^{4x-1}$

Full worked solutions online

Exam-style question

Given that $f(x) = \ln 2x - \dfrac{3}{e^{2x}}$, find the exact values of $f\left(\dfrac{1}{2}\right)$, $f'\left(\dfrac{1}{2}\right)$ and $f''\left(\dfrac{1}{2}\right)$.

Short answers on page 218

Full worked solutions online

The chain rule

REVISED

Key facts

1. The **chain rule** allows you to differentiate expressions that are a function of a function.
$$\frac{dy}{dx} = \frac{dy}{du} \times \frac{du}{dx}.$$

2. Differentiating **trigonometric functions**:
$$y = \sin kx \Rightarrow \frac{dy}{dx} = k\cos kx$$
$$y = \cos kx \Rightarrow \frac{dy}{dx} = -k\sin kx$$
$$y = \tan kx \Rightarrow \frac{dy}{dx} = k\sec^2 kx$$

3. Differentiating the **logarithm** of a function:
$$y = \ln|f(x)| \Rightarrow \frac{dy}{dx} = \frac{f'(x)}{f(x)}.$$

4. Differentiating the **exponential** of a function:
$$y = e^{f(x)} \Rightarrow \frac{dy}{dx} = f'(x) e^{f(x)}.$$

5. Differentiating a^{kx}:
$$y = a^{kx} \Rightarrow \frac{dy}{dx} = ka^{kx} \ln a$$

6. If y is a function of x then
$$\frac{dx}{dy} = \frac{1}{\frac{dy}{dx}}$$

7. You can use the chain rule to find the rate at which a quantity is changing with time.
For example, given $A = \pi r^2$
$$\frac{dA}{dt} = \frac{dA}{dr} \times \frac{dr}{dt} = 2\pi r \frac{dr}{dt}$$

Hint: You usually start by substituting u for the inside function, as in the worked examples below.

Common mistake: When you are using calculus with trigonometric functions the angles must always be in radians.

Make sure you remember this formula correctly. Alternatively, just use the chain rule as in Worked example 6.

Make sure you remember this formula correctly. Alternatively, just use the chain rule as in Worked example 4.

Make sure you remember this formula correctly. Alternatively, you can use the method shown in Worked example 7.

Common mistake: If you misremember any of these formulae you could lose quite a few marks. Consider learning fewer formulae and working out more differentials using the chain rule.

Worked examples

1 Differentiating composite functions

Differentiate:

i $y = \sqrt{6x + 2}$

ii $y = \dfrac{1}{(4x^3 - x^4)}$

Solution

i $y = \sqrt{6x + 2}$

Let $u = 6x + 2 \Rightarrow y = \sqrt{u} = u^{\frac{1}{2}}$

$\dfrac{du}{dx} = 6 \qquad \dfrac{dy}{du} = \dfrac{1}{2} u^{-\frac{1}{2}}$

u is always the 'inside' function.

Edexcel A Level Mathematics (Pure)

Using the chain rule: $\dfrac{dy}{dx} = \dfrac{dy}{du} \times \dfrac{du}{dx}$

$\dfrac{dy}{dx} = \dfrac{1}{2}u^{-\frac{1}{2}} \times 6$

$= 3 \times (6x+2)^{-\frac{1}{2}}$

$= \dfrac{3}{\sqrt{(6x+2)}}$

Common mistake: Don't forget to write your answer in terms of x.

ii $y = \dfrac{1}{(4x^3 - x)^4}$

$= (4x^3 - x)^{-4}$

Let $u = 4x^3 - x \quad \Rightarrow \quad y = u^{-4}$

$\dfrac{du}{dx} = 12x^2 - 1 \qquad \dfrac{dy}{du} = -4u^{-5}$

Hint: With experience you can answer questions like these in your head. That is fine, but only when you are certain you will get it right. Until then write it out in full, as here.

Chain rule: $\dfrac{dy}{dx} = \dfrac{dy}{du} \times \dfrac{du}{dx}$

$\dfrac{dy}{dx} = -4u^{-5} \times (12x^2 - 1)$

$= -4(4x^3 - x)^{-5} \times (12x^2 - 1)$

$= -\dfrac{4(12x^2 - 1)}{(4x^3 - x)^5} = \dfrac{4(1 - 12x^2)}{(4x^3 - x)^5}$

Make sure you tidy up the answer.

2 Using the chain rule to find the gradient at a point

Find the gradient of the curve $y = \dfrac{1}{(3x-1)^2}$ at the point $\left(1, \dfrac{1}{4}\right)$.

Solution

$y = \dfrac{1}{(3x-1)^2}$

$= (3x-1)^{-2}$

Let $u = 3x - 1 \quad \Rightarrow \quad y = u^{-2}$

$\dfrac{du}{dx} = 3 \qquad \dfrac{dy}{du} = -2u^{-3}$

Chain rule $\dfrac{dy}{dx} = \dfrac{dy}{du} \times \dfrac{du}{dx}$

$\dfrac{dy}{dx} = -2u^{-3} \times 3$

$= -6 \times (3x-1)^{-3}$

$= -\dfrac{6}{(3x-1)^3}$

Substituting $x = 1$ into $\dfrac{dy}{dx} = -\dfrac{6}{(3x-1)^3}$ gives:

$\dfrac{dy}{dx} = -\dfrac{6}{(3 \times 1 - 1)^3} = -\dfrac{3}{4}$

$\Rightarrow$ the gradient is $-\dfrac{3}{4}$

Now you need to find the value of $\dfrac{dy}{dx}$ (the gradient function) at the point $\left(1, \dfrac{1}{4}\right)$.

3 Differentiating trigonometric functions

Differentiate

i $y = 2\sin 3x - 4\cos 5x$

ii $y = x^5 - 3\tan 2x$

Solution

i $y = 2\sin 3x - 4\cos 5x$

$\Rightarrow \dfrac{dy}{dx} = 2 \times 3\cos 3x - 4 \times (-5)\sin 5x$

$= 6\cos 3x + 20\sin 5x$

ii $y = x^5 - 3\tan 2x$

$\dfrac{dy}{dx} = 5x^4 - 3 \times 2\sec^2 2x$

$= 5x^4 - 6\sec^2 2x$

> **Hint:** You can use the chain rule to do this but it is quicker to learn the standard results:
> $y = \sin kx \Rightarrow \dfrac{dy}{dx} = k\cos kx$,
> $y = \cos kx \Rightarrow \dfrac{dy}{dx} = -k\sin kx$.

4 Differentiating functions of e^x

Differentiate $y = e^{3x^2 - 2}$ with respect to x.

Solution

Let $u = 3x^2 - 2 \Rightarrow y = e^u$

$\dfrac{du}{dx} = 6x \qquad \dfrac{dy}{du} = e^u$

Chain rule $\dfrac{dy}{dx} = \dfrac{dy}{du} \times \dfrac{du}{dx}$

$\dfrac{dy}{dx} = e^u \times 6x$

$\dfrac{dy}{dx} = 6xe^{3x^2 - 2}$

> **Hint:** This result illustrates the general rule for differentiating exponentials.
> If $y = e^{f(x)}$ then $\dfrac{dy}{dx} = f'(x)e^{f(x)}$.

5 Using $\dfrac{dx}{dy}$

Show that the derivative of $y = \ln x$ is $\dfrac{1}{x}$.

Solution

$y = \ln x$

$\Rightarrow e^y = x$ ①

$\Rightarrow x = e^y$

$\Rightarrow \dfrac{dx}{dy} = e^y$

$\Rightarrow \dfrac{dx}{dy} = x$

$\Rightarrow \dfrac{dy}{dx} = \dfrac{1}{x}$ as required.

> $e^{\ln x} = x$ as $f(x) = e^x$ and $f(x) = \ln x$ are inverse functions.

> Differentiate x with respect to y. Remember the derivative of e^y is e^y.

> From ① $e^y = x$.

6 Differentiating functions of $\ln x$

Given that $y = \ln(x^3 - 4x)$ find $\dfrac{dy}{dx}$.

Solution

Let $u = x^3 - 4x \Rightarrow y = \ln u$

$\dfrac{du}{dx} = 3x^2 - 4 \qquad \dfrac{dy}{du} = \dfrac{1}{u}$

Edexcel A Level Mathematics (Pure)

Chain rule: $\dfrac{dy}{dx} = \dfrac{dy}{du} \times \dfrac{du}{dx}$

$\dfrac{dy}{dx} = \dfrac{1}{u} \times (3x^2 - 4)$

$= \dfrac{3x^2 - 4}{x^3 - 4x}$

Hint: This result illustrates the general rule for differentiating logarithms.
If $y = \ln[f(x)]$, then $\dfrac{dy}{dx} = \dfrac{f'(x)}{f(x)}$.

7 Differentiating a^x

Differentiate $y = 2^x$.

Solution

$y = 2^x$ ①

$\Rightarrow \ln y = \ln 2^x$

$\Rightarrow \ln y = x \ln 2$ ← Take natural logs of both sides.

$\Rightarrow x = \dfrac{1}{\ln 2} \times \ln y$ ← Make x the subject.

$\Rightarrow \dfrac{dx}{dy} = \dfrac{1}{\ln 2} \times \dfrac{1}{y}$ ← Differentiate x with respect to y. Remember that $\dfrac{1}{\ln 2}$ is just a number!

$\Rightarrow \dfrac{dy}{dx} = y \ln 2$ ← Don't forget to write your answer in terms of x. From ①, $y = 2^x$.

$\dfrac{dy}{dx} = 2^x \ln 2$

8 Solving problems involving rates of change

A culture of bacteria form a circle which increases in radius at a rate of 2 mm per hour.

Find the rate at which the area is increasing when the radius is 20 mm.

Hint: In general,
$y = a^x \Rightarrow \dfrac{dy}{dx} = a^x \ln a$ and
$y = a^{kx} \Rightarrow \dfrac{dy}{dx} = ka^{kx} \ln a$

Solution

Denote the radius of a circle at the time t hours by r mm.

Then the area $A = \pi r^2$ in mm².

You know that $\dfrac{dr}{dt} = 2$ mm per hour and you are asked to find $\dfrac{dA}{dt}$ when $r = 20$. ← This is the rate at which the radius is increasing.

Using the chain rule: $\dfrac{dA}{dt} = \dfrac{dA}{dr} \times \dfrac{dr}{dt}$

$A = \pi r^2 \Rightarrow \dfrac{dA}{dr} = 2\pi r$

So $\dfrac{dA}{dt} = 2\pi r \times \dfrac{dr}{dt}$

Substituting $r = 20$ and $\dfrac{dr}{dt} = 2$ gives

$\dfrac{dA}{dr} = 2 \times \pi \times 20 \times 2 = 80\pi$

So the area is increasing at 80π mm² per hour.

Test yourself

1. Find the derivative of $y = (2x^2 - 3)^3$.
   - A $3(2x^2 - 3)^2$
   - B $12x(2x^2 - 3)^3$
   - C $12x(2x^2 - 3)^2$
   - D $6x(2x^2 - 3)^2$

2. You are given that $y = \dfrac{1}{(1-2x)^3}$. Find $\dfrac{dy}{dx}$.
   - A $-\dfrac{3}{(1-2x)^4}$
   - B $\dfrac{6}{(1-2x)^4}$
   - C $\dfrac{6}{(1-2x)^2}$
   - D $-\dfrac{6}{(1-2x)^4}$
   - E $\dfrac{6}{(1-2x)^3}$

3. Three of the following statements about the function $y = \sqrt{x^2 - 4x + 3}$ are false and one is true. Which one is true?
   - A $\dfrac{dy}{dx} = \dfrac{2(x-2)}{\sqrt{x^2 - 4x + 3}}$.
   - B The curve with equation $y = \sqrt{x^2 - 4x + 3}$ crosses the x-axis at $(2, 0)$.
   - C $y = \sqrt{x^2 - 4x + 3}$ is an increasing function.
   - D Where the curve $y = \sqrt{x^2 - 4x + 3}$ crosses the y-axis its gradient is $-\dfrac{2\sqrt{3}}{3}$.

4. You are given that $f(x) = 3\ln(5x - 1)$. Find the exact value of $f'(2)$.
   - A $\dfrac{1}{3}$
   - B $15 \ln 9$
   - C $\dfrac{5}{3}$
   - D 15

5. Which of the following is the gradient function of $y = \sqrt{\sin 2x}$?
   - A $\dfrac{1}{2}(\sin 2x)^{-\frac{1}{2}}$
   - B $\dfrac{1}{2}(\sin 2x)^{-\frac{1}{2}} \cos 2x$
   - C $\dfrac{\cos 2x}{\sqrt{\sin 2x}}$
   - D $-\dfrac{\cos 2x}{\sqrt{\sin 2x}}$
   - E $\sqrt{\sin 2x} \cos 2x$

6. You are given the equation of a curve $y = \ln\left(\cos\dfrac{x}{2}\right)$. Which of the following is true?
   - A $\dfrac{dy}{dx} = \dfrac{1}{2}\tan\dfrac{x}{2}$
   - B $\dfrac{dy}{dx} = -\dfrac{1}{2}\tan\dfrac{x}{2}$
   - C $\dfrac{dy}{dx} = -\dfrac{1}{2}\cos\dfrac{x}{2}\sin\dfrac{x}{2}$
   - D $\dfrac{dy}{dx} = -\tan\dfrac{x}{2}$

7. The diagram shows a cone which contains water to a depth h.

   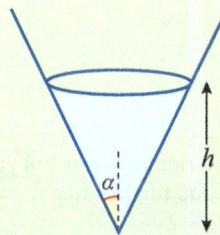

   The half angle, α, is $\arctan\dfrac{1}{2}$ and so the volume of water is given by $V = \dfrac{\pi}{12}h^3$.
   Water is poured into the cone at a steady rate of $5\,\text{cm}^3\,\text{s}^{-1}$.
   At what rate is the depth increasing when $h = 10\,\text{cm}$?
   - A $125\pi\,\text{cm}\,\text{s}^{-1}$
   - B $400\pi\,\text{cm}\,\text{s}^{-1}$
   - C $5\pi\,\text{cm}\,\text{s}^{-1}$
   - D $\dfrac{1}{5\pi}\,\text{cm}\,\text{s}^{-1}$

Full worked solutions online

Exam-style question

Find the exact gradient of the curve $y = \ln|1 + \sin 2x|$ at the point where $x = \dfrac{\pi}{3}$.
Give your answer in its simplest form.

Short answers on page 218

Full worked solutions online

The product and quotient rules

REVISED

Key facts

1. **Product rule:**

   If $y = uv$, where the variables u and v are both functions of x,

   then $\dfrac{dy}{dx} = u\dfrac{dv}{dx} + v\dfrac{du}{dx}$.

2. **Quotient rule:**

   If $y = \dfrac{u}{v}$, where the variables u and v are both functions of x,

   then $\dfrac{dy}{dx} = \dfrac{v\dfrac{du}{dx} - u\dfrac{dv}{dx}}{v^2}$.

Worked examples

1 Using the product rule

For $y = x\sqrt{3x - 2}$, find $\dfrac{dy}{dx}$.

Solution

$y = x\sqrt{3x - 2}$ — *y is the product of two functions x and $\sqrt{3x - 2}$. So you need to use the product rule.*

Let $u = x$ and $v = (3x - 2)^{\frac{1}{2}}$

$\dfrac{du}{dx} = 1 \qquad \dfrac{dv}{dx} = \dfrac{1}{2}(3x - 2)^{-\frac{1}{2}} \times 3$ — *Use the chain rule to differentiate $(3x - 2)^{\frac{1}{2}}$.*

$\qquad\qquad\qquad \dfrac{dv}{dx} = \dfrac{3}{2}(3x - 2)^{-\frac{1}{2}}$

The product rule states that $y = uv \Rightarrow \dfrac{dy}{dx} = u\dfrac{dv}{dx} + v\dfrac{du}{dx}$

Substituting gives $\dfrac{dy}{dx} = x \times \dfrac{3}{2}(3x - 2)^{-\frac{1}{2}} + (3x - 2)^{\frac{1}{2}} \times 1$

$\dfrac{dy}{dx} = \dfrac{3x}{2} \times \dfrac{1}{\sqrt{3x - 2}} + \sqrt{3x - 2}$

$\dfrac{dy}{dx} = \dfrac{3x}{2\sqrt{3x - 2}} + \sqrt{3x - 2}$. — *This is the right answer but you can tidy it up further to get $\dfrac{dy}{dx} = \dfrac{9x - 4}{2\sqrt{(3x - 2)}}$.*

2 Using the quotient rule

Differentiate $y = \dfrac{4x - 1}{4 + 3x}$. — *y is a quotient of two functions $(4x - 1)$ and $(4 + 3x)$ so you need to use the quotient rule.*

Solution

$y = \dfrac{4x - 1}{4 + 3x}$ — *This is of the form $y = \dfrac{u}{v}$.*

Let $u = 4x - 1$ and $v = 4 + 3x$

$\dfrac{du}{dx} = 4 \qquad \dfrac{dv}{dx} = 3$

The quotient rule: $y = \dfrac{u}{v} \Rightarrow \dfrac{dy}{dx} = \dfrac{v\dfrac{du}{dx} - u\dfrac{dv}{dx}}{v^2}$

Substituting gives $\dfrac{dy}{dx} = \dfrac{(4 + 3x) \times 4 - (4x - 1) \times 3}{(4 + 3x)^2}$

$\dfrac{dy}{dx} = \dfrac{16 + 12x - 12x + 3}{(4 + 3x)^2}$

$\dfrac{dy}{dx} = \dfrac{19}{(4 + 3x)^2}$

3 Finding the derivative of $y = \tan kx$

Use the quotient rule to prove that the derivative of $\tan 5x$ is $5\sec^2 5x$.

Solution

$y = \tan 5x \Rightarrow y = \dfrac{\sin 5x}{\cos 5x}$.

The quotient rule states that $y = \dfrac{u}{v} \Rightarrow \dfrac{dy}{dx} = \dfrac{v\dfrac{du}{dx} - u\dfrac{dv}{dx}}{v^2}$

Let $u = \sin 5x$ and $v = \cos 5x$

$\dfrac{du}{dx} = 5\cos 5x \qquad \dfrac{dv}{dx} = -5\sin 5x$

So $\dfrac{dy}{dx} = \dfrac{\cos 5x \times 5\cos 5x - \sin 5x \times (-5\sin 5x)}{(\cos 5x)^2}$

$= \dfrac{5\cos^2 5x + 5\sin^2 5x}{\cos^2 5x}$

$= \dfrac{5(\cos^2 5x + \sin^2 5x)}{\cos^2 5x}$ ← Since $\cos^2 \theta + \sin^2 \theta \equiv 1$ then $\cos^2 5x + \sin^2 5x \equiv 1$

$= \dfrac{5}{\cos^2 5x}$

$= 5\sec^2 5x$ ← You can quote this standard result, unless you have been asked to prove it!

4 Differentiating functions involving e^x and $\ln x$

Differentiate the following with respect to x:

i $y = x^2 e^{3x}$

ii $y = \dfrac{\ln x}{3x^2}$

Solution

i $y = x^2 e^{3x}$. ← This is of the form $y = uv$. So use the product rule to differentiate it.

Let $u = x^2$ and $v = e^{3x}$

$\dfrac{du}{dx} = 2x \qquad \dfrac{dv}{dx} = 3e^{3x}$ ← Remember that the derivative of e^{kx} is ke^{kx}.

The product rule: $y = uv \Rightarrow \dfrac{dy}{dx} = u\dfrac{dv}{dx} + v\dfrac{du}{dx}$

Substituting for u and v gives $\dfrac{dy}{dx} = x^2 \times 3e^{3x} + e^{3x} \times 2x$ ← Notice that x and e^{3x} are common factors.

$= e^{3x} x(3x + 2)$

ii $y = \dfrac{\ln x}{3x^2}$ ← This is of the form $y = \dfrac{u}{v}$. So use the quotient rule.

Let $u = \ln x$ and $v = 3x^2$

$\dfrac{du}{dx} = \dfrac{1}{x} \qquad \dfrac{dv}{dx} = 6x$

The quotient rule: $y = \dfrac{u}{v} \Rightarrow \dfrac{dy}{dx} = \dfrac{v\dfrac{du}{dx} - u\dfrac{dv}{dx}}{v^2}$

Substituting for u and v gives $\dfrac{dy}{dx} = \dfrac{3x^2 \times \dfrac{1}{x} - \ln x \times 6x}{(3x^2)^2}$

$\dfrac{dy}{dx} = \dfrac{3x - 6x \ln x}{9x^4}$

$\dfrac{dy}{dx} = \dfrac{3x(1 - 2\ln x)}{9x^4}$ ← Factorising.

$\dfrac{dy}{dx} = \dfrac{1 - 2\ln x}{3x^3}$

Test yourself

TESTED

1 The graph shows a part of the curve $y = x \sin 3x$. M is a maximum point.

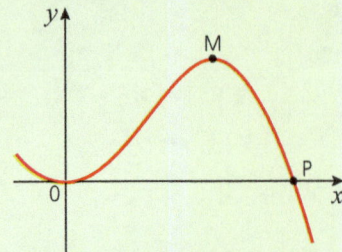

Four of the following statements are false and one is true. Which one is true?
- A The gradient at the point P is exactly -3.
- B The x coordinate of M is between 0.5 and 0.6
- C When $x = \frac{\pi}{6}$, the gradient of the curve is $\frac{\pi}{2}$.
- D The line $y = x$ touches the curve between $x = 0$ and $x = \frac{\pi}{2}$.
- E To the right of the point P, the gradient of $y = x \sin 3x$ is always negative.

2 Which of the following is the derivative of the function $y = x^3 \ln\left(\frac{1}{x}\right)$?

- A $3x^2 \ln\left(\frac{1}{x}\right) - x$
- B $x^2\left(3\ln\left(\frac{1}{x}\right) - 1\right)$
- C $x^2\left(3\ln\left(\frac{1}{x}\right) + 1\right)$
- D $-3x$

3 The equation of a curve is $y = \frac{x}{2x+1}$. Three of the following statements are false and one is true. Which one is true?
- A The gradient function is $\frac{dy}{dx} = \frac{1}{2}$.
- B The gradient function is positive for all values of x.
- C The gradient at $x = 2$ is $\frac{1}{25}$.
- D The gradient at $x = -2$ is $-\frac{1}{9}$.

4 Given that $y = \frac{x}{2 + 3\ln x}$, find the value of $\frac{dy}{dx}$ when $x = 1$.

- A $-\frac{1}{4}$
- B $\frac{1}{2}$
- C $\frac{1}{4}$
- D $\frac{1}{3}$

5 Find the gradient of the curve $y = \frac{\cos 2x}{x}$ at the point where $x = \pi$.

- A $-\frac{2}{\pi}$
- B $-\frac{1}{\pi^2}$
- C $\frac{1}{\pi^2}$
- D 1

Full worked solutions online CHECKED ANSWERS

Exam-style question

A curve has equation $y = (x + 2)e^{-x}$.
- i Find the coordinates of the points where the curve cuts the axes.
- ii Find the coordinates of the stationary point, S, on the curve.
- iii By evaluating $\frac{d^2y}{dx^2}$ at S, determine whether the stationary point is a maximum, a minimum, or a point of inflection.
- iv Sketch the curve.
  You may use the result that $xe^{-x} \to 0$ as $x \to \infty$.

Short answers on page 218

Full worked solutions online CHECKED ANSWERS

Implicit differentiation

REVISED

Key facts

1. An **explicit function** has one dependent variable (often y) as the subject and is in the form $y = f(x)$.

   An **implicit function** does not have the dependent variable as the subject, for example: $e^{2y} = x^2 + y$ or $x^2 + y^2 = 5$.

2. The **chain rule** is needed to differentiate implicit functions. Suppose you are asked to differentiate y^5 with respect to x. You know that if you differentiate it with respect to y, the answer is $5y^4$ but that is not what you were asked to do. To see how to do it, it is helpful to use a letter to denote y^5, say z.

   Let $z = y^5$ so $\frac{dz}{dy} = 5y^4$. Now apply the chain rule in the form

   $$\frac{dz}{dx} = \frac{dz}{dy} \times \frac{dy}{dx}$$

   $$\frac{dz}{dx} = 5y^4 \frac{dy}{dx}.$$

   This is called **implicit differentiation**.

   If you differentiate y^5 with respect to y, you get $5y^4$.
   If you differentiate y^5 with respect to x, you get $5y^4 \frac{dy}{dx}$.

> This is the answer. You will notice that you could have just written this answer down. There are two parts multiplied together:
> - $5y^4$ is the differential of y^5 with respect to y
> - $\frac{dy}{dx}$ allows for the fact that you are differentiating with respect to x not y.

Worked examples

1 Using implicit differentiation

In each case, differentiate the expression given with respect to x.

   i y^3 ii $\ln y$ iii $\sin(4y)$

Solution

> **Hint:** To differentiate with respect to x.... multiply by $\frac{dy}{dx}$.

i $\dfrac{d(y^3)}{dx} = \dfrac{d(y^3)}{dy} \dfrac{dy}{dx}$

      $= 3y^2 \dfrac{dy}{dx}$

ii $\dfrac{d(\ln y)}{dx} = \dfrac{d(\ln y)}{dy} \dfrac{dy}{dx}$

      $= \dfrac{1}{y} \dfrac{dy}{dx}$

iii $\dfrac{d(\sin(4y))}{dx} = \dfrac{d(\sin(4y))}{dy} \dfrac{dy}{dx}$

      $= 4\cos(4y) \dfrac{dy}{dx}$

2 Using the product rule

Differentiate $y^2 x^3$ with respect to x.

Solution

Let $u = y^2$ and $v = x^3$

so that $\dfrac{du}{dx} = 2y \dfrac{dy}{dx}$ and $\dfrac{dv}{dx} = 3x^2$

Edexcel A Level Mathematics (Pure)

Using the product rule

$$u\frac{dv}{dx} + v\frac{du}{dx} = y^2 \times 3x^2 + x^3 \times 2y\frac{dy}{dx}$$

$$= x^2 y\left(3y + 2x\frac{dy}{dx}\right)$$

3 Using the quotient rule

Differentiate $\dfrac{\sin y}{e^x}$ with respect to x.

Solution

Let $u = \sin y$ and $v = e^x$

so that $\dfrac{du}{dx} = \cos y \dfrac{dy}{dx}$ and $\dfrac{du}{dx} = e^x$

$u = \sin y$ so $\dfrac{du}{dx} = \dfrac{d(\sin y)}{dx}$
$= \dfrac{d(\sin y)}{dy}\dfrac{dy}{dx}$
$= \cos y \dfrac{dy}{dx}$

Using the quotient rule

$$\frac{v\frac{du}{dx} - u\frac{dv}{dx}}{v^2} = \frac{e^x \times \cos y \frac{dy}{dx} - \sin y \times e^x}{(e^x)^2}$$

$$= \frac{\cos y \frac{dy}{dx} - \sin y}{e^x}$$

4 When more than one term contains the dependent variable

You are given $e^{2y} = x^2 + y$.

i Find $\dfrac{dy}{dx}$.

ii By differentiating with respect to y, find $\dfrac{dx}{dy}$.

iii Comment on these results.

Hint: In this example, $\dfrac{dy}{dx}$ occurs more than once after you have differentiated. In such cases you need to collect all the $\dfrac{dy}{dx}$ terms on one side and then factorise it.

Solution

i Differentiating both sides of $e^{2y} = x^2 + y$ with respect to x gives

$$2e^{2y}\frac{dy}{dx} = 2x + \frac{dy}{dx}$$

$$\Rightarrow 2e^{2y}\frac{dy}{dx} - \frac{dy}{dx} = 2x$$

$$\Rightarrow \frac{dy}{dx}(2e^{2y} - 1) = 2x$$

$$\Rightarrow \frac{dy}{dx} = \frac{2x}{2e^{2y} - 1}$$

$\dfrac{d(e^{2y})}{dx} = \dfrac{d(e^{2y})}{dy}\dfrac{dy}{dx}$
$= 2e^{2y}\dfrac{dy}{dx}$

$\dfrac{d(y)}{dx} = \dfrac{d(y)}{dy}\dfrac{dy}{dx}$
$= 1 \times \dfrac{dy}{dx} = \dfrac{dy}{dx}$

Make $\dfrac{dy}{dx}$ the subject of this equation.

ii Differentiating both sides of $e^{2y} = x^2 + y$ with respect to y gives

$$2e^{2y} = 2x\frac{dx}{dy} + 1$$

$$\Rightarrow 2e^{2y} - 1 = 2x\frac{dx}{dy}$$

$$\Rightarrow \frac{dx}{dy} = \frac{2e^{2y} - 1}{2x}$$

Here you are differentiating x^2 with respect to y. Differentiate it with respect to x and then multiply by $\dfrac{dx}{dy}$ to give $2x\dfrac{dx}{dy}$.

Make $\dfrac{dx}{dy}$ the subject of this equation.

iii $\dfrac{dy}{dx} = \dfrac{2x}{2e^{2y}-1}$ and $\dfrac{dx}{dy} = \dfrac{2e^{2y}-1}{2x} \Rightarrow \dfrac{dx}{dy} = \dfrac{1}{\frac{dy}{dx}}$

5 Finding coordinates of turning points

A curve is defined by the equation $3y^2 + y = 7x^2 + 2$.

  i Find $\dfrac{dy}{dx}$ and hence find the gradient of the curve at the point A(2, 3).

  ii Find the coordinates of the points on the curve where $\dfrac{dy}{dx} = 0$.

Solution

i Differentiating both sides of $3y^2 + y = 7x^2 + 2$ with respect to x gives

$$6y\dfrac{dy}{dx} + \dfrac{dy}{dx} = 14x$$

$$\Rightarrow \dfrac{dy}{dx}(6y + 1) = 14x$$

$$\Rightarrow \dfrac{dy}{dx} = \dfrac{14x}{6y+1}$$

Make $\dfrac{dy}{dx}$ the subject of this equation.

For the point (2, 3), substitute $x = 2$ and $y = 3$:

$$\dfrac{dy}{dx} = \dfrac{28}{19}$$

ii $\dfrac{dy}{dx} = 0 \Rightarrow \dfrac{14x}{6y+1} = 0 \Rightarrow x = 0$

$\dfrac{14x}{6y+1} = 0$ only when $14x = 0$, i.e. when $x = 0$.

Substitute $x = 0$ into $3y^2 + y = 7x^2 + 2$:

$$\Rightarrow 3y^2 + y - 2 = 0$$
$$\Rightarrow (3y - 2)(y + 1) = 0$$
$$y = \dfrac{2}{3} \text{ or } y = -1.$$

Remember to find the y coordinates.

The points on the curve where $\dfrac{dy}{dx} = 0$ are $\left(0, \dfrac{2}{3}\right)$ and $(0, -1)$.

Test yourself

TESTED

1. Differentiate y^2 with respect to x.
   - A $2y$
   - B $2x$
   - C $y^2 \dfrac{dy}{dx}$
   - D $2y \dfrac{dy}{dx}$

2. Differentiate $\cos y$ with respect to x.
   - A $-\sin y$
   - B $-\sin y \dfrac{dy}{dx}$
   - C $\sin y \dfrac{dy}{dx}$
   - D $-\sin x$

3. Find $\dfrac{dy}{dx}$ when $y^2 = e^x + \sin y$.
   - A $\dfrac{e^x}{2y - \cos y}$
   - B $\dfrac{e^x}{2y + \cos y}$
   - C $\dfrac{e^x + \cos y}{2y}$
   - D $\dfrac{2y - e^x}{\cos y}$

4. The curve shown is defined implicitly by the equation $y^2 + 2 = x^3 + y$. Find the gradient of the curve at the point (2, 3).

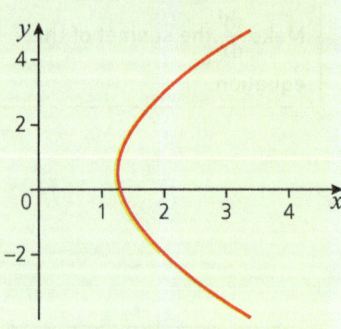

   - A $\dfrac{12}{5}$
   - B 9
   - C 2
   - D -6

5. Find $\dfrac{dy}{dx}$ when $y^2 = ye^x + x^2$.
   - A $\dfrac{ye^x + 2x}{2y + e^x}$
   - B $\dfrac{2x}{2y - e^x}$
   - C $\dfrac{ye^x + 2x}{2y - e^x}$
   - D $\dfrac{2x}{2y + e^x}$

Full worked solutions online   CHECKED ANSWERS

Exam-style question

The diagram shows the curve that is defined implicitly by the equation $y^3 = 2xy + x^2$.

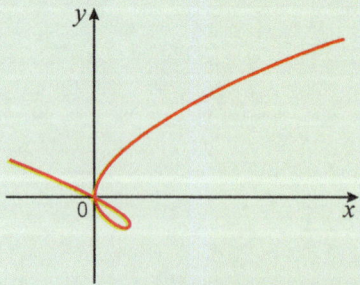

i   Show that $\dfrac{dy}{dx} = \dfrac{2(x+y)}{3y^2 - 2x}$

ii Write down $\dfrac{dx}{dy}$ in terms of x and y.
    Show that A(3, 3) is a point on the curve and find the value of $\dfrac{dx}{dy}$ at A.

iii State, with a reason, whether the line $y = x$ is a tangent to the curve at (3, 3).

Short answers on page 219

Full worked solutions online   CHECKED ANSWERS

Chapter 8 Integration

About this topic

Integration is the opposite process to differentiation. You can use integration to find the equation of a curve if you know its derivative, $\frac{dy}{dx}$, and a point that it passes through. Integration also gives the area under a curve and you can use it to find the area between a curve and the x- or y-axis and the area between two curves. Not every function can be integrated (see Chapter 14 for more on numerical integration), but integration by parts and integration by substitution are two powerful techniques which increase the number of functions you can integrate.

Before you start, remember

How to differentiate
- powers of x
- exponentials and logs
- trigonometric functions.

Integration as the reverse of differentiation

REVISED

Key facts

1. The rule for integrating a power of x:
$$\int ax^n \, dx = \frac{ax^{n+1}}{n+1} + c$$

   Think of this as 'add one to the power and divide by the new power'.

   This is true when n is any value except for -1.

2. A **definite integral** has **limits** which you substitute into the integrated function.
$$\int_a^b x^n \, dx = \left[\frac{x^{n+1}}{n+1}\right]_a^b = \frac{b^{n+1}}{n+1} - \frac{a^{n+1}}{n+1}$$

   An **indefinite** integral has no limits, for example:
   $$\int x^3 \, dx.$$

3. If you know a derivative, $\frac{dy}{dx}$, you can find the equation of a family of curves with that derivative by integrating.
   If you know one point on the curve, you can substitute this in to find the constant of integration and hence the equation of the curve that goes through that particular point.

   Don't forget c, the constant of integration!

4. $\int e^x \, dx = e^x + c$ and $\int e^{kx} \, dx = \frac{1}{k} e^{kx} + c$

   $\frac{d}{dx}(e^{kx}) = ke^{kx} \Rightarrow \int e^{kx} \, dx = \frac{1}{k} e^{kx} + c.$

5. $\int \frac{1}{x} \, dx = \ln|x| + c$

   The modulus signs are used because you can only find the log of a positive number.

Worked examples

1 Finding indefinite integrals with any power of x

Find:

i $\quad \int \left(x - \dfrac{1}{x^2} + \sqrt{x} - \dfrac{1}{x} \right) dx$

ii $\quad \int \left(\dfrac{\sqrt{x} + 4}{x} \right) dx$

> Rewrite using negative and fractional powers of x. Remember that $\dfrac{1}{x^n} = x^{-n}$.

Solution

i
$$\int \left(x - \frac{1}{x^2} + \sqrt{x} - \frac{1}{x} \right) dx$$
$$= \int \left(x - x^{-2} + x^{\frac{1}{2}} - \frac{1}{x} \right) dx$$
$$= \frac{1}{2}x^2 + x^{-1} + \frac{2}{3}x^{\frac{3}{2}} - \ln|x| + c$$
$$= \frac{x^2}{2} + \frac{1}{x} + \frac{2}{3}x^{\frac{3}{2}} - \ln|x| + c$$

> Remember that $\int \dfrac{1}{x}\,dx = \ln|x| + c$.

> Don't forget '$+c$'.

> Dividing by $\dfrac{3}{2}$ is the same as multiplying by $\dfrac{2}{3}$.

ii
$$\int \left(\frac{\sqrt{x} + 4}{x} \right) dx = \int \left(\frac{\sqrt{x}}{x} + \frac{4}{x} \right) dx$$
$$= \int \left(x^{-\frac{1}{2}} + \frac{4}{x} \right) dx$$
$$= 2x^{\frac{1}{2}} + 4\ln|x| + c$$
$$= 2\sqrt{x} + 4\ln|x| + c$$

> You can't integrate this as it stands, you need to write it so each term is a single power of x.

> Remember the laws of indices: $\dfrac{x^a}{x^b} = x^{a-b}$

2 Integrating e^x

Given that $\dfrac{dy}{dx} = e^x - 6e^{-3x} + e^2$, find y.

Solution

$$\frac{dy}{dx} = e^x - 6e^{-3x} + e^2$$
$$\Rightarrow y = \int \left(e^x - 6e^{-3x} + e^2 \right) dx$$
$$\Rightarrow y = e^x - \frac{6e^{-3x}}{-3} + e^2 x + c$$
$$\Rightarrow y = e^x + 2e^{-3x} + e^2 x + c$$

> **Common mistake**: Watch out for constant terms like $\ln 5$, e^2 or 2π when you integrate. These integrate to $x\ln 5$, xe^2 and $2\pi x$ respectively.

> Integrate both sides.

> Remember that $\int e^{kx}\,dx = \dfrac{1}{k}e^{kx} + c$.

3 Evaluating definite integrals

Find the exact value of $\int_{-\frac{1}{2}}^{\frac{1}{2}} \left(e^{2x} + \dfrac{1}{e^{2x}} \right) dx$.

Solution

$$\int_{-\frac{1}{2}}^{\frac{1}{2}} \left(e^{2x} + \frac{1}{e^{2x}} \right) dx = \int_{-\frac{1}{2}}^{\frac{1}{2}} \left(e^{2x} + e^{-2x} \right) dx$$
$$= \left[\frac{1}{2}e^{2x} - \frac{1}{2}e^{-2x} \right]_{-\frac{1}{2}}^{\frac{1}{2}}$$
$$= \left(\frac{1}{2}e^{2 \times \frac{1}{2}} - \frac{1}{2}e^{-2 \times \frac{1}{2}} \right) - \left(\frac{1}{2}e^{2 \times \left(-\frac{1}{2}\right)} - \frac{1}{2}e^{-2 \times \left(-\frac{1}{2}\right)} \right)$$

> **Common mistake**: Remember that an exact answer may be an integer, or contain a fraction, a surd, or a number like π or e. $\sqrt{3}$ is exact; 1.732 is not. $\dfrac{2}{3}$ is exact; 0.666 is not.

> Remember that $\dfrac{1}{e^{2x}} = e^{-2x}$.

$$= \left(\frac{1}{2}e^1 - \frac{1}{2}e^{-1}\right) - \left(\frac{1}{2}e^{-1} - \frac{1}{2}e^1\right)$$
$$= e^1 - e^{-1} = e - \frac{1}{e}$$

Remember that $e^1 = e$.

4 Finding the equation of a curve given its gradient function, $\frac{dy}{dx}$

The gradient of a curve is given by $\frac{dy}{dx} = \frac{3}{x}$. The curve passes through (e, 2). Find the equation of the curve.

If you know $\frac{dy}{dx}$ then integrating will give you y.

Solution

$y = \int \frac{3}{x} dx \Rightarrow y = 3\ln|x| + c$

This is the **general equation** of the curve.

When $x = e$, $y = 2$, so $2 = 3 \ln e + c$
$= 3 + c$, so $c = -1$

Use the coordinates of the point you have been given to find the value of c.

So the equation of the curve is $y = 3\ln|x| - 1$

This is the **particular solution** that passes through the point (e, 2).

Test yourself

TESTED

1 Find $\int \frac{3}{x^2} dx$.

A $3x^{-2} + c$ B $-\frac{6}{x^3} + c$ C $-\frac{3}{x} + c$ D $2x^{\frac{3}{2}} + c$

2 Find $\int \sqrt[3]{x} \, dx$.

A $-\frac{1}{2x^2} + c$ B $x^{\frac{1}{3}} + c$ C $\frac{1}{3\sqrt[3]{x^2}} + c$ D $\frac{3(\sqrt[3]{x})^4}{4} + c$

3 Find $\int_1^4 \sqrt{x} \, dx$.

A $\frac{21}{2}$ B $\frac{3}{4}$ C 1 D $\frac{14}{3}$

4 Find $\int_2^4 \left(x^3 + \frac{1}{x^3}\right) dx$.

A $56\frac{9}{64}$ B $60\frac{3}{32}$ C $60\frac{15}{1024}$ D $59\frac{29}{32}$

5 The gradient of a curve is given by $\frac{dy}{dx} = \frac{3}{x^4}$. The curve passes through (1, 2). Find the equation of the curve.

A $y = 3 - \frac{1}{x^3}$ B $y = -\frac{12}{x^5}$ C $y = 1\frac{1}{8} - \frac{1}{x^3}$ D $y = 2\frac{3}{5} - \frac{3}{5x^5}$

Full worked solutions online

CHECKED ANSWERS

Edexcel A Level Mathematics (Pure)

Exam-style question

A curve has gradient given by $\dfrac{dy}{dx} = \dfrac{e^{3x} - 2}{e^{2x}}$.

The curve passes through the point (0, 6).

Find the equation of the curve.

Short answers on page 219

Full worked solutions online

CHECKED ANSWERS

Finding areas

REVISED

Key facts

1. You can find the area under the graph of a function by dividing it up into narrow rectangles.

   The height of each rectangle is f(x).
   The width of each rectangle is δx.
   Area of one rectangle is $f(x) \times \delta x$.
   Total area is the sum of all the rectangles = $\sum f(x)\, \delta x$.

   The narrower the rectangles, the better the approximation.

   So as $\delta x \to 0$, $\sum_{a}^{b} f(x)\, \delta x \to \int_{a}^{b} f(x)\, dx$

   > To find the height of each rectangle you substitute the x coordinate into f(x).

   > The **fundamental theorem of calculus** says that integration is the reverse of differentiation.

2. **Area under a curve**

   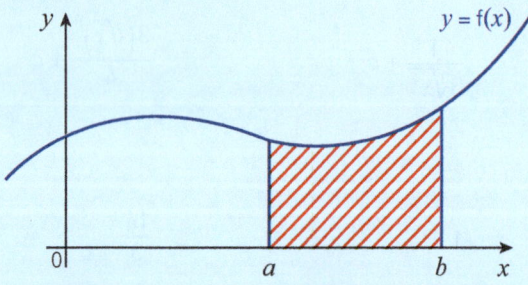

   The shaded area between the curve y = f(x) and the x-axis, and between the values x = a and x = b is given by

   $A = \int_{a}^{b} f(x)\, dx = \left[F(x) \right]_{a}^{b} = F(b) - F(a)$

   > F(x) is obtained by integrating f(x).

3. Integration gives **areas below the x-axis** as negative.

   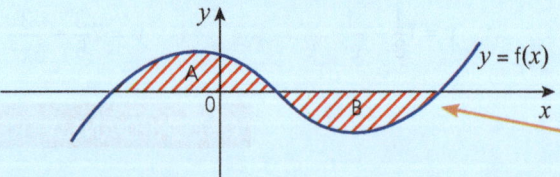

   If y = f(x) crosses the x-axis in the required region then calculate the area in two parts A and B and add them together.
   Total area is A + |B|.

   > Be careful when finding areas below the axis.
   > When you calculate the area of Region B the answer will have a negative sign.

   > Take the positive value of the area B.

4. An alternative method for finding the **area between two curves** is to integrate the difference between the curves. The limits are the x coordinates of the points of intersection.

$$A = \int (\text{top curve} - \text{bottom curve})\,dx$$

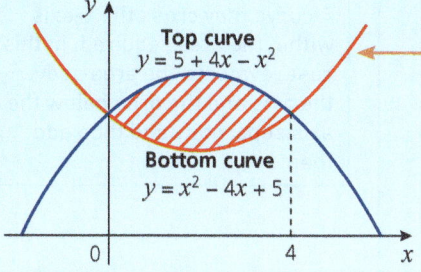

In this case,
$$\text{area} = \int_0^4 \left([5 + 4x - x^2] - [x^2 - 4x + 5] \right) dx$$
$$= \int_0^4 (8x - 2x^2)\,dx$$
$$= \left[\frac{8x^2}{2} - \frac{2x^3}{3} \right]_0^4$$
$$= \left(64 - \frac{128}{3} \right) - (0)$$
$$= 21\frac{1}{3} \text{ units}^2$$

5. **The area between a curve and the y-axis** is $\int_a^b x\,dy$

To find an area between the curve and the y-axis:

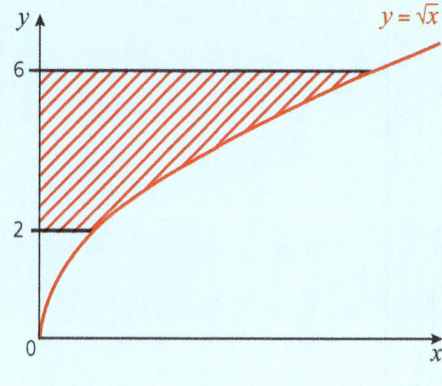

Step 1: Make x the subject
$y = \sqrt{x} \Rightarrow x = y^2$

Step 2: Integrate $x = f(y)$ with respect to y
$$\int_a^b x\,dy = \int_a^b y^2\,dy$$
$$= \left[\frac{1}{3} y^3 \right]_2^6$$
$$= \frac{1}{3} \times 6^3 - \frac{1}{3} \times 2^3$$
$$= 69\frac{1}{3}.$$

Worked examples

1 Finding the area under a curve

The shaded region in the diagram is bounded by the curve $y = \frac{1}{2}e^{3x}$, the lines $x = 0.4$ and $x = 0.8$ and the x-axis.

Find the area of the shaded region.

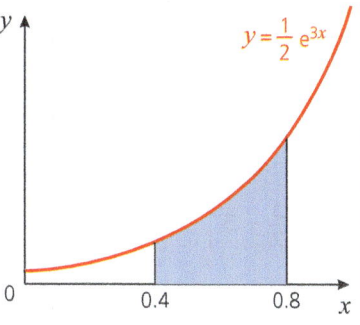

Solution

The area is
$$\int_{0.4}^{0.8} \frac{1}{2} e^{3x}\,dx = \left[\frac{1}{2} \times \frac{1}{3} e^{3x} \right]_{0.4}^{0.8}$$
$$= \frac{1}{6} e^{2.4} - \frac{1}{6} e^{1.2}$$
$$= 1.2838\ldots$$
$$= 1.28 \text{ (to 3 s.f.)}$$

To find an area you need to integrate the function between the two limits.

Edexcel A Level Mathematics (Pure)

2 Finding areas below the x-axis

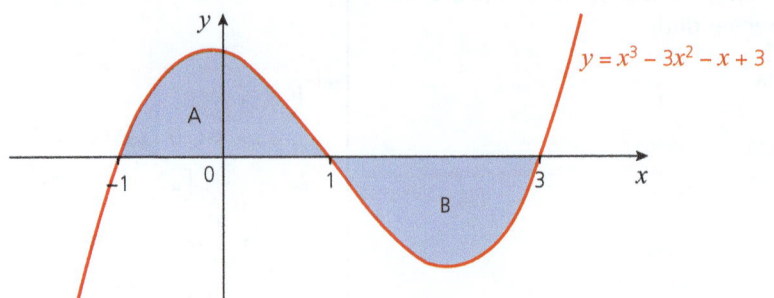

Hint: Remember that if an area is below the x-axis it has a negative sign.

A curve may cross the x-axis within the area required. In this case, evaluate the area above the axis and the area below the axis separately and then add them.

The graph shows the curve $y = x^3 - 3x^2 - x + 3$.

  i Find the area of the shaded region.
  ii Evaluate $\int_{-1}^{3}(x^3 - 3x^2 - x + 3)\,dx$.
  iii Explain why your answers to parts **i** and **ii** are not the same.

Solution

i $A = \int_{-1}^{1}(x^3 - 3x^2 - x + 3)\,dx = \left[\dfrac{x^4}{4} - x^3 - \dfrac{x^2}{2} + 3x\right]_{-1}^{1}$

$= \left(\dfrac{1}{4} - 1 - \dfrac{1}{2} + 3\right) - \left(\dfrac{1}{4} + 1 - \dfrac{1}{2} - 3\right)$

$= 4$

and $B = \int_{1}^{3}(x^3 - 3x^2 - x + 3)\,dx = \left[\dfrac{x^4}{4} - x^3 - \dfrac{x^2}{2} + 3x\right]_{1}^{3}$

$= \left(\dfrac{81}{4} - 27 - \dfrac{9}{2} + 9\right) - \left(\dfrac{1}{4} - 1 - \dfrac{1}{2} + 3\right)$

$= -4$

This area is negative because it is below the x-axis.

Total shaded area = $A + B = 4 + 4 = 8$ square units.

ii $\int_{-1}^{3}(x^3 - 3x^2 - x + 3)\,dx$

$= \left[\dfrac{x^4}{4} - x^3 - \dfrac{x^2}{2} + 3x\right]_{-1}^{3}$

$= \left(\dfrac{81}{4} - 27 - \dfrac{9}{2} + 9\right) - \left(\dfrac{1}{4} + 1 - \dfrac{1}{2} - 3\right)$

$= 0$

$= \left(-\dfrac{9}{4}\right) - \left(-\dfrac{9}{4}\right)$

Common mistake: If you are just asked to evaluate a definite integral you don't need to worry about whether the curve is above or below the x-axis.

iii The answers to **i** and **ii** are different because in **i** the fractions are added without regard to sign but in **ii** the minus sign means that the fractions were subtracted.

The answer to **ii** is the net area above the x-axis.

3 Finding the area between a curve and a line

The diagram shows the curve $y = 2x^2 + 6x$ and the line $y = x + 3$.

Find the exact area of the shaded region.

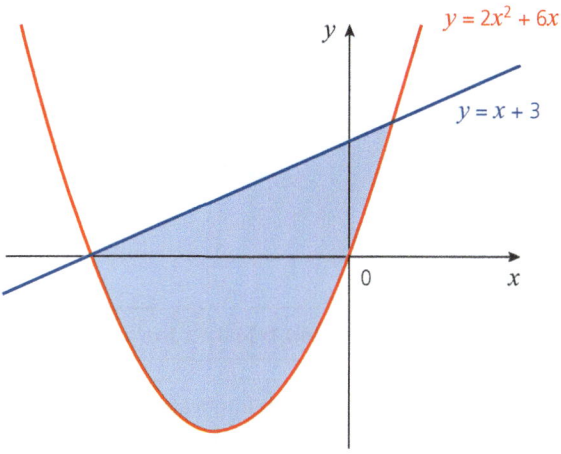

Solution

The limits are the x coordinates of the points where the curve and line intersect. ◄── Solve $y = 2x^2 + 6x$ and $y = x + 3$.

$\Rightarrow \quad 2x^2 + 6x = x + 3$

$\Rightarrow \quad 2x^2 + 5x - 3 = 0$

$\Rightarrow (2x - 1)(x + 3) = 0$

$\Rightarrow \quad x = \frac{1}{2}$ or $x = -3$ ◄── So the limits are -3 and $\frac{1}{2}$.

$\text{Area} = \int_{-3}^{\frac{1}{2}} ((x + 3) - (2x^2 + 6x)) \, dx$ ◄── Area = $\int$ (top curve − bottom curve) dx

$= \int_{-3}^{\frac{1}{2}} (-2x^2 - 5x + 3) \, dx$ ◄── Simplify before you integrate.

$= \left[-\frac{2}{3}x^3 - \frac{5}{2}x^2 + 3x \right]_{-3}^{\frac{1}{2}}$

$= \left(-\frac{2}{3} \times \left(\frac{1}{2}\right)^3 - \frac{5}{2} \times \left(\frac{1}{2}\right)^2 + 3 \times \left(\frac{1}{2}\right) \right)$

$\qquad - \left(-\frac{2}{3} \times (-3)^3 - \frac{5}{2} \times (-3)^2 + 3 \times (-3) \right)$

$= \left(\frac{19}{24}\right) - \left(-\frac{27}{2}\right)$

$= \frac{343}{24}$ ◄── Leave your answer as a fraction as the question asks for the **exact** area.

4 Finding the area between a curve and the y-axis

The diagram shows the curve $y = \ln x$.
Find the exact area of the shaded region.

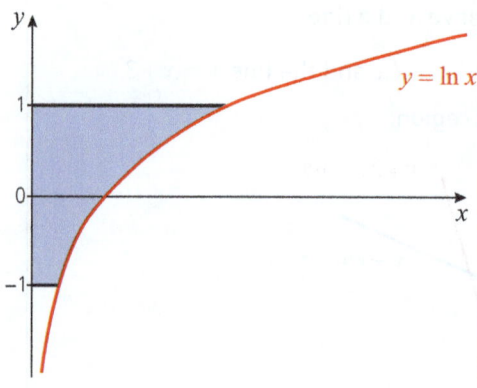

Solution

$y = \ln x \Rightarrow x = e^y$ ← **Step 1:** Make x the subject.

$\int_{-1}^{1} x \, dy = \int_{-1}^{1} e^y \, dy$ ← **Step 2:** Integrate $x = f(y)$ with respect to y.

$= \left[e^y \right]_{-1}^{1}$

$= e^1 - e^{-1}$

$= e - \dfrac{1}{e}$ ← Leave the answer in exact form.

Test yourself

TESTED

1 Here is the graph of $y = 4 - x^2$

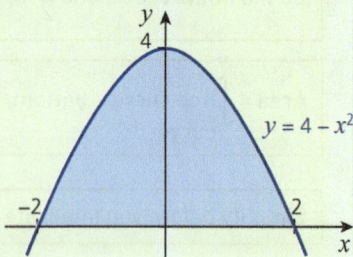

You are asked to find the area between the curve $y = 4 - x^2$ and the x-axis.

Three of the following expressions will give the correct answer.

Which of them **cannot** lead to the correct answer?

A $\int_{-2}^{2} (4 - x^2) \, dx$ B $\int_{0}^{4} (4 - x^2) \, dx$ C $\left[4x - \dfrac{x^3}{3} \right]_{-2}^{2}$ D $2 \int_{0}^{2} (4 - x^2) \, dx$

2 Here is the graph of $y = x^3 - x$.

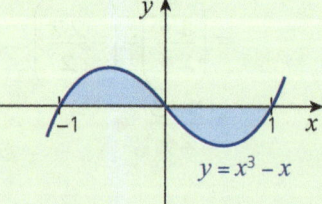

You are asked to find the shaded area between the curve $y = x^3 - x$ and the x-axis. One of the following expressions will give the correct answer.

Which one leads to the correct answer?

A $-\int_{-1}^{0} (x^3 - x) \, dx + \int_{0}^{1} (x^3 - x) \, dx$ B $\left[\dfrac{x^4}{4} - \dfrac{x^2}{2} \right]_{-1}^{1}$

C $\int_{-1}^{0} (x^3 - x) \, dx + \int_{0}^{1} (x^3 - x) \, dx$ D $\int_{-1}^{0} (x^3 - x) \, dx - \int_{0}^{1} (x^3 - x) \, dx$

3 The diagram shows the curves $y = x^2 - 2x$ and $y = x^3 + 3x^2 - 5x$ which intersect at P, Q and R.

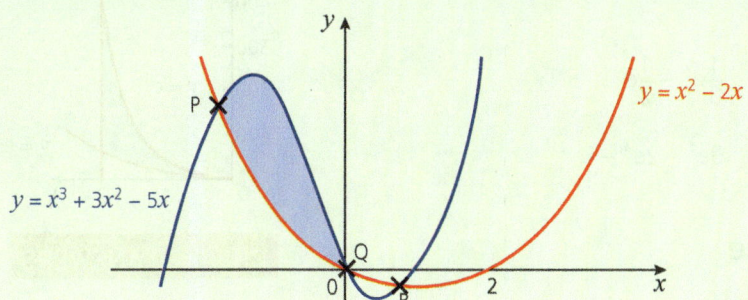

You are required to find the area between the curves, point P and point Q.
The working is given below but there is a mistake in it. At which line does the mistake occur?

Line A   At points of intersection: $x^3 + 3x^2 - 5x = x^2 - 2x$
$$x^3 + 2x^2 - 3x = 0$$

Line B $$x(x+3)(x-1) = 0$$

Line C   So P is (−3, 3), Q is (0, 0), and R is (1, −1)
$$\text{Area} = \int_{-3}^{0} (x^3 + 3x^2 - 5x)\,dx - \int_{-3}^{0} (x^2 - 2x)\,dx$$

Line D $$= \int_{-3}^{0} (x^3 + 2x^2 - 3x)\,dx$$

$$= \left[\frac{x^4}{4} + \frac{2x^3}{3} - \frac{3x^2}{2}\right]_{-3}^{0}$$

Line E $$= 0 - \left(\frac{81}{4} - 18 - \frac{27}{2}\right) = -11\frac{1}{4}$$

4 The diagram shows the curves $y = x^5$ and $y = x^6$ which intersect at (0, 0) and (1, 1).

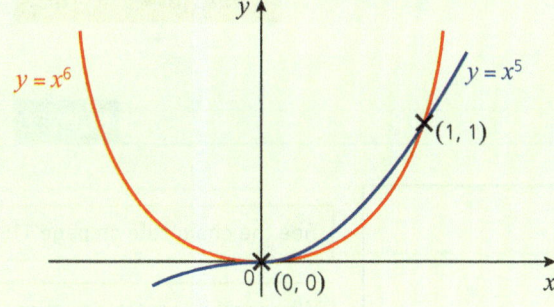

Find the area enclosed between the curves.

A $\frac{1}{6}$ B $\frac{13}{42}$ C $\frac{1}{42}$ D $\frac{1}{7}$

5 Which of the following is the area of the shaded region?

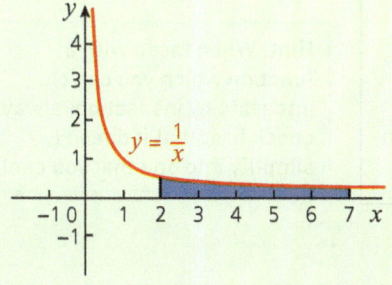

A $\ln 5$ B $\ln 3.5$ C $\frac{1}{4} - \frac{1}{49}$ D $\frac{\ln 7}{\ln 2}$ E 0.544

6 The shaded region in this graph is formed by the curves $y = e^{3x}$ and $y = e^{2x}$ and the line $x = 2$.
Find the area of the region.

A $\frac{1}{3}e^6 - \frac{1}{2}e^4 + \frac{1}{6}$

B $\frac{1}{3}e^6 - \frac{1}{2}e^4$

C $e^2 - 1$

D $3e^6 - 2e^4 - 1$

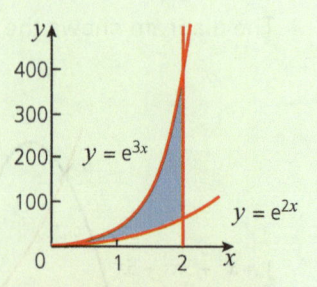

Full worked solutions online CHECKED ANSWERS

Exam-style question

The graph shows the curve $y = e^x$ and the line $y = p$.

It is drawn with equal scales along the x-axis and y-axis.

The line $y = p$ cuts the curve at the point P.

The region between the curve, the y-axis and the line $y = p$ is shaded.
i   Find the coordinates of P.
ii Find the area of the region bounded by the curve, the axes and the line through P parallel to the y-axis.
    Hence find the area of the shaded region.
iii Copy the graph and then draw on it a sketch of $y = \ln x$.
    Shade the region bounded by the curve $y = \ln x$, the x-axis and the line $x = p$.
iv Write down, in terms of p, the value of $\int_1^p \ln x \, dx$.

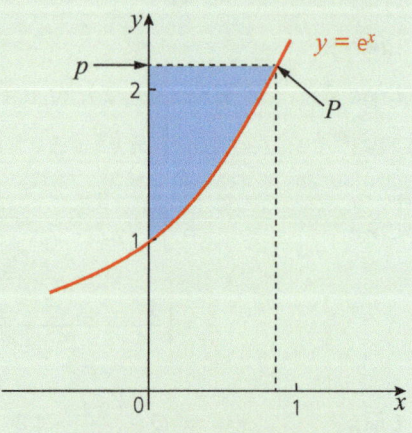

Short answers on page 219

Full worked solutions online CHECKED ANSWERS

Integration by substitution   REVISED

Key facts

1 Since $y = \ln|f(x)| \Rightarrow \dfrac{dy}{dx} = \dfrac{f'(x)}{f(x)}$.

   then $\int \dfrac{f'(x)}{f(x)} dx = \ln|f(x)| + c$.

   See the chain rule on page 118.

   When you integrate a quotient, check if the top line is the derivative of the bottom line. If it is, the answer is $\ln|\text{bottom line}| + c$.

2 Some integrals which cannot be integrated directly can be transformed into a simpler form by using a **substitution**. You need to write one of the functions as u and then rewrite the whole integral (including the 'dx') in terms of u.

   When an integral contains a 'function of a function' it usually helps to use a substitution for the 'inside' function.

   For example, for $\int \sqrt{x-3} \, dx$, $\sqrt{x-3}$ is a function ($\sqrt{}$) of a function $(x-3)$ so use $u = x - 3$.

   Hint: When faced with a function which you can't integrate by inspection, always check first that it doesn't simplify into one that you can!

Worked examples

1 Indefinite integration by substitution

Find $\int \sqrt{(2x+5)}\,dx$.

> $\sqrt{(2x+5)}$ is a function of a function. You can't integrate it directly but you can use the substitution $u = 2x+5$ to transform this integral into one you can do.

Solution

Let $u = 2x + 5 \Rightarrow \dfrac{du}{dx} = 2$

$\Rightarrow du = 2\,dx$

$\Rightarrow \dfrac{1}{2}du = dx$

> You can't find $\int \sqrt{u}\,dx$ as you can't integrate u with respect to x so you need to convert the dx to du.

Now substitute into the original integral.

$\int \sqrt{(2x+5)}\,dx = \int \sqrt{u}\,\dfrac{1}{2}du$

$= \int \dfrac{1}{2} u^{\frac{1}{2}}\,du$

$= \dfrac{1}{2} \times \dfrac{2}{3} u^{\frac{3}{2}} + c$

$= \dfrac{1}{3} u^{\frac{3}{2}} + c$

So $\int \sqrt{(2x+5)}\,dx = \dfrac{1}{3}(2x+5)^{\frac{3}{2}} + c$.

> Replace $(2x+5)$ with u and dx with $\dfrac{1}{2}du$.

> Don't forget the '+ c'.

> **Common mistake:** Don't forget to substitute back for x. Never leave your final answer in terms of u.

> **Hint:** This process is like the chain rule for differentiation in reverse. With practice you can integrate functions like this example by inspection.

2 Definite integration by substitution

Evaluate $\int_0^2 x(x^2 - 3)^5\,dx$ by using the substitution $u = x^2 - 3$.

Solution

Let $u = x^2 - 3$.

Convert dx to du:

$\dfrac{du}{dx} = 2x \Rightarrow du = 2x\,dx \Rightarrow \dfrac{1}{2}du = x\,dx$

Convert the limits:

$x = 2 \Rightarrow u = 2^2 - 3 = 1$

$x = 0 \Rightarrow u = 0 - 3 = -3$

> **Common mistake:** When you have a definite integral you need to rewrite the limits before substituting them in.

> You have an 'x' in the integral – so leave the 'x' with the 'dx' when you rearrange.

> Rewriting the integral like this makes it easier to see how to make the substitution.

Now substitute into the original integral.

$\int_0^2 x(x^2-3)^5\,dx = \int_0^2 (x^2-3)^5\, x\,dx$

$= \int_{-3}^{1} u^5 \times \dfrac{1}{2}\,du$

$= \left[\dfrac{1}{2} \times \dfrac{1}{6} u^6 \right]_{-3}^{1}$

$= \left(\dfrac{1}{12} \times 1^6 \right) - \left(\dfrac{1}{12} \times (-3)^6 \right)$

$= -60\dfrac{2}{3}$

> Replace $(x^2 - 3)$ with u and $x\,dx$ with $\dfrac{1}{2}du$ and then change the limits.

> This is what you are aiming for: a function you can integrate directly.

Edexcel A Level Mathematics (Pure)

3 Finding areas

Here is a sketch of the curve $y = 3x^2 e^{(1-x^3)}$.

Find the exact area of the shaded region.

Use the substitution $u = (1 - x^3)$.

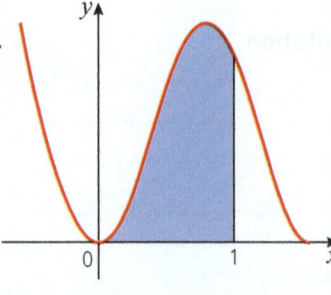

Solution

To find $\int_0^1 3x^2 e^{(1-x^3)} \, dx$:

Let $u = (1 - x^3)$.

Convert dx to du:

$\frac{du}{dx} = -3x^2 \Rightarrow du - 3x^2 \, dx \Rightarrow -du = 3x^2 \, dx$.

> You might find it easier to think of this as $-1 du = 3x^2 dx$.

Convert the limits:

$x = 1 \Rightarrow u = 1 - 1^3 = 0$

$x = 0 \Rightarrow u = 1 - 0 = 1$

Now substitute into the original integral.

$$\int_0^1 3x^2 e^{(1-x^3)} \, dx = \int_0^1 e^{(1-x^3)} \times 3x^2 \, dx$$
$$= \int_1^0 e^u \times -1 \, du$$
$$= \int_1^0 -e^u \, du$$
$$= \left[-e^u\right]_1^0$$
$$= (-e^0) - (-e^1)$$
$$= -1 + e$$
$$= e - 1$$

> **Common mistake:** Take extra care with the limits. The upper limit isn't always the larger number.

> **Common mistake:** You were asked for the **exact** value, so don't work this out.

4 Integrating a quotient

Find $\int \frac{2x}{x^2 + 4} dx$.

Solution

$\int \frac{2x}{x^2 + 4} dx = \int \frac{1}{x^2 + 4} \times 2x \, dx$

$\frac{1}{x^2 + 4}$ is a function of a function so use the substitution $u = x^2 + 4$

So let $u = x^2 + 4$

Convert dx to du:

$\frac{du}{dx} = 2x \Rightarrow du = 2x \, dx$

$\int \frac{2x}{x^2 + 4} dx = \int \frac{1}{u} du$
$= \ln|u| + c$

So $\int \frac{2x}{x^2 + 4} dx = \ln(x^2 + 4) + c$

> **Hint:** Notice in $\int \frac{2x}{x^2+4} dx$ the top line ($2x$) is the derivative of the bottom line ($x^2 + 4$) and when you integrate you get an answer of $\ln|\text{bottom line}| + c$.
>
> So you could have used Key fact 1 instead
>
> $\int \frac{f'(x)}{f(x)} dx = \ln|f(x)| + c$.
>
> With practice, you can use inspection to integrate quotients which are in this form.

> To get the final answer replace u with $x^2 + 4$

> You don't need the modulus signs as $x^2 + 4$ is always positive.

5 Using the result $\int \frac{f'(x)}{f(x)} dx = \ln|f(x)| + c$

Evaluate $\int_{-2}^{0} \frac{6}{1-2x} dx$

> You could use the substitution $u = 1 - 2x$ instead, but it is quicker if you recognise this type of integral.

Solution

The derivative of $1 - 2x$ is -2.

$$\int_{-2}^{0} \frac{6}{1-2x} dx = \int_{-2}^{0} \frac{-3 \times -2}{1-2x} dx = -3\int_{-2}^{0} \frac{-2}{1-2x} dx$$

> You can adjust the top line to make it the derivative of the bottom line. Since $-3 \times -2 = 6$ the integral is unchanged.

Using the rule:

$$\int \frac{f'(x)}{f(x)} dx = \ln|f(x)| + c$$

$$-3\int_{-2}^{0} \frac{-2}{1-2x} dx = \Big[-3\ln|1-2x|\Big]_{-2}^{0}$$

$$= (-3\ln 1) - (-3\ln 5)$$

$$= 3\ln 5$$

> You may be able to go straight to this line but to avoid mistakes it is best to show all of your working.

$-3 \ln 1 = 0$

Test yourself

TESTED

1 Find $\int \frac{x}{\sqrt{1+x^2}} dx$. You may wish to use the substitution $u = 1 + x^2$.

A $2\sqrt{1+x^2} + c$ B $\sqrt{1+x^2} + c$ C $u^{\frac{1}{2}} + c$ D $\frac{1}{2}\ln(1+x^2) + c$

2 Find $\int \frac{6x^2}{1+4x^3} dx$.

Two of these answers are equivalent and both are correct. Look for both of them.

A $\frac{1}{2}\ln|1+4x^3| + c$ B $2\ln|1+4x^3| + c$ C $\frac{2x^3}{x+x^4} + c$

D $2x^3 + \frac{3}{2}\ln|x| + c$ E $\frac{1}{2}\ln|2(1+4x^3)| + c$

3 The diagram shows a sketch of the curve $y = 10x(2x-1)^3$.

Find the area of the shaded region using the substitution $u = 2x - 1$.

A $\frac{1}{4}$ B 0.055

C $\frac{1}{8}$ D $-\frac{1}{8}$

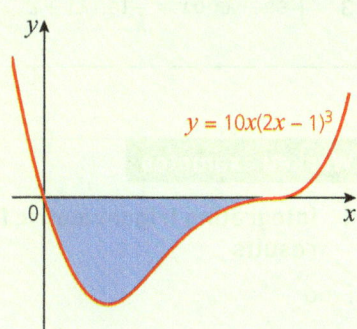

4 The diagram shows a sketch of part of the curve $y = xe^{1-x^2}$.
Find the exact area of the shaded region which is between the curve, the x-axis and the line $x = 1$.

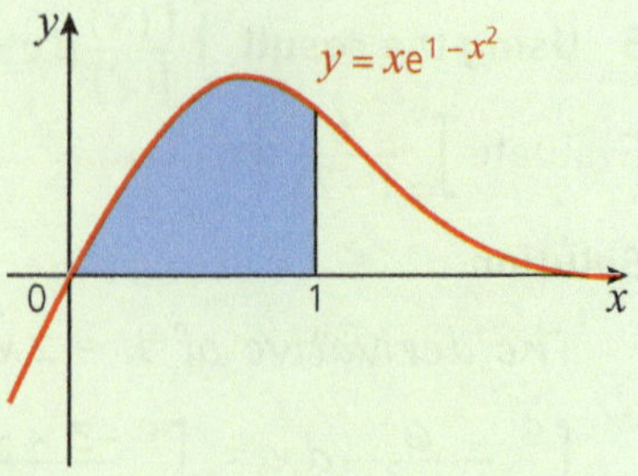

A $\dfrac{1-e}{2}$ B $\dfrac{e-1}{2}$

C $2(e-1)$ D $\dfrac{1}{2}e^{\frac{2}{3}}$

5 Find the exact value of $\int_0^1 \dfrac{2e^x}{1+e^x} dx$. You may wish to use the substitution $u = 1 + e^x$.

A $2 - 2\ln 2$ B $\dfrac{2e}{1+e} - 2$ C $\dfrac{1}{2}\ln\left(\dfrac{1+e}{2}\right)$ D $2\ln\left(\dfrac{1+e}{2}\right)$

Full worked solutions online   CHECKED ANSWERS

Exam-style question

i Show that the substitution $u = 2x - 1$ can be used to transform the integral $\int \dfrac{x}{(2x-1)^2} dx$ into $\dfrac{1}{4}\int \dfrac{u+1}{u^2} du$.

ii Hence show that the exact value of $\int_1^2 \dfrac{x}{(2x-1)^2} dx$ is $\dfrac{2 + 3\ln 3}{12}$.

Short answers on page 219

Full worked solutions online   CHECKED ANSWERS

Integrating trigonometric functions   REVISED

Key facts

1 $\int \cos kx \, dx = \dfrac{1}{k}\sin kx + c$ Using the facts that integration is the reverse of differentiation and $\dfrac{d}{dx}(\sin kx) = k\cos kx$.

2 $\int \sin kx \, dx = -\dfrac{1}{k}\cos kx + c$

3 $\int \sec^2 kx \, dx = \dfrac{1}{k}\tan kx + c$ Using $\dfrac{d}{dx}(\tan kx) = k\sec^2 kx$.

Worked examples

1 **Integrating trigonometric functions using the standard results**

Find

i $\int (\cos 5x + \sin 2x) \, dx$ ii $\int \sec^2 3x \, dx$.

Solution

i $\int (\cos 5x + \sin 2x) \, dx = \dfrac{1}{5}\sin 5x - \dfrac{1}{2}\cos 2x + c$ Use the standard results given in the Key facts.

ii $\int \sec^2 3x \, dx = \dfrac{1}{3}\tan 3x + c$

2 Finding areas

The sketch shows the curve $y = \sin\left(3x + \frac{\pi}{4}\right)$.

Find the exact area of the shaded region.

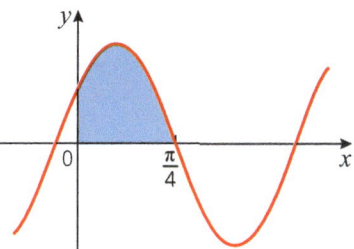

Hint: With practice you may find you can integrate a lot of trigonometric functions by inspection. But if you are in any doubt it is best to show all of your working.

Solution

$$\int_0^{\frac{\pi}{4}} \sin\left(3x + \frac{\pi}{4}\right) dx$$

Let $u = 3x + \frac{\pi}{4}$

Convert dx to du: $\frac{du}{dx} = 3 \Rightarrow \frac{1}{3}du = dx$

Change the limits:

$$x = \frac{\pi}{4} \Rightarrow u = 3 \times \frac{\pi}{4} + \frac{\pi}{4} = \pi$$

$$x = 0 \Rightarrow u = 3 \times 0 + \frac{\pi}{4} = \frac{\pi}{4}$$

So $\int_0^{\frac{\pi}{4}} \sin\left(3x + \frac{\pi}{4}\right) dx = \int_{\frac{\pi}{4}}^{\pi} \sin u \times \frac{1}{3} du$

$$= \left[-\frac{1}{3}\cos u\right]_{\frac{\pi}{4}}^{\pi}$$

$$= \left(-\frac{1}{3} \times -1\right) - \left(-\frac{1}{3} \times \frac{1}{\sqrt{2}}\right)$$

$$= \frac{1}{3} + \frac{1}{3\sqrt{2}}$$

$$= \frac{\sqrt{2} + 1}{3\sqrt{2}}$$

$$= \frac{2 + \sqrt{2}}{6}$$

Remember that $\frac{\pi}{4}$ is just a number, so when you differentiate it you get '0'.

Replace $3x + \frac{\pi}{4}$ with u, dx with $\frac{1}{3}du$ and change the limits.

This can be written as $\frac{\sqrt{2}}{3\sqrt{2}} + \frac{1}{3\sqrt{2}} = \frac{\sqrt{2} + 1}{3\sqrt{2}}$.

Rationalise the denominator by multiplying by $\frac{\sqrt{2}}{\sqrt{2}}$.

3 Using integration by substitution

Find the exact value of $\int_0^{\frac{\pi}{3}} \frac{\sin\theta}{1 + \cos\theta} d\theta$.

Hint: $\frac{1}{1 + \cos\theta}$ is a 'function of a function' and the derivative of $1 + \cos\theta$ is $-\sin\theta$ so use the substitution $u = 1 + \cos\theta$.

Solution

Let $u = 1 + \cos\theta$

Convert $d\theta$ to du:

$\frac{du}{d\theta} = -\sin\theta \Rightarrow -du = \sin\theta\, d\theta$

Change the limits:

$\theta = \frac{\pi}{3} \Rightarrow u = 1 + \cos\frac{\pi}{3} = 1 + \frac{1}{2} = \frac{3}{2}$

$\theta = 0 \Rightarrow u = 1 + \cos 0 = 1 + 1 = 2$

$-du$ means $-1\, du$.

Edexcel A Level Mathematics (Pure)

So:

$$\int_0^{\frac{\pi}{3}} \frac{\sin\theta}{1+\cos\theta} d\theta = \int_2^{\frac{3}{2}} -\frac{1}{u} du$$

$$= \left[-\ln u\right]_2^{\frac{3}{2}}$$

$$= \left(-\ln \frac{3}{2}\right) - (-\ln 2)$$

$$= \ln \frac{2}{3} + \ln 2$$

$$= \ln \frac{4}{3}$$

Common mistake: Take care that you get the new limits the right way round.

Remember the rules of logs: $n \ln a = \ln a^n$.
So $-\ln \frac{3}{2} = \ln \left(\frac{3}{2}\right)^{-1} = \ln \frac{2}{3}$.

Also: $\ln a + \ln b = \ln ab$.

Common mistake: You were asked for the **exact** value, so don't work this out.

4 Using trigonometric identities

Find:

i $\int \tan\theta \, d\theta$ **ii** $\int \sin^2\theta \, d\theta$.

Solution

i $\int \frac{\sin\theta}{\cos\theta} d\theta = -\ln|\cos\theta| + c$

$= \ln\left|\frac{1}{\cos\theta}\right| + c$

$= \ln|\sec\theta| + c$

ii $\int \sin^2\theta \, d\theta = \int \frac{1}{2}(1 - \cos 2\theta) d\theta$

$= \int \left(\frac{1}{2} - \frac{1}{2}\cos 2\theta\right) d\theta$

$= \frac{1}{2}\theta - \frac{1}{4}\sin 2\theta + c$

The 'top' is the negative of the derivative of the 'bottom'.

Use the rules of logs:
$-\ln a = \ln a^{-1} = \ln\left(\frac{1}{a}\right)$.

Use the identity

$\cos 2\theta \equiv 1 - 2\sin^2\theta$

$\Rightarrow \sin^2\theta \equiv \frac{1}{2}(1 - \cos 2\theta)$

See page 95 for a reminder of this.

Now you can just carry out the integration by inspection using the standard results.

Test yourself   TESTED

1 Find $\int \left(\cos\frac{x}{2} - 3\sin 3x\right) dx$.

    A $-2\sin\frac{x}{2} - \cos 3x + c$ **B** $-\frac{1}{2}\sin\frac{x}{2} - 9\cos 3x + c$

    C $2\sin\frac{x}{2} + \cos 3x + c$ **D** $\frac{1}{2}\sin\frac{x}{2} + 9\cos 3x + c$

2 Find the exact value of $\int_0^{\frac{\pi}{3}} \cos(2\theta - \pi) d\theta$

    A $-\frac{\sqrt{3}}{4}$ **B** $\frac{\sqrt{3}}{4}$ **C** $\frac{\sqrt{3}}{4} + \frac{\pi}{3}$ **D** $-\sqrt{3}$

3 Find $\int x^2 \cos(x^3) dx$

A $3\sin(x^3)+c$ B $-\frac{1}{3}\sin(x^3)+c$ C $\frac{1}{3}\sin u+c$ D $\frac{1}{3}\sin(x^3)+c$

4 The diagram shows a sketch of part of the curve $y=\frac{\cos x}{\sin x}$.
Calculate the exact area of the shaded region which is enclosed by the curve, the line $x=\frac{\pi}{3}$ and the x-axis.

A 0.144 B $-\ln\left(\frac{\sqrt{3}}{2}\right)$

C $\ln\left(\frac{\sqrt{3}}{2}\right)$ D $\ln\frac{3}{2}$

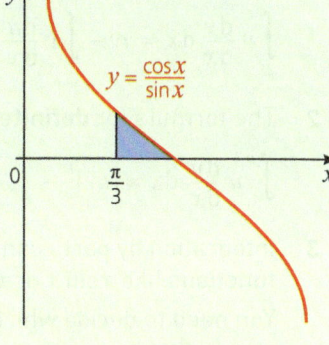

5 Find $\int (\sin\theta\, e^{\cos\theta} - 1)d\theta$

A $-e^{\cos\theta} - \theta + c$ B $-\cos\theta\, e^{\sin\theta} - \theta + c$ C $-e^{\cos\theta - 1} + c$ D $e^{\cos\theta} - \theta + c$

Full worked solutions online   CHECKED ANSWERS

Exam-style question

i Find $\int \frac{\cos x}{e^{\sin x}} dx$.

ii Show that $\int_0^{\frac{\pi}{6}} \frac{\sin 2x}{1+\cos 2x} dx = \ln\left(\frac{2\sqrt{3}}{3}\right)$.

Short answers on page 219
Full worked solutions online   CHECKED ANSWERS

Integration by parts

REVISED

Key facts

1. The **formula** for integrating by parts is:

   $$\int u \frac{dv}{dx} dx = uv - \int v \frac{du}{dx} dx$$

2. The formula for **definite** integration by parts is:

   $$\int_a^b u \frac{dv}{dx} dx = [uv]_a^b - \int_a^b v \frac{du}{dx} dx$$

3. Integration by parts can be used to integrate the product of two functions like $x \sin x$ or $x^2 e^x$.

   You need to decide which function to use as u and which function to use as $\frac{dv}{dx}$.
   - Choose u to be the function which becomes **simpler** when you **differentiate** it.
   - The other function will be $\frac{dv}{dx}$.

> **Hint:** Always check you can **integrate** your choice of $\frac{dv}{dx}$. If this function can't be integrated then you need to change your choice.

You can use integration by parts to integrate products of...	
a small polynomial in x like...	and a function of...
x	$\sin x$
$(x^2 - x)$	$\cos x$
x^2	e^x
...choose as u	...choose as $\frac{dv}{dx}$

4.

You can use integration by parts to integrate products of...	
A function of...	...and a small polynomial in x like...
	$(x + 2)$
$\ln x$	x^2
	1
...choose as u	...choose as $\frac{dv}{dx}$

> To integrate $\ln x$ write it as $1 \times \ln x$ and use parts.

Worked examples

1 Indefinite integration by parts

Find $\int 2x e^{3x} dx$.

> **Hint:** The derivative of $2x$ is 2 which is a simpler function. The derivative of e^{3x} is $3e^{3x}$ which is not simpler. So choose u to be $2x$.

Solution

Let $u = 2x$ $\Rightarrow$ $\frac{du}{dx} = 2$

> Differentiate u to find $\frac{du}{dx}$.

and $\frac{dv}{dx} = e^{3x}$ $\Rightarrow$ $v = \frac{1}{3} e^{3x}$

> **Common mistake:** Integrate $\frac{dv}{dx}$ to find v. Don't write '+ c' here otherwise you will end up with incorrect extra terms in your final answer.

Now substitute into the formula:

$$\int u \frac{dv}{dx} dx = uv - \int v \frac{du}{dx} dx$$

Chapter 8 Integration

$$\int 2x e^{3x} dx = 2x \times \tfrac{1}{3} e^{3x} - \int \tfrac{1}{3} e^{3x} \times 2 \, dx$$
$$= \tfrac{2}{3} x e^{3x} - \int \tfrac{2}{3} e^{3x} \, dx$$
$$= \tfrac{2}{3} x e^{3x} - \tfrac{2}{3} \times \tfrac{1}{3} e^{3x} + c$$
$$= \tfrac{2}{3} x e^{3x} - \tfrac{2}{9} e^{3x} + c$$

Don't forget the '+ c'.

2 Using integration by parts twice

Find $\int x^2 \sin 2x \, dx$.

Solution

Let $u = x^2$ $\Rightarrow$ $\dfrac{du}{dx} = 2x$

and $\dfrac{dv}{dx} = \sin 2x$ $\Rightarrow$ $v = -\dfrac{1}{2} \cos 2x$

Hint: The order of the polynomial can tell you how many times you need to integrate by parts. In this case you have an x^2 so you have to use parts twice.

Hint: Set out your work like this – then it is easier to use the formula.

Now substitute into the formula:

$$\int u \dfrac{dv}{dx} dx = uv - \int v \dfrac{du}{dx} dx$$

$$\int x^2 \sin 2x \, dx = x^2 (-\tfrac{1}{2} \cos 2x) - \int (-\tfrac{1}{2} \cos 2x) \times 2x \, dx$$

$$= -\tfrac{1}{2} x^2 \cos 2x + \int x \cos 2x \, dx \quad (1)$$

Common mistake: Take care with your signs! You will make fewer mistakes if you tidy up at this stage.

You can't integrate this directly so use parts again.

Using parts again on $\int x \cos 2x \, dx$.

Now let $u = x$ $\Rightarrow$ $\dfrac{du}{dx} = 1$

and $\dfrac{dv}{dx} = \cos 2x$ $\Rightarrow$ $v = \tfrac{1}{2} \sin 2x$

Now substitute into the parts formula to give:

$$\int x \cos 2x \, dx = x \times \tfrac{1}{2} \sin 2x - \int \tfrac{1}{2} \sin 2x \, dx$$

$$= \tfrac{1}{2} x \sin 2x - \tfrac{1}{2} \times (-\tfrac{1}{2} \cos 2x) + c$$

$$= \tfrac{1}{2} x \sin 2x + \tfrac{1}{4} \cos 2x + c \quad (2)$$

You haven't finished yet – remember you are trying to find $\int x^2 \sin 2x \, dx$.

Substitute (2) into (1)

$$\int x^2 \sin 2x \, dx = -\tfrac{1}{2} x^2 \cos 2x + \int x \cos 2x \, dx$$

$$= -\tfrac{1}{2} x^2 \cos 2x + \tfrac{1}{2} x \sin 2x + \tfrac{1}{4} \cos 2x + c$$

Don't forget '+ c'.

3 Integrating ln x

Find $\int \ln x \, dx$.

Hint: You can't integrate $\ln x$ by inspection but you can rewrite $\ln x$ as $1 \times \ln x$ so you have a product. Now that you have a product you can use integration by parts.

Edexcel A Level Mathematics (Pure)

Solution

$$\int \ln x \, dx = \int 1 \times \ln x \, dx$$

Let $u = \ln x \Rightarrow \dfrac{du}{dx} = \dfrac{1}{x}$

and $\dfrac{dv}{dx} = 1 \Rightarrow v = x$

Now substitute into the formula:

$$\int u \dfrac{dv}{dx} dx = uv - \int v \dfrac{du}{dx} dx$$

$$\int 1 \times \ln x \, dx = \ln x \times x - \int x \times \dfrac{1}{x} dx$$

$$= x \ln x - \int 1 \, dx$$

$$= x \ln x - x + c$$

Common mistakes: You can't let $\dfrac{dv}{dx} = \ln x$ as you can't integrate it to find an expression for v.

So you have to choose u to be $\ln x$ and $\dfrac{dv}{dx}$ to be 1.

A common mistake is to think that the answer is $\dfrac{1}{x}$, but that's when you differentiate.

Notice that $\ln x$ does become simpler when you differentiate it.

Simplify before you try to integrate.

4 Definite integration by parts

The diagram shows a sketch of part of the curve $y = -x \cos 2x$.

Find the area of the shaded region which lies between the curve, the axes and the line $x = \dfrac{\pi}{4}$.

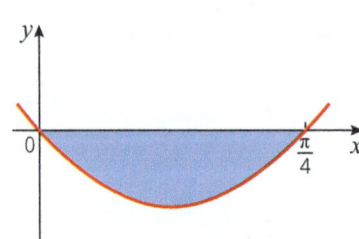

Common mistake: You must make sure your calculator is in radians mode when you integrate a trig function.

Solution

Work out $\int_0^{\frac{\pi}{4}} -x \cos 2x \, dx$.

Let $u = -x \Rightarrow \dfrac{du}{dx} = -1$

and $\dfrac{dv}{dx} = \cos 2x \Rightarrow v = \dfrac{1}{2} \sin 2x$

Now substitute into the formula:

$$\int_a^b u \dfrac{dv}{dx} dx = \left[uv \right]_a^b - \int_a^b v \dfrac{du}{dx} dx$$

$$\int_0^{\frac{\pi}{4}} -x \cos 2x \, dx = \left[-x \times \dfrac{1}{2} \sin 2x \right]_0^{\frac{\pi}{4}} - \int_0^{\frac{\pi}{4}} \dfrac{1}{2} \sin 2x \times -1 \, dx$$

$$= \left[-\dfrac{1}{2} x \sin 2x \right]_0^{\frac{\pi}{4}} + \int_0^{\frac{\pi}{4}} \dfrac{1}{2} \sin 2x \, dx$$

Hint: Take care with your signs! It is easier to tidy up as you go.

$$= -\dfrac{1}{2} \left[x \sin 2x \right]_0^{\frac{\pi}{4}} + \left[\dfrac{1}{2} \times -\dfrac{1}{2} \cos 2x \right]_0^{\frac{\pi}{4}}$$

$$= -\dfrac{1}{2} \left[x \sin 2x \right]_0^{\frac{\pi}{4}} - \dfrac{1}{4} \left[\cos 2x \right]_0^{\frac{\pi}{4}}$$

$$= -\dfrac{1}{2} \left(\dfrac{\pi}{4} \times 1 - 0 \right) - \dfrac{1}{4} (0 - 1)$$

Remember:
$\sin\left(2 \times \dfrac{\pi}{4}\right) = 1$

$\cos\left(2 \times \dfrac{\pi}{4}\right) = 0$

and $\cos 0 = 1$.

$$= -\dfrac{1}{2} \times \dfrac{\pi}{4} - \dfrac{1}{4} \times -1$$

$$= -\left(\dfrac{\pi}{8} - \dfrac{1}{4} \right)$$

This integral is negative as the curve is below the x-axis here.

$$= -\dfrac{(\pi - 2)}{8}$$

So the area is $\dfrac{\pi - 2}{8}$ square units.

You are looking for an area so you need to multiply the answer by '-1' to make it positive.

Test yourself

1. Find $\int (x+1)\sin x\, dx$.

   A $-\left(\frac{1}{2}x^2 + x\right)\cos x + c$
   B $(x+1)\cos x + \sin x + c$
   C $-(x+1)\cos x + \sin x + c$
   D $(x+1)\cos x - \sin x + c$

2. Find $\int x\ln 2x\, dx$

   A $\frac{1}{2}x^2\ln 2x - \frac{1}{4}x^2 + c$
   B $\frac{1}{2}x^2\ln 2x - \frac{1}{8}x^2 + c$
   C $\ln 2$
   D $1 - \ln x + c$

3. Find $\int x^2 \cos x\, dx$

   A $x^2\sin x + 2x\cos x - 2\sin x + c$
   B $-x^2\sin x + 2x\cos x - 2\sin x + c$
   C $\frac{1}{3}x^3\sin x + c$
   D $-x^2\sin x + 2x\cos x + 2\sin x + c$
   E $x^2\sin x - 2x\cos x + 2\sin x + c$

4. Find $\int_0^1 xe^{2x}\, dx$.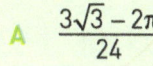

   A $\frac{e^2}{4}$
   B $2(2 - e^2)$
   C $\frac{1}{2}xe^{2x} - \frac{1}{4}e^2 + \frac{1}{4}$
   D $\frac{1}{4}(e^2 + 1)$

5. The diagram shows part of the curve $y = x\sin 2x$. Find the exact area of the shaded region which lies between the curve, the line $x = \frac{\pi}{3}$ and the x-axis.

   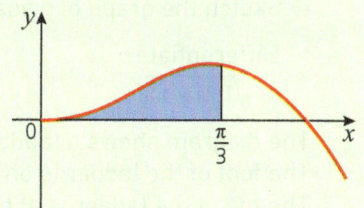

   A $\frac{3\sqrt{3} - 2\pi}{24}$
   B $\frac{\pi + 6\sqrt{3}}{2}$
   C $\frac{2\pi + 3\sqrt{3}}{24}$
   D $\frac{\pi^2}{72}$

Full worked solutions online   CHECKED ANSWERS

Exam-style question

i Show that $\int xe^{-x}\, dx = -(x+1)e^{-x} + c$.

The diagram shows part of the curve $y = \frac{(1-x^2)}{e^x}$.

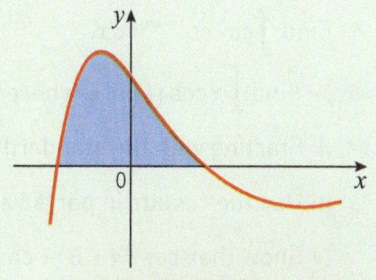

ii Find the coordinates of the points where the curve cuts the x-axis.

iii Find the exact area of the shaded region which is between the curve $y = \frac{(1-x^2)}{e^x}$ and the x-axis.

Short answers on page 219

Full worked solutions online   CHECKED ANSWERS

Review questions (Chapters 6–8)

1 Given that $\cos\theta = \frac{1}{3}$ and θ is acute, find the exact value of $\cot\theta$.

2 Solve $2\tan^2\theta = 3\sec\theta$ for $0° \leq \theta \leq 360°$.

3 *In this question you must show detailed reasoning.*
Determine a sequence of two transformations which maps the graph of $y = \sin\theta$ onto the graph of $y = 3\sin\theta + 4\cos\theta$ where θ is in degrees.

4 A cylindrical waste paper bin is made from a thin sheet of metal. The bin has no lid.
The bin needs to hold 30 litres and has a height h cm and radius r cm.
Find the minimum surface area of metal required to make the bin and prove that this value is the minimum.

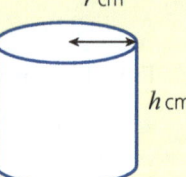

Hint: 30 litres = 30 000 cm³

5 You are given that $y = 5\sin\theta - 4\cos\theta$.

  i Find $\frac{dy}{d\theta}$.

  ii Find the coordinates of the stationary points for $-\pi \leq \theta \leq \pi$.

  iii Find $\frac{d^2y}{d\theta^2}$ and hence determine whether each of the stationary points is a maximum, a minimum or point of inflection.

  iv Sketch the graph of y against θ for $-\pi \leq \theta \leq \pi$.

6 i Differentiate
    a $\sqrt{100 - x^2}$ b $x\sqrt{100 - x^2}$

The diagram shows a ladder of length 10 m leaning against a vertical wall.
The foot of the ladder is on horizontal ground at a distance x m from the wall.
The top of the ladder is at height h m.

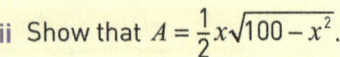

The area of the triangle formed by the ladder, the wall and the ground is A m².

  ii Show that $A = \frac{1}{2}x\sqrt{100 - x^2}$.

  iii Show that the maximum possible value of A is 25 m².

  iv The foot of the ladder slides away from the wall at a steady speed of 0.01 m s⁻¹.
Find the rate at which the area is changing when $x = 6$.

7 Find $\int \sin x \, e^{1-\cos x} \, dx$.

8 i Find $\int x\cos(kx) \, dx$ where k is a non-zero constant.

  ii Starting with the standard formula for $\cos(A + B)$, show that $\cos^2 x = \frac{1}{2}(\cos 2x + 1)$.

  iii Use the results in parts **i** and **ii** to find $\int x\cos^2 x \, dx$.

  iv Show that $\cos(A + B) + \cos(A - B) = 2\cos A \cos B$.
Hence show that $2\cos 3x \cos x = \cos 4x + \cos 2x$.

  v Use the results of parts **i** and **iv** to show that $\int_0^{\frac{\pi}{4}} x\cos x \cos(3x) \, dx = \frac{\pi - 3}{16}$.

Short answers on page 219

Full worked solutions online

SECTION 3

Target your revision (Chapters 9–14)

1 Use sigma notation

A sequence is defined by $a_n = (-1)^n x^n$ for $n = 1, 2, 3, \ldots$

i Write down the terms a_1, a_2, a_3, a_4.

A series is defined by $S_n = \sum_{k=1}^{n} (-1)^k x^k$.

ii Write down the series S_6.

(see page 156)

2 Solve problems involving geometric sequences and series

A geometric series has first term $\frac{1}{2}$.

The sum to infinity of the series is $\frac{3}{4}$. Find

i the common ratio
ii the least number of terms for the sum of the series to be greater than 90% of the sum to infinity.

(see page 161)

3 Solve problems involving arithmetic sequences and series

The fifth term of an arithmetic sequence is –3 and the tenth term is 12.

i Find the first term.
ii Find the nth term.
iii Find the sum of the first 50 terms.

(see page 160)

4 Use partial fractions

Express $\dfrac{x^2 - 2}{x(2x - 1)(3x + 2)}$ in partial fractions.

(see page 163)

5 Recognise when to use partial fractions in integration

Find the exact values of

i $\displaystyle\int_{1}^{2} \dfrac{4x-1}{6x^2 - 3x} \, dx$ ii $\displaystyle\int_{2}^{3} \dfrac{1}{6x^2 - 3x} \, dx$.

(see page 163)

6 Use the binomial expansion for positive integer powers

a Expand $\left(4 - \dfrac{x}{2}\right)^5$.

b Find the term in x^5 in the expansion $\left(4 - \dfrac{x}{2}\right)^9$.

(see page 167)

7 Use the general binomial expansion

a Write down the first three terms in the binomial expansion of $\sqrt{1 + 3x}$ in ascending powers of x. For what range of values of x is this expansion valid?

b Find a quadratic approximation for $(5 - x)^{-2}$, stating the range of values of x for which this expansion is valid.

(see page 167)

8 Using parametric equations to sketch a curve

A curve is given by $x = 1 - t$, $y = t^2$.

i Plot the curve for values of t between –3 and +3.
ii Give the equation of the line of symmetry of the curve in terms of t.

(see page 172)

9 Convert between parametric and Cartesian equations

i Find the Cartesian equation of the curve defined by the parametric equations

a $x = \dfrac{1}{1+t}$, $y = \dfrac{2t}{1-t}$

b $x = e^{3t}$, $y = \dfrac{3}{1+t}$.

ii The equation of a curve is $x^2 + 4y^2 = 16$. In each case, complete the pair of parametric equations to describe this curve.

a $x = 4\cos\theta$ and $y = $ _____.
b $y = 2\cos 2\theta$ and $x = $ _____.

(see page 173)

10 Use the equation of a circle in parametric form

A circle is described by the parametric equations $x = -3 + 5\cos\theta$, $y = 2 + 5\sin\theta$. Find

i the coordinates of the centre, and
ii the radius of the circle.

(see page 172)

11 Find the gradient of a curve defined parametrically

The parametric equations of a curve, C, are $x = 2t + \ln t$, $y = t - \ln 2t$ for $t > 0$

Find the gradient of the curve at the point where $t = 4$.

(see page 176)

Edexcel A Level Mathematics (Pure)   153

12 Find the equations of tangents and normals for a parametric curve

The parametric equations of a curve are

$x = 2\sec\theta$, $y = 3\tan\theta$ for $0 < \theta < \frac{\pi}{2}$.

i Show that $\frac{dy}{dx} = \frac{3}{2}\csc\theta$.

ii Find the equation of
   a the tangent and   b the normal
   to the curve at the point with parameter $\theta = \frac{\pi}{4}$.

iii Show that the equation of the curve is $4y^2 = 9(x^2 - 4)$.

(see page 177)

13 Find the stationary points of a curve defined parametrically

A curve is defined by the parametric equations $x = \cos 2\theta$, $y = \sin 3\theta$ for $0 < \theta < \frac{\pi}{2}$.

The curve has a stationary point at P.

i Find $\frac{dy}{dx}$.

ii Find the coordinates of P.

(see page 177)

14 Use direct integration to solve a differential equation

i Solve the differential equation $\frac{dx}{dt} = 6\cos 2t$.

ii Given that $x = 6$ when $t = 0$, find the particular solution.

iii Sketch this particular solution for $0 \leq t \leq 2\pi$.

(see page 181)

15 Use separation of variables to solve a differential equation

Solve $\frac{dy}{dx} = 2xy$ for $y > 0$.

(see page 181)

16 Add and subtract vectors and multiply a vector by a scalar

You are given $\mathbf{a} = x\mathbf{i} - \mathbf{j} + 2\mathbf{k}$ and $\mathbf{b} = 3\mathbf{i} + y\mathbf{j} + z\mathbf{k}$.

i The resultant of $\mathbf{a}$ and $\mathbf{b}$ is $2\mathbf{i} + \mathbf{j} - 2\mathbf{k}$. Find the values of x, y and z.

ii You are given that $\mathbf{c} = 3\mathbf{a} + \mathbf{b}$. Find the vector $\mathbf{c}$ and the angle between $\mathbf{c}$ and the positive $\mathbf{j}$ direction.

(see page 188)

17 Find the magnitude of a vector

You are given that $\mathbf{q} = \begin{pmatrix} 4 \\ -3 \\ -5 \end{pmatrix}$.

The vector $\mathbf{p}$ has a magnitude of $10\sqrt{2}$ and is in the same direction as $\mathbf{q}$.
Find the vector $\mathbf{p}$.

(see page 189)

18 Solve problems involving vectors

The points A, B and C have position vectors $3\mathbf{i} + \mathbf{j} - 2\mathbf{k}$, $2\mathbf{i} - \mathbf{j} - 3\mathbf{k}$ and $5\mathbf{i} + 2\mathbf{j} + 3\mathbf{k}$ respectively.

i Find $|\overrightarrow{AB}|$.

ii The point M divides $\overrightarrow{BC}$ in the ratio $1:2$. Find the position vector of M.

iii Point D is such that $\overrightarrow{AD} = 2\overrightarrow{CD}$. Find the position vector of D.

(see page 189)

19 Use change of sign method to find an interval for the root of an equation

Show that the equation $e^x - x = x^2 + 2$ has a root between $x = 2$ and $x = 3$.

(see page 192)

20 Use an iterative formula

i Show that the equation $2x^3 - 3x - 8 = 0$ can be rearranged into the form $x = \sqrt[3]{\frac{3x+8}{2}}$.

ii Use the iteration $x_{n+1} = \sqrt[3]{\frac{3x_n + 8}{2}}$ with $x_1 = 1$ to find the values of x_2, x_3, x_4 and x_5, giving your answers to 5 decimal places.

(see page 193)

21 Use the Newton–Raphson method to solve an equation

The equation $x^5 - 3x^2 + 4x - 1 = 0$ has a single root, α, which lies between $x = 0$ and $x = 1$. Use the Newton–Raphson method together with $x_1 = 0$ to find the value of α correct to 5 decimal places.

(see page 193)

22 Know the problems that can arise when numerical methods are used to solve an equation

Explain fully why each of the following methods may fail to find a root of $f(x) = 0$.

i Change of sign method.

ii Newton–Raphson

iii An iteration formula derived from rearrangement of $f(x) = 0$ to $x = g(x)$.

(see page 193)

23 Know whether the trapezium rule results in an overestimate or underestimate

The diagram shows part of the graph of $y = 3e^{-x^2}$ for $-3 \leqslant x \leqslant 3$.

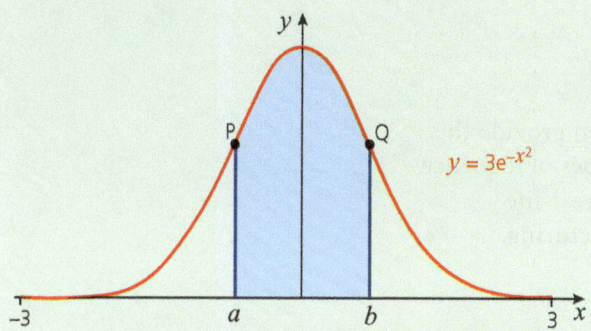

The trapezium rule is used to find $\int_a^b 3e^{-x^2} dx$.

P and Q are non-stationary points of inflection.

i Does the trapezium rule give an underestimate, an overestimate or is it not possible to tell, for the value of

  a $\int_a^b 3e^{-x^2} dx$?

  b $\int_b^3 3e^{-x^2} dx$?

  c $\int_a^3 3e^{-x^2} dx$?

  Give a reason for each answer.

ii Find the exact values of a and b.

(see page 196)

24 Use the trapezium rule

The diagram shows the graph of $y = \sin\sqrt{x}$. The shaded area between the curve, the x-axis and the lines $x = 1$ and $x = 2$ has been divided into 4 strips.

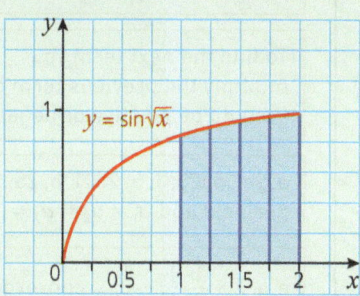

i Use the trapezium rule with 4 strips to estimate the value of $\int_1^2 \sin\sqrt{x}\, dx$.

Write down the first 6 significant figures of the answer from your calculator display. State, with a reason, whether your answer is an underestimate or an overestimate.

ii Use the trapezium rule again with

  a 8 strips and
  b 16 strips

  to obtain an estimate for the value of

  $\int_1^2 \sin\sqrt{x}\, dx$.

iii Comment on the accuracy to which you can give the integral.

(see page 196)

25 Use the sum of a series of rectangles to find upper and lower bounds for a definite integral

The diagram shows the curve $y = \sqrt{\cos x}$ for $0 \leqslant x \leqslant \frac{\pi}{2}$.

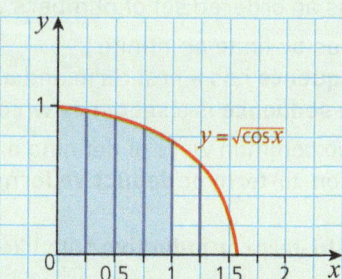

By finding the sum of the area of a series of five rectangles find

i an upper bound and
ii a lower bound

for $\int_0^1 \sqrt{\cos x}\, dx$.

Give your answers to 5 decimal places.

Hence write down an inequality for $\int_0^1 \sqrt{\cos x}\, dx$.

(see page 196)

Short answers on pages 219–220

Full worked solutions online

Chapter 9 Sequences and series

About this topic
You meet sequences and series in everyday life; they often provide the patterns of events around us. The two most common types of sequence are arithmetic and geometric and they have a variety of real-life applications including in banking, medicine and manufacturing.

Before you start, remember
- number patterns from GCSE
- how to solve equations.

Definitions and notation

REVISED

Key facts

1. A **sequence** is an ordered set of numbers $a_1, a_2, \ldots\ldots a_k, \ldots\ldots a_n$ ← e.g. 1, 4, 7, 10...

2. Sequences can be finite or infinite
   - a **finite sequence** has a first term and a last term
   - an **infinite sequence** is a sequence which continues forever.

   Each term can be worked out from its position in the sequence.

3. There are two common ways of defining a sequence
   - as a position-to-term or **deductive** formula like $a_k = k^2 - 1$ for $k = 1, 2, 3, \ldots$
   - as a term-to-term or **inductive** rule like $a_1 = 4; a_{k+1} = 2a_k + 3$. ←

   This is also called a **recurrence relation**.

4. In an **arithmetic sequence** (also called an **arithmetic progression**) the difference between two consecutive terms is constant.
   So $a_{k+1} = a_k + d$, where d is a fixed number called the **common difference**. ←

   To find the next term you add d to the previous term (d could be positive or negative).

5. In a **geometric sequence** (also called a **geometric progression**) the ratio between two consecutive terms is constant.
   So $a_{k+1} = ra_k$, where r is a fixed number called the **common ratio**. ←

   To find the next term you multiply the previous term by r (r could be positive or negative).

6. In an **increasing sequence** each term is greater than the term before.
   In a **decreasing sequence** each term is less than the term before.

7. In a **periodic sequence** $a_{k+p} = a_k$ for a fixed integer p, called the **period**. ←

   e.g. 1, 5, 25, 1, 5, 25, 1, 5, 25... has period 3 and $a_1 = a_4 = a_7 = \ldots$

8. In a **convergent sequence** the differences between successive terms is smaller each time. ←
   A convergent sequence tends to a **limit** as $n \to \infty$ and $a_k \to a_{k+1}$.

   e.g. $1, -\frac{1}{2}, \frac{1}{4}, -\frac{1}{8}, \ldots$ converges to zero.

9. A sequence which is not convergent is said to be a **divergent** sequence. The terms may tend to $+\infty$ or $-\infty$ or to no particular value.

   The terms get progressively closer to a fixed number called the limit of the sequence.

   $-1, -4, -7, -10, \ldots$

10. A **series** is the sum of the terms of a sequence.
    S_n denotes the sum to n terms of the series.

    $$S_n = \sum_{k=1}^{k=n} a_k = a_1 + a_2 + \ldots + a_n$$

    e.g. 1, 2, 4, 8, 16, ...

    e.g. $S_5 = \sum_{k=1}^{k=5} k^2 = 1 + 4 + 9 + 16 + 25 = 55.$

Worked examples

1 Recognising types of sequence

Describe the sequence 2, 4, 6, 8, 2, 4, 6, 8, 2, 4, 6, 8, …

Solution

This sequence repeats itself every 4th term so it is periodic with period 4.

It is infinite and divergent. ← *There is no last term and the terms do not converge.*

2 Finding the terms in a sequence from a deductive formula

A sequence is defined by $a_k = k^3 - k^2 + 1$ for $k = 1, 2, 3, …$

Write down the first four terms of the sequence.

Solution

$a_k = k^3 - k^2 + 1$

When $k = 1$, $a_1 = 1^3 - 1^2 + 1 = 1$

When $k = 2$, $a_2 = 2^3 - 2^2 + 1 = 8 - 4 + 1 = 5$

When $k = 3$, $a_3 = 3^3 - 3^2 + 1 = 27 - 9 + 1 = 19$

When $k = 4$, $a_4 = 4^3 - 4^2 + 1 = 64 - 16 + 1 = 49$

3 Using an inductive formula or recurrence relation

A sequence is defined by $a_1 = 2$; $a_{k+1} = 0.8a_k + 3$.
  i Calculate the value of a_3.
  ii What is the smallest value of n for which $a_n \geq 10$?

Solution

i $a_1 = 2$,

$a_2 = 0.8a_1 + 3 = 0.8 \times 2 + 3 = 4.6$,

$a_3 = 0.8a_2 + 3 = 0.8 \times 4.6 + 3 = 6.68$ ← *Notice that the difference between successive terms is growing smaller each time. So the sequence is convergent. It converges to 15 (use your calculator to check this).*

ii To find the first term that is over 10, find the terms one by one

$a_4 = 0.8a_3 + 3 = 0.8 \times 6.68 + 3 = 8.344$,

$a_5 = 0.8a_4 + 3 = 0.8 \times 8.344 + 3 = 9.6752$,

$a_6 = 0.8a_5 + 3 = 0.8 \times 9.6752 + 3 = 10.74016$, so $a_6 \geq 10$

So 6 is the smallest value of n for which $a_n \geq 10$.

4 Solving problems involving sequences

A sequence is defined by $a_{k+1} = pa_k + q$ where $a_1 = 48$.

Given that $a_2 = 20$ and $a_3 = 13$, find the values of p and q.

Solution

$a_2 = 20 \Rightarrow 48p + q = 20$ (1)

$a_3 = 13 \Rightarrow 20p + q = 13$ (2) ← *There are 2 unknowns so you need to form 2 equations.*

Subtracting (2) from (1) gives $28p = 7 \Rightarrow p = \frac{1}{4}$.

Substituting $p = \frac{1}{4}$ in (1) $\Rightarrow 48 \times \frac{1}{4} + q = 20 \Rightarrow q = 8$.

So $p = \frac{1}{4}$, $q = 8$.

Edexcel A Level Mathematics (Pure)

5 Using series

The sum of n terms of a series is given by $S_n = \dfrac{n^2(n+1)^2}{4}$.

i Write down the first four terms of the series.

ii Find an expression for the nth term of the series.

Solution

i $S_1 = a_1 = \dfrac{1^2 \times 2^2}{4} = \dfrac{4}{4} = 1 \Rightarrow a_1 = 1$

$S_2 = a_1 + a_2 = \dfrac{2^2 \times 3^2}{4} = 9 \Rightarrow a_2 = 8$

$S_3 = a_1 + a_2 + a_3 = \dfrac{3^2 \times 4^2}{4} = 36 \Rightarrow a_3 = 27$

$S_4 = a_1 + a_2 + a_3 + a_4 = \dfrac{4^2 \times 5^2}{4} = 100 \Rightarrow a_4 = 64$

> Notice that these are all cube numbers. You will prove this result in part ii.

So the series is $1 + 8 + 27 + 64 \ldots$

ii In general, $a_n = S_n - S_{n-1}$.

> For example, $a_2 = S_2 - S_1$ and $a_3 = S_3 - S_2$.

$a_n = S_n - S_{n-1} = \dfrac{n^2(n+1)^2}{4} - \dfrac{(n-1)^2 n^2}{4}$

$= \dfrac{n^2(n^2 + 2n + 1) - (n^2 - 2n + 1)n^2}{4}$

$= \dfrac{n^4 + 2n^3 + n^2 - (n^4 - 2n^3 + n^2)}{4}$

$= \dfrac{4n^3}{4}$

So the nth term of the series is n^3.

> Check the first four terms of the series to see that this is correct.

6 Using sigma notation

A sequence is defined by $a_k = (k+1)2^k$.

Write out the series $\displaystyle\sum_{k=2}^{5} a_k$ without simplifying the terms.

Solution

Substituting $k = 2$ into $(k+1)2^k \Rightarrow a_2 = 3 \times 2^2$

Substituting $k = 3$ into $(k+1)2^k \Rightarrow a_3 = 4 \times 2^3$

Substituting $k = 4$ into $(k+1)2^k \Rightarrow a_4 = 5 \times 2^4$

Substituting $k = 5$ into $(k+1)2^k \Rightarrow a_5 = 6 \times 2^5$

So $\displaystyle\sum_{k=2}^{5}(k+1)2^k = 3 \times 2^2 + 4 \times 2^3 + 5 \times 2^4 + 6 \times 2^5$

Test yourself

1 Which of the following is the best description of the sequence whose nth term is $\cos(n \times 60°)$?
   A divergent and geometric
   B periodic with period 6
   C both divergent and periodic with period 6
   D convergent and geometric

2 The sum of n terms of a series is given by $S_n = \dfrac{n}{n+1}$.
   Which of the following is the correct series?
   A $\dfrac{1}{2} + \dfrac{2}{3} + \dfrac{3}{4} + \dfrac{4}{5}...$
   B $\dfrac{1}{2} + \dfrac{1}{6} + \dfrac{1}{12} + \dfrac{1}{20}...$
   C $\dfrac{1}{2} + \dfrac{1}{6} + \dfrac{1}{12} + \dfrac{1}{72}...$
   D $\dfrac{1}{2} + \dfrac{7}{6} + \dfrac{17}{12} + \dfrac{31}{20}...$

3 Which one of the following statements is true?
   A $\sum_{r=3}^{7} r^2 = 140$
   B The sequence $1, -1, 1, -1, 1...$ converges
   C The sequence $1, 3, 5, 7,...$ is defined by $a_{k+1} = a_k + 2$
   D The sum to n terms of the series $1 + \dfrac{1}{2} + \dfrac{1}{4} + \dfrac{1}{8}...$ is $2 - \left(\dfrac{1}{2}\right)^{n-1}$

4 Which of the following series is the same as $1 - x + 2x^2 - 3x^3 + ...$?
   A $\sum_{r=1}^{\infty} (-1)^r rx^r$
   B $1 + \sum_{r=1}^{\infty} (-1)^r rx^r$
   C $1 - \sum_{r=1}^{\infty} (-1)^r rx^r$
   D $\sum_{r=1}^{\infty} (-1)^{r+1} rx^r$

Full worked solutions online

Exam-style question

Jan is being treated with a special drug.

At 0100 hours she is given 120 units of the drug.

Each hour the amount in her body reduces by 10 units.

At 0600, 1100, and 1600 she is given subsequent doses, in each case enough to bring the amount in her body up to 120 units.

The amounts in her body every hour are denoted by a_1 at 0100, a_2 at 0200 and so on.
   i   Write down the sequence $a_1, a_2, ..., a_{20}$.
   ii Describe this sequence.

Jan is not given any more of the drug, (she is recovering well).
   iii Write down the value of a_{21}.
   iv   What is the mean amount of drug in Jan's body from 0100 to 2100?
   v   What is the total amount of the drug given to Jan?
   vi   When is there no drug left in Jan's body?

Short answers on pages 220–221

Full worked solutions online

Sequences and series

REVISED

Key facts

1. In an **arithmetic** sequence (progression) with first term a, common difference d and n terms:
   - the kth term is given by $a_k = a + (k-1)d$
   - the last term, $l = a_n = a + (n-1)d$
   - the sum of n terms is $S_n = \frac{1}{2}n(a+l) = \frac{1}{2}n[2a + (n-1)d]$.

2. In a **geometric** sequence (progression) with first term a, common ratio r and n terms:
   - the kth term is given by $a_k = ar^{k-1}$
   - the last term, $a_n = ar^{n-1}$
   - the sum of n terms is $S_n = \frac{a(r^n - 1)}{(r-1)} = \frac{a(1-r^n)}{(1-r)}$
   - for an **infinite** geometric series to converge the common ratio must be between -1 and 1, so $-1 < r < 1$ which is sometimes written $|r| < 1$
   - the **sum to infinity** of a convergent G.P. is $S = S_\infty = \frac{a}{1-r}$.

Worked examples

1 Finding arithmetic sequences

The fourth term of an arithmetic sequence is 13 and the seventh term is 19.
  i Find the first term.
  ii Find the nth term.

Solution

  i Using $a_k = a + (k-1)d$ gives
     4th term $a_4 = a + 3d = 13$
     7th term $a_7 = a + 6d = 19$
     By subtraction $3d = 6$ ← Find $a_7 - a_4$.
     Hence $d = 2$ and $a = 7$.
     So the first term is 7.

  ii The nth term is $7 + (n-1) \times 2 = 5 + 2n$ ← Use $l = a_n = a + (n-1)d$.

  Check: when $n = 7$ then $5 + 2 \times 7 = 19$ ✓

2 Solving problems involving arithmetic sequences

  i Show that the series whose kth term is given by $a_k = 3k + 1$ is an arithmetic series.
  ii Find the 20th term and the sum to 20 terms.

Solution

  i By substitution
     $a_1 = 3 \times 1 + 1 = 4$
     $a_2 = 3 \times 2 + 1 = 7$
     $a_3 = 3 \times 3 + 1 = 10$ ← You can see that so far there is a common difference of 3. Now you need to prove that this is true for any pair of terms.
     In general, $(k+1)$th term, $a_{k+1} = 3(k+1) + 1 = 3k + 4$
     kth term: $a_k = 3k + 1$
     Difference: $a_{k+1} - a_k = 3$ ← $(3k+4) - (3k+1) = 4 - 1 = 3$.
     So the sequence is an arithmetic progression with first term 4 and common difference 3.

  ii The 20th term is $3 \times 20 + 1 = 61$.
     Using $S_n = \frac{1}{2}n(a+l)$,
     the sum to 20 terms is $\frac{1}{2} \times 20(4 + 61) = 650$.

3 Finding geometric sequences

A geometric sequence has second term 3 and fifth term 24.
  i Find the first term and the common ratio.
  ii Find the 8th term and the sum to 8 terms.

Solution

  i Using $a_k = ar^{k-1}$ gives:

     2nd term = 3 = ar^2 ⇒ $ar = 3$,

     5th term = 24 = ar^{5-1} ⇒ $ar^4 = 24$

     So $\dfrac{ar^4}{ar} = \dfrac{24}{3}$ ← *This is a very useful technique: dividing one equation by the other cancels a.*

     ⇒ $r^3 = 8$

     Hence $r = 2$ and $a = 1.5$ ← *Using $ar = 3$.*

  ii 8th term: $a_8 = ar^{8-1} = 1.5 \times 2^7 = 192$

     Sum to 8 terms: $S_8 = \dfrac{1.5 \times (2^8 - 1)}{2 - 1} = 382.5$ ← *Using $S_n = \dfrac{a(r^n - 1)}{(r - 1)}$.*

4 Finding the sum to infinity

  i Show that the geometric series

     $5 + \dfrac{5}{2} + \dfrac{5}{4} + \dfrac{5}{8} \ldots$

     has a sum to infinity.

  ii Find the sum to infinity.

Solution

  i The first term is $a = 5$. ← *Remember that a geometric series can only have a sum to infinity if it converges, so you need to show the common ratio is between −1 and 1.*

     Each successive term is half of its predecessor so $r = \tfrac{1}{2}$.
     Since $-1 < r < 1$, the geometric sequence is convergent and has a sum to infinity.

  ii Sum to infinity $S = \dfrac{a}{1-r} = \dfrac{5}{1-\tfrac{1}{2}} = \dfrac{5}{\tfrac{1}{2}} = 10$.

5 Solving problems involving geometric series

  i State the common ratio of the geometric series

     $3 + \dfrac{3x}{2} + \dfrac{3x^2}{4} + \dfrac{3x^3}{8} \ldots$

  ii State the restrictions on x for the series to have a sum to infinity.
  iii State the sum to infinity in terms of x.
  iv Find x if the sum to infinity is 15.

Solution

  i The first term is $a = 3$.
     The common ratio is $r = \dfrac{x}{2}$ ← *The terms in the sequence involve x, so the common ratio is in terms of x.*

  ii For sum to infinity to exist $-1 < r < 1$ ← *The sequence must converge in order to have a sum to infinity.*
     so, in this case, $-1 < \dfrac{x}{2} < 1$.

     Multiplying through by 2 gives $-2 < x < 2$.

  iii $S = \dfrac{a}{1-r} = \dfrac{3}{1-\tfrac{x}{2}} = \dfrac{6}{2(1-\tfrac{x}{2})} = \dfrac{6}{2-x}$ ← *Multiply both the numerator and denominator by 2 to clear the fraction, $\tfrac{x}{2}$, in the denominator.*

  iv Given that $\dfrac{6}{2-x} = 15$

     ⇒ $6 = 15(2-x)$
     $= 30 - 15x$

     ⇒ $15x = 24$ so $x = 1.6$ or $1\tfrac{3}{5}$. ← *Notice that this satisfies $-2 < x < 2$.*

Test yourself

TESTED

1. The numbers p, 4, q, form a geometric sequence.
   Which of the following values of p and q are possible?

   A $p = 0, q = 8$ B $p = 0, q = 16$ C $p = 1, q = 16$ D $p = 6, q = 2$

2. The first three terms of an arithmetic sequence are −3, 2 and 7.
   What is the sum of the first 12 terms?

   A 52   B 294   C 324   D 648

3. The 2nd term of an arithmetic sequence is 7 and the 6th term is −5.
   Three of the following statements are false and one is true. Which one is true?

   A The common difference is 3   B The first term is 13

   C $a_k < -10$ if $k \geq 8$ D $\sum_{k=2}^{8} a_k = -4$

4. In which of the following cases does the series $1 - 2x + 4x^2 - 8x^3 + \ldots$ have the given sum to infinity, S?

   A $x = 3, S = \frac{1}{7}$ B $x = -\frac{1}{5}, S = \frac{5}{7}$

   C $x = \frac{1}{3}, S = \frac{5}{3}$ D $x = \frac{1}{4}, S = \frac{2}{3}$

Full worked solutions online   CHECKED ANSWERS

Exam-style question

i   On Ian's first birthday his parents put £25 into a money box. On his second birthday they give him £28, on his third birthday £31 and so on.

   a How much money does he receive on his 18th birthday?
   b How much money is in his money box at the end of his 18th birthday if none has been spent?

ii On Ian's eleventh birthday his grandparents put £200 into a bank account for him.
   Compound interest at 7% is added to this account every following year on his birthday.
   His grandparents add another £200 on each birthday.
   How much is in his account the day after Ian is eighteen?

Short answers on page 221

Full worked solutions online   CHECKED ANSWERS

Chapter 10 Further algebra

About this topic

Partial fractions are a way of changing the form of an algebraic fraction to one that may make it easier to apply a required technique, for example binomial expansion or integration.

The binomial theorem can be used to expand expressions to a positive integer power, like $(x + 2)^6$. You can also use the binomial theorem to find the first few terms in the expansion of expressions like $(1 + x)^{-2}$ or $(2 - 3x)^{\frac{1}{2}}$.

Before you start, remember

- how to simplify algebraic fractions
- how to integrate an expression
- polynomials from Chapter 3.

Partial fractions

REVISED

Key facts

1. For a polynomial identity $p(x) \equiv q(x)$
   - you can **substitute for any value of** x, say $x = a$, to give $p(a) = q(a)$.
   - you can compare coefficients as the coefficients of corresponding powers of x in the polynomials are equal.

   For example, if $ax^2 + bx + c \equiv 3x^2 + 2$, then $a = 3$, $b = 0$ and $c = 2$.

2. **Forms of partial fractions**
   - $\dfrac{px + q}{(ax + b)(cx + d)} \equiv \dfrac{A}{ax + b} + \dfrac{B}{cx + d}$
   - $\dfrac{px + q}{(ax + b)^2} \equiv \dfrac{A}{ax + b} + \dfrac{B}{(ax + b)^2}$
   - $\dfrac{px^2 + qx + r}{(ax + b)(cx + d)^2} \equiv \dfrac{A}{ax + b} + \dfrac{B}{cx + d} + \dfrac{C}{(cx + d)^2}$

   Note that any of A, B or C may be zero.

3. $\int \dfrac{f'(x)}{f(x)} dx = \ln|f(x)| + c$

   In Chapter 8, you found that when you integrate a quotient, if the top line is the derivative of the bottom line then the answer is $\ln|\text{bottom line}| + c$.

   Take care, not every quotient is in this form.

   For example:
   - $\int \dfrac{7}{3x + 5} dx \equiv \dfrac{7}{3} \int \dfrac{3}{3x + 5} dx = \dfrac{7}{3} \ln|3x + 5| + c$

   Rewrite the integral so that the top line is the derivative of the bottom line.

   - $\int \dfrac{7}{(3x + 5)^2} dx = -\dfrac{7}{3(3x + 5)} + c$

   Use the chain rule in reverse or use the substitution $u = 3x + 5$.

Worked examples

1 Working with a denominator in the form $(ax+b)(cx+d)$

Express $\dfrac{x-1}{x^2+x}$ in terms of the sum of partial fractions.

Hint: Each partial fraction question involves the following steps.

Solution

Step 1: Fully factorise the bottom line (denominator).

The denominator is $x^2 + x$.

Factorising gives $x(x + 1)$.

Step 2: Select the matching form of partial fraction. *(See Key fact 2.)*

So $\dfrac{x-1}{x(x+1)} \equiv \dfrac{A}{x} + \dfrac{B}{x+1}$

Note that A and B are constants to be determined. On the left hand side, the bottom line is written in factorised form.

Step 3: Multiply through by the denominator of the left-hand side to clear all fractions.

This gives $\dfrac{x-1}{x(x+1)} \times x(x+1) \equiv \dfrac{A}{x} \times x(x+1) + \dfrac{B}{(x+1)} \times x(x+1)$

So $x - 1 \equiv A(x+1) + Bx$.

Multiply each term by $x(x+1)$.

Step 4: Use one or both of the methods of **substitution** or **comparing coefficients** to find the values of the constants.

$x - 1 \equiv A(x+1) + Bx$

Substituting $x = 0$: $0 - 1 = A \times (0+1) + B \times 0$
$\Rightarrow \quad -1 = A$
$\Rightarrow \quad A = -1.$

Hint: You can substitute in any value for x as this is an identity, but it is best to choose a value that will make one of the terms equal to 0. In this example, you don't need to compare coefficients.

Substituting $x = -1$: $(-1) - 1 = A \times ((-1)+1) + B \times (-1)$
$\Rightarrow \quad -2 = -B$
$\Rightarrow \quad B = 2.$

Note that this is an equation not an identity.

Hence $\dfrac{x-1}{x(x+1)} \equiv -\dfrac{1}{x} + \dfrac{2}{x+1}.$

Make sure you write down the final answer.

2 Working with a denominator in the form $(ax+b)(cx+d)^2$

Express $\dfrac{x-2}{x^3 - 2x^2 + x}$ as the sum of partial fractions.

Solution

Step 1: Factorising gives

$x^3 - 2x^2 + x \equiv x(x^2 - 2x + 1) \equiv x(x-1)^2.$

Fully factorise the bottom line.

There is a repeated factor in the denominator, so this is the right form to use.

Step 2: $\dfrac{x-2}{x(x-1)^2} \equiv \dfrac{A}{x} + \dfrac{B}{x-1} + \dfrac{C}{(x-1)^2}$

Multiply each term by $x(x-1)^2$.

Step 3: $x - 2 \equiv A(x-1)^2 + Bx(x-1) + Cx$

You can find A and C using the substitution method. Use the comparing coefficients method to find B.

Step 4: $x - 2 \equiv A(x-1)^2 + Bx(x-1) + Cx$

Substituting $x = 0$: $0 - 2 = A(0-1)^2 + B \times 0 + C \times 0$
$\Rightarrow \quad -2 = A + 0 + 0$
$\Rightarrow \quad A = -2.$

The 2nd and 3rd terms both contain an x factor and so must be zero when $x = 0$.

Substituting $x = 1$: $1 - 2 = A \times 0 + B \times 0 + C \times 1$
$\Rightarrow \quad -1 = 0 + 0 + C$
$\Rightarrow \quad C = -1.$

The 1st and 2nd terms both contain an $x - 1$ factor and so must be zero when $x = 1$.

Now find B by equating coefficients of one of the terms involving B.

Comparing the coefficient of x^2 in
$x - 2 \equiv A(x-1)^2 + Bx(x-1) + Cx$

gives $0 = A + B$.

Since $A = -2$ then $B = 2$.

So $\dfrac{x-2}{x^3 - 2x^2 + x} \equiv \dfrac{x-2}{x(x-1)^2} \equiv -\dfrac{2}{x} + \dfrac{2}{(x-1)} - \dfrac{1}{(x-1)^2}$.

> This can be the x^2 or the x term.

> You can see by inspection that B is in the coefficient of x^2 when you expand $Bx(x-1)$.

> **Hint:** All the partial fractions may be found using **only** the equating coefficients method for solving identities. However, there is often less working required if substitution is used.
>
> Substituting $x = 0$ into an identity is always easy and is equivalent to comparing the constant terms. There is no point in substituting $x = 0$ **and** comparing the constant terms as you will obtain the same equations.

3 Using partial fractions in integration

Find the exact value of $\displaystyle\int_2^4 \dfrac{6x^2 + 32}{(3x-1)(x+2)^2}\,dx$.

> **Common mistake:** You cannot integrate this as it stands, so you need to rewrite it using partial fractions first.

Solution

Start by expressing $\dfrac{6x^2 + 32}{(3x-1)(x+2)^2}$ in partial fractions.

Step 1: The denominator is already factorised.

Step 2: $\dfrac{6x^2 + 32}{(3x-1)(x+2)^2} \equiv \dfrac{A}{(3x-1)} + \dfrac{B}{(x+2)} + \dfrac{C}{(x+2)^2}$

Step 3: $6x^2 + 32 \equiv A(x+2)^2 + B(3x-1)(x+2) + C(3x-1)$

Step 4: Substituting $x = -2$: $56 = C \times (-7)$

$\Rightarrow \qquad C = -8$

Substituting $x = \tfrac{1}{3}$: $6 \times \left(\tfrac{1}{3}\right)^2 + 32 = A\left(\tfrac{1}{3} + 2\right)^2$

$\Rightarrow \qquad \dfrac{98}{3} = \dfrac{49}{9} A$

$\Rightarrow \qquad A = 6$

Comparing the coefficients of x^2 in
$6x^2 + 32 \equiv A(x+2)^2 + B(3x-1)(x+2) + C(3x-1)$

gives $6 = A + 3B$

Since $A = 6$ then $B = 0$.

$\Rightarrow \dfrac{6x^2 + 32}{(3x-1)(x+2)^2} \equiv \dfrac{6}{(3x-1)} - \dfrac{8}{(x+2)^2}$

So $\displaystyle\int_2^4 \dfrac{6x^2 + 32}{(3x-1)(x+2)^2}\,dx$

$= \displaystyle\int_2^4 \dfrac{6}{(3x-1)}\,dx - \int_2^4 \dfrac{8}{(x+2)^2}\,dx$

$= 2\displaystyle\int_2^4 \dfrac{3}{(3x-1)}\,dx - 8\int_2^4 (x+2)^{-2}\,dx$

$= 2\Big[\ln|3x-1|\Big]_2^4 - 8\Big[-(x+2)^{-1}\Big]_2^4$

$= 2\Big[\ln|3x-1|\Big]_2^4 + 8\left[\dfrac{1}{(x+2)}\right]_2^4$

$= 2(\ln 11 - \ln 5) + 8\left(\dfrac{1}{6} - \dfrac{1}{4}\right)$

$= 2\ln\dfrac{11}{5} - \dfrac{2}{3}$

> Do not forget the repeated factor.

> Multiply each term by $(3x-1)(x+2)^2$.

> You could also find B by comparing the constant term by inspection or by substituting $x = 0$.

> So there in no $\dfrac{B}{(x+2)}$ term.

> Now you have a form you can integrate.

> The top line is not the derivative of the bottom. You need to use the reverse chain rule or substitution to integrate this.

> Rewrite the first integral so that the top line is the derivative of the bottom.

> Remember the rules of logs: $\ln a - \ln b = \ln \dfrac{a}{b}$.

Edexcel A Level Mathematics (Pure)

Test yourself

TESTED

1 Simplify $\dfrac{6x^3}{(x+1)^2} \times \dfrac{3x+3}{2x}$ as far as possible.

A $9x$
B $\dfrac{6x^3(x+3)}{(x+1)^2}$
C $\dfrac{3x^2(x+3)}{(x+1)^2}$
D $\dfrac{12x^2}{x+1}$
E $\dfrac{9x^2}{x+1}$

2 Which of the following is the correct form of partial fractions for the expression $\dfrac{x}{x^2-3x-4}$?

A $\dfrac{A}{x+4}+\dfrac{B}{x-1}$
B $\dfrac{Ax}{x+1}+\dfrac{Bx}{x-4}$
C $\dfrac{Ax+B}{x^2-3x-4}$
D $\dfrac{A}{x+1}+\dfrac{B}{x-4}$

3 Express $\dfrac{4+6x-x^2}{(x-1)(x+2)^2}$ as the sum of partial fractions.

A $\dfrac{1}{x-1}+\dfrac{4}{(x+2)^2}$
B $\dfrac{1}{x-1}-\dfrac{2}{x+2}+\dfrac{4}{(x+2)^2}$
C $\dfrac{1}{x-1}-\dfrac{2}{x+2}+\dfrac{4}{3(x+2)^2}$
D $\dfrac{6}{x+2}+\dfrac{4}{(x+2)^2}-\dfrac{1}{x-1}$

4 Express $\dfrac{5-x}{x^2-x-2}$ in partial fractions and use these to find $\displaystyle\int\left(\dfrac{5-x}{x^2-x-2}\right)dx$.

A $\ln\left(\dfrac{|x-2|}{(x+1)^4}\right)+c$
B $\ln|x-2|+2\ln|x+1|+c$
C $\ln\left(\dfrac{x-2}{(x+1)^2}\right)+c$
D $\ln|x-2|-2\ln|x+1|+c$

5 Express $\dfrac{5-2x}{(1-x)^2}$ in partial fractions and use these to evaluate $\displaystyle\int_0^{\frac{1}{2}}\left(\dfrac{5-2x}{(1-x)^2}\right)dx$.

A $7\ln 2 - 2$
B $2\ln 2 + 5$
C $2\ln 2 + 3$
D $2\ln 2 - 3$

Full worked solutions online

CHECKED ANSWERS

Exam-style question

i Express $\dfrac{3-2x^2}{(2x-3)(x-1)^2}$ in partial fractions.

ii Hence find $\displaystyle\int_2^3 \dfrac{3-2x^2}{(2x-3)(x-1)^2}\,dx$.

Short answers on page 221

Full worked solutions online

CHECKED ANSWERS

The binomial theorem

REVISED

Key facts

1. $n! = n(n-1)(n-2)\ldots\times 3 \times 2 \times 1$.
   $0! = 1$ and $1! = 1$

   $3! = 3 \times 2 \times 1 = 6$.

2. The number of ways of arranging n unlike objects in line is $n!$

3. The number of possible selections (combinations) of r objects from n unlike objects is $^nC_r = \dfrac{n!}{r!(n-r)!}$

   $^{10}C_3 = \dfrac{10!}{3! \times (10-3)!} = \dfrac{10!}{3! \times 7!} = 120$.

   The number of ways of choosing 3 students from a group of 10 to go on a school trip is $^{10}C_3 = 120$.

   You should use ^nC_r when the order in which the objects are selected doesn't matter.

   - You may see ^nC_r written as $_nC_r$ or $\binom{n}{r}$.
   - $^nC_0 = {^nC_n} = 1$.

   There is 1 way to choose none of the students so $^{10}C_0 = 1$ and 1 way to choose all of them so $^{10}C_{10} = 1$.

4. A **binomial expression** is an expression with two terms. For example: $(x+2)$ or $(3x-y)$.

   Here are some binomial expressions raised to a power and their expansions.

   $(x+a)^2 = 1x^2 + 2ax + 1a^2$
   $(x+a)^3 = 1x^3 + 3ax^2 + 3a^2x + 1a^3$
   $(x+a)^4 = 1x^4 + 4ax^3 + 6a^2x^2 + 4a^3x + 1a^4$

   - The powers of the terms in x reduce by 1 each time and the powers of a increase by 1 each time.
   - The sum of the two powers is always the same as the power of the bracket.
   - The number of terms is $n+1$.

   The coefficients in the expansions above are called **binomial coefficients**.

   Polynomials produced by expanding binomials in this way are called **binomial expansions.**
   - The powers of the terms in x reduce by one each term and the powers of a increase by one each term.
   - The sum of the two powers is always the same as the power of the bracket.

5. You can use Pascal's triangle to find binomial coefficients.

   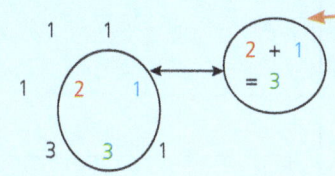

   Each number in Pascal's triangle is found by adding the two above it.

   Hint: Remember the binomial coefficients are symmetrical and so $^nC_r = {^nC_{n-r}}$. For example, $^{10}C_3 = \dfrac{10!}{3! \times 7!}$ and $^{10}C_7 = \dfrac{10!}{7! \times 3!}$ so both equal 120.

6. You can also use the formula $^nC_r = \dfrac{n!}{r!(n-r)!}$ to calculate **binomial coefficients.**

7. The **binomial theorem** states that when $n \in \mathbb{N}$,

   $(ax+by)^n = {^nC_0}(ax)^n + {^nC_1}(ax)^{n-1}(by)^1 + {^nC_2}(ax)^{n-2}(by)^2$
   $+ {^nC_3}(ax)^{n-3}(by)^3 + \ldots\ldots + {^nC_n}(by)^n$

   Choosing 3 students out of 10 to go on a trip is the same as choosing which 7 out of 10 will stay behind.

8. You can extend the binomial theorem to expand binomials in the form $(1+ax)^n$ when n is negative or a fraction.
   The general binomial theorem states:

   $\mathbb{N}$ is the set of natural numbers $\{1, 2, 3,\ldots\}$.

   $(1+x)^n = 1 + nx + \dfrac{n(n-1)}{2!}x^2 + \dfrac{n(n-1)(n-2)}{3!}x^3 + \ldots$

   The dots show that the expansion continues like this forever.

   When $n \in \mathbb{N}$, x may take any value but when $n \notin \mathbb{N}$, $|x| < 1$. As the expansion is infinite for $n \notin \mathbb{N}$ you will only be asked to find the first few terms.

   The expansion is only valid for values of x between -1 and 1.

9. You can rewrite $(a+x)^n$ as $a^n\left(1+\dfrac{x}{a}\right)^n$.

Edexcel A Level Mathematics (Pure)

Worked examples

1 Finding a binomial expansion when n is a positive integer

Expand $(2x - 3)^4$.

Solution

$(2x - 3)^4 = {}^4C_0(2x)^4 + {}^4C_1(2x)^3(-3)^1 + {}^4C_2(2x)^2(-3)^2$
$\qquad\qquad + {}^4C_3(2x)^1(-3)^3 + {}^4C_4(-3)^4$

$\qquad\quad = 1 \times 16x^4 + 4 \times 8x^3 \times (-3) + 6 \times 4x^2 \times 9$
$\qquad\qquad + 4 \times 2x \times (-27) + 1 \times 81$

$\qquad\quad = 16x^4 - 96x^3 + 216x^2 - 216x + 81$

> Remember that ${}^nC_0 = {}^nC_n = 1$. You can use the 4th row of Pascal's triangle to generate the binomial coefficients: 1, 4, 6, 4, 1.

> Take care with your signs! The −3 results in alternating positive and negative terms.

2 Using the formula ${}^nC_r = \dfrac{n!}{r!(n-r)!}$ to calculate single terms in a binomial expansion

Find the coefficient of x^5 in the expansion $(4 - 3x)^7$.

Solution

The term will be

${}^7C_5(4)^{7-5}(-3x)^5 = 21 \times 16 \times (-243x^5)$
$\qquad\qquad\qquad\quad = -81648x^5$.

So the coefficient is −81 648.

> Make sure you use brackets.

> Use your calculator to find ${}^7C_5$.

3 Use the general binomial theorem with a negative power

Find the first four terms in the expansion of $(1 + x)^{-5}$ in ascending powers of x.

State the values of x for which the expansion is valid.

Solution

Use $(1+x)^n = 1 + nx + \dfrac{n(n-1)}{2!}x^2 + \dfrac{n(n-1)(n-2)}{3!}x^3 + \ldots$

$\qquad\qquad$ when $|x| < 1$

Replace n with −5

$(1+x)^{-5} = 1 + (-5)x + \dfrac{(-5)(-6)}{2!}x^2 + \dfrac{(-5)(-6)(-7)}{3!}x^3 + \ldots$

$\qquad\quad = 1 - 5x + 15x^2 - 35x^3 + \ldots$

So $(1+x)^{-5} \approx 1 - 5x + 15x^2 - 35x^3$ when $|x| < 1$.

> Watch your signs and take care when you subtract 1 from a negative number.

> Remember that 1 is a term too!

> **Common mistake:** Notice the use of the ≈ sign. When you use 3 dots (...) you are saying that the expansion carries on further so the expression on the LHS has exactly the same value as the one on the RHS. However, when you just give the first few terms of the expansion (without the 3 dots) then the two expressions are no longer exactly equal and so you need to write '≈' instead of '='.

Chapter 10 Further algebra

4 Use the general binomial theorem with a fractional power

Write down the first three terms in the binomial expansion of $\dfrac{1}{\sqrt{1-2x}}$ in ascending powers of x.

For what range of values of x is this expansion valid?

Solution

You can write $\dfrac{1}{\sqrt{1-2x}}$ as $(1-2x)^{-\frac{1}{2}}$

Use $(1+x)^n = 1 + nx + \dfrac{n(n-1)}{2!}x^2 + \dfrac{n(n-1)(n-2)}{3!}x^3 + \ldots$

when $|x| < 1$.

Replace x with $(-2x)$ and n with $-\dfrac{1}{2}$:

$$(1-2x)^{-\frac{1}{2}} = 1 + \left(-\dfrac{1}{2}\right)(-2x) + \dfrac{\left(-\frac{1}{2}\right)\left(-\frac{3}{2}\right)}{2!}(-2x)^2 + \ldots$$

$\Rightarrow (1-2x)^{-\frac{1}{2}} = 1 + x + \dfrac{3}{2}x^2 + \ldots$ when $|2x| < 1$

So $\dfrac{1}{\sqrt{1-2x}} \approx 1 + x + \dfrac{3}{2}x^2$ when $|x| < \dfrac{1}{2}$.

Common mistake: You need to divide both sides of the inequality by 2 as you are asked for the range of values of x not $2x$.

This means that $-\dfrac{1}{2} < x < \dfrac{1}{2}$.

5 Use the binomial theorem for $(a + x)^n$

i Find a quadratic approximation for $\sqrt{(16 - x^2)}$, stating the range of values of x for which this expansion is valid.

ii Use your expansion to find an approximation for $\sqrt{15.9}$.

Solution

i You can write $\sqrt{(16 - x^2)}$ as $\left(16 - \dfrac{16x^2}{16}\right)^{\frac{1}{2}}$

So $(16 - x^2)^{\frac{1}{2}} = 16^{\frac{1}{2}}\left(1 - \dfrac{x^2}{16}\right)^{\frac{1}{2}}$

$= 4\left(1 - \dfrac{x^2}{16}\right)^{\frac{1}{2}}$

Hint: Sometimes you are asked to expand an expression which is not in the form $(1+x)^n$. When this happens, you need to rearrange the expression so that the first term inside the brackets is 1 **before** you expand it.

Now you have '16' in both terms and so you can take out $16^{\frac{1}{2}}$ as a common factor.

In general: $(a+x)^n = a^n\left(1 + \dfrac{x}{a}\right)^n$.

Use $(1+x)^n = 1 + nx + \dfrac{n(n-1)}{2!}x^2 + \dfrac{n(n-1)(n-2)}{3!}x^3 + \ldots$

when $|x| < 1$.

Replace x with $\left(-\dfrac{x^2}{16}\right)$ and n with $\dfrac{1}{2}$:

$$4\left(1 - \dfrac{x^2}{16}\right)^{\frac{1}{2}} = 4\left(1 + \dfrac{1}{2}\left(-\dfrac{x^2}{16}\right) + \dfrac{\frac{1}{2}\left(-\frac{1}{2}\right)}{2!}\left(-\dfrac{x^2}{16}\right)^2 + \ldots\right)$$

You are asked for a quadratic so you don't need this term since $(x^2)^2 = x^4$.

when $\left|\dfrac{x^2}{16}\right| < 1$.

Edexcel A Level Mathematics (Pure)

Tidy this up to get:

$$4\left(1-\frac{x^2}{16}\right)^{\frac{1}{2}} = 4\left(1-\frac{1}{32}x^2-\ldots\right) \text{ when } |x^2| < 16.$$

> Multiply both sides of the inequality by 16 and then square root to find the range of values of x.

So $\sqrt{(16-x^2)} \approx 4 - \frac{1}{8}x^2$ when $|x| < 4$.

ii $\sqrt{15.9} = \sqrt{(16-x^2)}$ when $x^2 = 0.1$

So $\sqrt{(16-0.1)} \approx 4 - \frac{1}{8} \times 0.1$

≈ 3.9875

> Notice that the correct value for $\sqrt{15.9}$ is 3.9874804… . So this approximation is very close.

6 Working with two binomials

Find a, b and c such that $\dfrac{1}{(1+3x)(1-4x)^2} \approx a + bx + cx^2$.

State the values of x for which the expansion is valid.

Solution

$$\frac{1}{(1+3x)(1-4x)^2} = (1+3x)^{-1}(1-4x)^{-2}$$

Using the binomial expansion:

$(1+3x)^{-1} = 1 + (-1)(3x) + \dfrac{(-1)(-2)}{2!}(3x)^2 + \ldots$ when $|3x| < 1$

$\approx 1 - 3x + 9x^2$ when $|x| < \dfrac{1}{3}$

$(1-4x)^{-2} = 1 + (-2)(-4x) + \dfrac{(-2)(-3)}{2!}(-4x)^2 + \ldots$ when $|4x| < 1$

$\approx 1 + 8x + 48x^2$ when $|x| < \dfrac{1}{4}$

> $-\dfrac{1}{3} < x < \dfrac{1}{3}.$

> $-\dfrac{1}{4} < x < \dfrac{1}{4}.$

So $\dfrac{1}{(1+3x)(1-4x)^2} \approx (1-3x+9x^2)(1+8x+48x^2)$

$\approx 1 + 8x + 48x^2$

$ - 3x - 24x^2$

$ + 9x^2$

$\approx 1 + 5x + 33x^2$

> Multiply the second bracket by 1.

> Multiply the second bracket by $-3x$. You are only interested in powers up to x^2 so ignore $-3x \times 48x^2$.

> Multiply the second bracket by $9x^2$. Ignore any powers higher than x^2.

So $a = 1$, $b = 5$ and $c = 33$.

The two expansions are valid when $|x| < \dfrac{1}{3}$ and $|x| < \dfrac{1}{4}$.

The tighter restriction is $|x| < \dfrac{1}{4}$ so this is the one that the overall expansion is valid for.

> If, say, $x = 0.3$ then the first expansion would be valid as $0.3 < \dfrac{1}{3}$ but the second one would not be valid as $0.3 > \dfrac{1}{4}$.

> **Hint:** Sometimes you will be asked to rewrite an expression as two or more partial fractions before using the binomial expansion.
>
> The advantage of using partial fractions is that you can add/subtract the resulting expansions rather than multiplying them. It also makes it easier to find the coefficient of one particular term.
>
> See question 3 of the Review section on page 199 for an example of this.

Test yourself

1. Find the coefficient of x^5 in the expansion of $(2-x)^8$.
   A $-{}^8C_5$ B $-{}^8C_5 \times 2^5$ C ${}^8C_5 \times 2^3$ D $-{}^8C_5 \times 2^3$ E $-{}^8C_3 \times 2^5$

2. Write out the binomial expansion of $(1-3x)^4$.
   A $1+12x+54x^2+108x^3+81x^4$
   B $1-12x+54x^2-108x^3+81x^4$
   C $1-12x-18x^2-12x^3-3x^4$
   D $1-12x+108x^2-684x^3+1944x^4$
   E $1-3x+9x^2-27x^3+81x^4$

3. Find the first four terms in the binomial expansion of $(1-5x)^{-2}$.
   A $1+10x+75x^2+500x^3$
   B $1-10x+75x^2-500x^3$
   C $1-2x+3x^2-4x^3$
   D $1+10x+15x^2+20x^3$

4. Use the first three terms in the expansion of $\sqrt{1-x}$ to find an approximation for $\sqrt{0.95}$.
   Write down all the numbers on your calculator display.
   A 1.0246875 B 0.9746875 C 0.9746796875 D 0.9753125

5. Find a, b and c such that $\dfrac{1}{(1+3x)^3} \approx 1+ax+bx^2+cx^3$
   A $a=9$, $b=27$ and $c=27$
   B $a=-3$, $b=6$ and $c=-10$
   C $a=-9$, $b=18$ and $c=-30$
   D $a=-9$, $b=54$ and $c=-270$

6. Find a quadratic approximation for $\dfrac{1}{\sqrt{4+x}}$.
   A $1-\dfrac{1}{2}x+\dfrac{3}{8}x^2$ B $4-\dfrac{1}{2}x+\dfrac{3}{32}x^2$ C $\dfrac{1}{2}-\dfrac{1}{16}x+\dfrac{3}{256}x^2$ D $\dfrac{1}{2}-\dfrac{1}{4}x+\dfrac{3}{16}x^2$

7. State the values of x for which the expansion of $\dfrac{1}{\left(1-\frac{x}{2}\right)(4+3x)}$ is valid.
   A $|x|<\dfrac{4}{3}$ and $|x|<2$
   B $|x|<\dfrac{4}{3}$
   C $|x|<1$
   D $|x|<\dfrac{1}{3}$

Full worked solutions online

Exam-style question

i Find the first three terms in the binomial expansion of $(1+y)^{\frac{1}{2}}$ in ascending powers of y.

ii Explain why you cannot substitute $y=4$ into your expansion to provide an estimate for $\sqrt{5}$.

iii Find the first three terms in the binomial expansion of $(4+x)^{\frac{1}{2}}$.
    State the range of values of x for which the expansion is valid.

iv Use your expansion of $(4+x)^{\frac{1}{2}}$ to find an estimate for $\sqrt{5}$.
    Using the value of $\sqrt{5}$ on your calculator, find the percentage error in your answer.

Short answers on page 221

Full worked solutions online

Chapter 11 Parametric equations

About this topic

Until now you have considered equations of lines and curves given in the form $y = f(x)$. However, in some cases, these equations are quite complicated and some curves just cannot be written in this way. It is sometimes easier to write the x and y coordinates in terms of some intermediate variable called a parameter.

Before you start, remember

- curve sketching
- coordinate geometry
- trigonometric identities
- differentiation and the chain rule.

Parametric equations

REVISED

Key facts

1. The equation of a curve is often written in Cartesian form, for example $y = \frac{3}{4}x^2$ or $x^2 + y^2 = 1$.

   A **Cartesian equation** describes the curve with a single equation linking the x and y coordinates.

   A curve may also be written in **parametric form**.
   For example, the curve $y = \frac{3}{4}x^2$ is the same as $x = 2t, y = 3t^2$ where t is a parameter.

2. The Cartesian equation of the curve is obtained from the parametric form by **eliminating the parameter** between the two equations for x and y.

3. When **plotting a curve** given in parametric form, the x and y coordinates of individual points lying on the curve can be found by substituting different values for the parameter into the two equations.

4. The same information is required when **sketching a curve** given in parametric form as when sketching a curve given in Cartesian form:
   - the points of intersection with the axes
   - any restrictions on the values that x and y can take
   - the behaviour of the curve as x and y tend to ∞.

5. The parametric equations for a **circle** with radius r and
   - centre $(0, 0)$ are $x = r\cos\theta, y = r\sin\theta$
   - centre (a, b) are $x = a + r\cos\theta, y = b + r\sin\theta$.

Worked examples

1 Using parametric equations to sketch a curve

A curve has parametric equations $x = t + 2$, $y = \frac{1}{t}$.

i Find the coordinates of points on the curve for the following values of t: $-3, -2, -1, -0.5, 0.5, 1, 2, 3$.

ii Are there any values of x for which the curve is undefined?

iii Plot the points you have found and join them to give the curve.

Solution

i

t	-3	-2	-1	-0.5	0.5	1	2	3
$x = t + 2$	-1	0	1	1.5	2.5	3	4	5
$y = \frac{1}{t}$	$-\frac{1}{3}$	$-\frac{1}{2}$	-1	-2	2	1	$\frac{1}{2}$	$\frac{1}{3}$

ii The curve is not defined when $t = 0$, since $y = \frac{1}{t}$ and so y is undefined there.

When $t = 0$, $x = 2$ and this is the equation of an asymptote to the curve.

iii

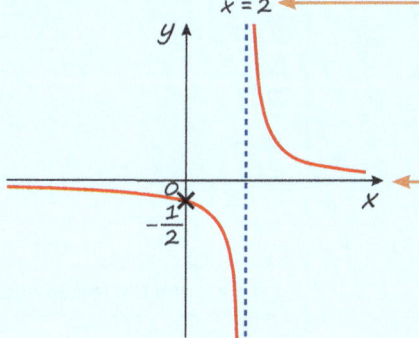

> The line $x = 2$ is an asymptote.

> The x-axis is also an asymptote since y can never equal 0.

2 Converting between parametric and Cartesian equations

> You will often be asked to find the Cartesian equation of a curve given in parametric form. You need to eliminate the parameter between the two equations.
>
> You may also be asked to give the parametric equations to describe a curve. The parametric equations are not unique and so you will be given a suitable substitution to use.

i Find the Cartesian equations of the curve with parametric equations $x = t + 2$, $y = \frac{1}{t-1}$.

ii Find the parametric equations of the curve $x^2 + y^2 = 5$ given $x = \sqrt{5} \cos t$.

Solution

i Rearrange the expression for x to make t the subject.
$x = t + 2 \Rightarrow t = x - 2$.
Substituting this into $y = \frac{1}{t-1}$ gives
$y = \frac{1}{(x-2)-1}$
$\Rightarrow y = \frac{1}{x-3}$

> Always use the coordinate with the linear equation (if there is one) to get the substitution for the parameter.

Edexcel A Level Mathematics (Pure)

ii $x^2 + y^2 = 5$

Substitute $x = \sqrt{5}\cos t$: $5\cos^2 t + y^2 = 5$.

Make y^2 the subject $\Rightarrow$
$$y^2 = 5 - 5\cos^2 t$$
$$y^2 = 5(1 - \cos^2 t)$$ ← Use $\sin^2\theta + \cos^2\theta \equiv 1$.
$$y^2 = 5\sin^2 t$$
$$y = \sqrt{5}\sin t.$$ ← You need to find y, not y^2.

So the parametric equations are $x = \sqrt{5}\cos t, y = \sqrt{5}\sin t$.

3 Using trigonometric identities

Find the Cartesian equation of the curve given by the parametric equations

$$x = \cos\theta + \sin\theta, \quad y = 2\cos\theta + \sin\theta$$

Solution

Subtract the expressions for x and y to obtain expressions for $\cos\theta$ and $\sin\theta$.

$$\begin{array}{ll} y = 2\cos\theta + \sin\theta & 2x = 2\cos\theta + 2\sin\theta \\ x = \cos\theta + \sin\theta & y = 2\cos\theta + \sin\theta \\ \hline y - x = \cos\theta & 2x - y = \sin\theta \end{array}$$

Now use the identity $\sin^2\theta + \cos^2\theta = 1$
$$(2x - y)^2 + (y - x)^2 = 1$$

Expand the brackets: $(4x^2 - 4xy + y^2) + (y^2 - 2xy + x^2) = 1$

Collect like terms: $5x^2 - 6xy + 2y^2 = 1$ ← This is the Cartesian equation.

4 The equation of a circle

Find the Cartesian equation of the curve described by the parametric equations $x = 2 + 3\cos\theta, y = -1 + 3\sin\theta$

Solution

$x = 2 + 3\cos\theta \Rightarrow \cos\theta = \dfrac{x-2}{3}$

$y = -1 + 3\sin\theta \Rightarrow \sin\theta = \dfrac{y+1}{3}$

$\cos^2\theta + \sin^2\theta \equiv 1 \Rightarrow \left(\dfrac{x-2}{3}\right)^2 + \left(\dfrac{y+1}{3}\right)^2 = 1$ ← Multiply each term by $3^2 = 9$.

Hint: This is the equation of the circle with centre $(2, -1)$ and radius 3. You need to be able to recognise the parametric equations of a circle. See Key fact 5.

5 Solving problems involving parametric equations

Find the coordinates of the points where the line $y = 2x$ cuts the curve $x = t^2$, $y = t^3$.

Solution

To find the points of intersection, substitute $x = t^2$, $y = t^3$ into the equation of the line.

$$y = 2x$$
$$t^3 = 2t^2$$
$$t^3 - 2t^2 = 0$$

Factorising: $t^2(t - 2) = 0$

$t = 0$ or 2

When $t = 0$, $x = 0$ and $y = 0$.

When $t = 2$, $x = 4$ and $y = 8$.

The points of intersection of the line and the curve are $(0, 0)$ and $(4, 8)$.

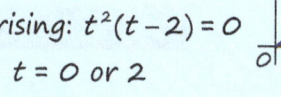

Test yourself

TESTED

1. Find the Cartesian equation of the curve given by the parametric equations $x = 2t^2$, $y = 4(t-1)$.

   A $y = \pm\sqrt{8x - 16}$ B $y = -4 \pm \sqrt{8x}$ C $y = -4 \pm \sqrt{2x}$ D $y = -1 \pm \sqrt{8x}$

2. Find the Cartesian equation of the curve given by the parametric equations $x = \cos\theta - \sin\theta$, $y = \sin\theta + 2\cos\theta$.

   A $5x^2 - 2xy + 2y^2 = 9$
   B $37x^2 - 34xy + 10y^2 = 9$
   C $-3x^2 - 2xy + 2y^2 = 9$
   D $5x^2 - 2xy + 2y^2 = 3$

3. A curve is defined parametrically by $x = 3 + 5\cos\theta$, $y = 5\sin\theta - 2$.
   Three of the following statements are false and one is true. Which statement is true?
   A The curve is a circle centre (3, 2) and radius 5.
   B The curve cuts the y-axis at the point where $\theta = 0.41$ radians.
   C There is only one point on the curve with x coordinate 3.
   D At the point (8, −2), $\theta = 0$.

4. A line has equation $y = 2x + 3$ and a curve has equations $x = 3t$, $y = \dfrac{3}{t}$.
   Three of the following statements are false and one is true. Which statement is true?
   A The line and curve intersect at points (−3, −3) and $\left(1\tfrac{1}{2}, 6\right)$.
   B The line and curve intersect at (−3, −3) and $\left(1\tfrac{1}{2}, 1\tfrac{1}{2}\right)$.
   C The line and curve intersect at (3, 3) only.
   D The line and curve do not intersect at any point.

Full worked solutions online CHECKED ANSWERS

Exam-style question

A curve C has parametric equations $x = 2t^2, y = 4t$.

A line p passes through the point (2, 0) and has gradient $\frac{4}{3}$.

i   Find the equation of line p.
ii Find the values of the parameter t for which p intersects C and hence write down the coordinates of the point(s) of intersection.

A second line q has equation $3y = 4x + 12$

iii Prove that the line q does not intersect the curve C.
iv From the equations of p and q what can you deduce about the lines?
v   Work out the Cartesian equation of C and draw a sketch showing C, p and q on the same set of axes.

Short answers on page 221

Full worked solutions online

CHECKED ANSWERS

Calculus with parametric equations

REVISED

Key fact

1. You can use differentiation and the chain rule to find the gradient of a curve defined by parametric equations with parameter t,

$$\frac{dy}{dx} = \frac{\frac{dy}{dt}}{\frac{dx}{dt}} \text{ or } \frac{dy}{dx} = \frac{dy}{dt} \times \frac{dt}{dx}.$$

2. To find the area under a curve defined parametrically, you need to integrate with respect to the parameter.
   Area under a parametric curve
   $$= \int_{t_a}^{t_b} y \frac{dx}{dt} \, dt$$
   where t_a is the value of the parameter t at a and t_b is the value of the parameter t at b.

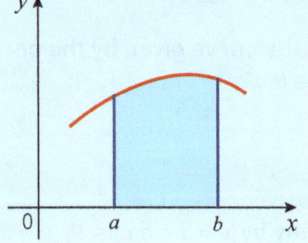

Worked examples

1 Finding the gradient

A curve has parametric equations $x = 4t^2 + 5, y = 2t$.

Find $\frac{dy}{dx}$ in terms of the parameter t.

Solution

$y = 2t \Rightarrow \frac{dy}{dt} = 2$

$x = 4t^2 + 5 \Rightarrow \frac{dx}{dt} = 8t$

Differentiate x and y with respect to t.

$\frac{dy}{dx} = \frac{\frac{dy}{dt}}{\frac{dx}{dt}}$

Therefore $\frac{dy}{dx} = \frac{2}{8t} = \frac{1}{4t}$.

2 Finding the equation of tangents and normals

A curve has parametric equations $x = 2\sin t$, $y = \cos 2t$.

i Find the coordinates of the points where $t = -\frac{\pi}{2}, -\frac{\pi}{6}, 0, \frac{\pi}{6}, \frac{\pi}{2}$.

Draw the curve for $-\frac{\pi}{2} \leq t \leq \frac{\pi}{2}$, using equal scales for x and y.

ii Show that $\frac{dy}{dx} = 2\sin t$.

iii Find the equation of the tangents and normals at the points where $t = -\frac{\pi}{6}$ and $\frac{\pi}{6}$.

iv Add these tangents and normals to your graph. What shape is formed by the four lines?

Solution

i

t	$-\frac{\pi}{2}$	$-\frac{\pi}{6}$	0	$\frac{\pi}{6}$	$\frac{\pi}{2}$
$(x, y) = (2\sin t, \cos 2t)$	$(-2, -1)$	$(-1, \frac{1}{2})$	$(0, 1)$	$(1, \frac{1}{2})$	$(2, -1)$

See the red curve in part **iv**.

ii $x = 2\sin t \quad \Rightarrow \quad \frac{dx}{dt} = 2\cos t$

$y = \cos 2t \quad \Rightarrow \quad \frac{dy}{dt} = -2\sin 2t$

$$\frac{dy}{dx} = \frac{\frac{dy}{dt}}{\frac{dx}{dt}}$$

$\Rightarrow \quad \frac{dy}{dx} = \frac{-2\sin 2t}{2\cos t} = \frac{-4\sin t \cos t}{2\cos t}$ ← Remember that $\sin 2t = 2\sin t \cos t$.

$\Rightarrow \quad \frac{dy}{dx} = -2\sin t$ as required.

iii When $t = -\frac{\pi}{6}, (x, y) = \left(-1, \frac{1}{2}\right), \frac{dy}{dx} = -2 \times -\frac{1}{2} = +1.$

Gradients: tangent at $\left(-1, \frac{1}{2}\right)$ is 1, normal is $\frac{1}{-1} = -1$. ← For perpendicular lines $m_1 m_2 = -1$ $m_2 = -\frac{1}{m_1}$.

Equations: tangent: $y - \frac{1}{2} = 1(x-(-1))$

$\Rightarrow y = x + \frac{3}{2}$

normal: $y - \frac{1}{2} = -1(x-(-1))$

$\Rightarrow y = -x - \frac{1}{2}$

When $t = \frac{\pi}{6}: (x, y) = \left(1, \frac{1}{2}\right)$, and $\frac{dy}{dx} = -2 \times \frac{1}{2} = -1.$ ← So the gradient of the tangent is −1 and the gradient of the normal is 1.

Equations: tangent: $y - \frac{1}{2} = -1(x-1)$

Edexcel A Level Mathematics (Pure) 177

$$\Rightarrow y = -x + \frac{3}{2}$$

normal: $y - \frac{1}{2} = 1(x - 1)$

$$\Rightarrow y = x - \frac{1}{2}$$

iv The four lines form a square.

> You can see this on the diagram.

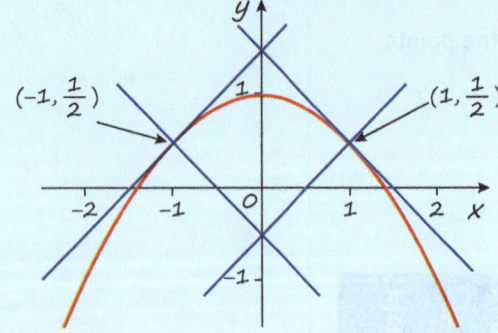

3 Using calculus to solve problems

The graph shows the curve with parametric equations $x = 4 - t^2$, $y = 4t - 3t^3$.

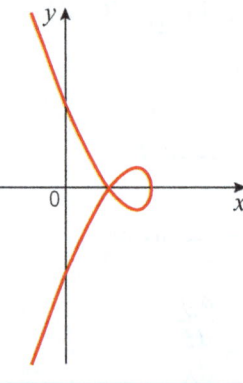

i Find the values of t at the points where the curve crosses the axes. Hence find the coordinates of these points.

ii Find $\dfrac{dy}{dx}$ in terms of t and hence find the coordinates of the turning points.

iii Find the area of the loop.

Solution

i When the curve crosses the y-axis, $x = 0$

$4 - t^2 = 0 \quad \Rightarrow \quad t^2 = 4$

$t = 2 \quad \text{or} \quad t = -2$

When $t = 2$, $y = 4 \times 2 - 3 \times 2^3 = 8 - 24 = -16$.

The point is $(0, -16)$.

When $t = -2$, $y = 4 \times -2 - 3 \times (-2^3) = -8 + 24 = 16$.

The point is $(0, 16)$.

The curve cuts the y-axis at the points $(0, -16)$ and $(0, 16)$.

When the curve cuts the x-axis, $y = 0$

$4t - 3t^3 = 0$

$\Rightarrow \quad t(4 - 3t^2) = 0$

$\Rightarrow \quad 3t\left(t^2 - \frac{4}{3}\right) = 0$

$\Rightarrow \quad t = 0, \sqrt{\frac{4}{3}} \text{ or } -\sqrt{\frac{4}{3}}$

> Don't forget the negative square root.

So $t = 0$, $\dfrac{2}{\sqrt{3}}$ or $-\dfrac{2}{\sqrt{3}}$

When $t = 0$, $x = 4 - 0^2 = 4$. The point is $(4, 0)$.

When $t = \dfrac{2}{\sqrt{3}}$, $x = 4 - \dfrac{4}{3} = \dfrac{8}{3}$. The point is $\left(\dfrac{8}{3}, 0\right)$.

When $t = -\frac{2}{\sqrt{3}}$, $x = 4 - \frac{4}{3} = \frac{8}{3}$. The point is $\left(\frac{8}{3}, 0\right)$.

The curve crosses the x-axis at $(4, 0)$ and twice at $\left(\frac{8}{3}, 0\right)$.

ii $x = 4 - t^2 \Rightarrow \frac{dx}{dt} = -2t$

$y = 4t - 3t^3 \Rightarrow \frac{dy}{dt} = 4 - 9t^2$

$$\frac{dy}{dx} = \frac{\frac{dy}{dt}}{\frac{dx}{dt}}$$

$\Rightarrow \frac{dy}{dx} = \frac{4 - 9t^2}{-2t}$

$\frac{dy}{dx} = 0$ when $\frac{4 - 9t^2}{-2t} = 0$

$\Rightarrow 9t^2 = 4$

$\Rightarrow t = \pm\sqrt{\frac{4}{9}} = \pm\frac{2}{3}$

When $t = \frac{2}{3}$, $x = 4 - \frac{4}{9} = 3\frac{5}{9}$ and $y = \frac{8}{3} - \frac{8}{9} = \frac{16}{9} = 1\frac{7}{9}$.

One turning point is $\left(3\frac{5}{9}, 1\frac{7}{9}\right)$.

When $t = -\frac{2}{3}$, $x = 3\frac{5}{9}$ and $y = -\frac{16}{9} = -1\frac{7}{9}$.

The other turning point is $\left(3\frac{5}{9}, -1\frac{7}{9}\right)$.

iii $x = 4 - t^2 \Rightarrow \frac{dx}{dt} = -2t$

$y = 4t - 3t^3$

Area under a curve $= \int_{t_a}^{t_b} y \frac{dx}{dt} dt$

$= \int_{\frac{2}{\sqrt{3}}}^{0} (4t - 3t^3)(-2t) dt$

$= \int_{\frac{2}{\sqrt{3}}}^{0} (-8t + 6t^4) dt$

$= \left[-\frac{8}{3}t^3 + \frac{6}{5}t^5\right]_{\frac{2}{\sqrt{3}}}^{0}$

$= (0) - \left(-\frac{8}{3} \times \left(\frac{2}{\sqrt{3}}\right)^3 + \frac{6}{5} \times \left(\frac{2}{\sqrt{3}}\right)^5\right)$

$= 1.642\ldots$

So the total area of the loop is

$2 \times 1.642\ldots = 3.28$ (to 3 s.f.)

> The only way a fraction can equal zero is if the top line is zero.

> **Common mistake:** The curve crosses the x-axis at $t = 0$, $t = \frac{2}{\sqrt{3}}$ and $t = -\frac{2}{\sqrt{3}}$. The upper part of the loop goes from $t = \frac{2}{\sqrt{3}}$ (at $x = \frac{8}{3}$) to $t = 0$ (at $x = 4$) and the lower part from $t = -\frac{2}{\sqrt{3}}$ to $t = 0$. If you use $t = -\frac{2}{\sqrt{3}}$ and $t = \frac{2}{\sqrt{3}}$ the areas will cancel out and you will get an answer of zero, so you should find the area of half a loop and then double it.

> **Hint:** Note the upper limit is not always going to be the larger number. If you get the limits the wrong way round the answer will be negative.

Test yourself

TESTED

1. A curve has parametric equations $x = 4t$, $y = 1 - \frac{1}{t}$. Find the value of $\frac{dy}{dx}$ when $t = 3$.

   A $\frac{1}{36}$ B 36   C $\frac{4}{9}$ D $-\frac{1}{36}$

2. Find the gradient of the curve given parametrically by $x = 2\cos^3 t$, $y = 2\sin^3 t$.

   A $\tan t$ B $-\cot^3 t$ C $-\tan t$ D $-\tan^2 t$

3. A curve is given by $x = t^2 + 1$, $y = t(t-3)^2$.

   Three of the following statements are false and one is true. Which statement is true?

   A The curve has no stationary points.
   B The curve has one stationary point only at (10, 0).
   C The curve has stationary points at (10, 0) and (2, 4).
   D The curve has stationary points at (1, 0) and (10, 0).

4. The parametric equations of a curve are given by $x = \cos 2t$, $y = 4\sin t$.

   At the point where $t = \frac{\pi}{2}$, three of the following statements are false and one is true. Which statement is true?

   A The equation of the normal at the point where $t = \frac{\pi}{2}$ is $y = x + 3$.

   B The equation of the tangent at the point where $t = \frac{\pi}{2}$ is $x + y = 3$.

   C When $t = 0$, $\frac{dy}{dx} = 0$.

   D Both the x-axis and the y-axis are normals to the curve.

Full worked solutions online   CHECKED ANSWERS

Exam-style question

The graph shows the curve with parametric equations $x = 2t$, $y = \frac{2}{t}$.

P and Q are points on the same branch of the curve with parameters p and q respectively.

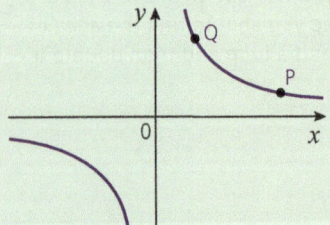

i Find the gradient of the chord PQ. Hence find the equation of the chord PQ.

ii Explain why replacing q by p in the equation of the chord gives the equation of the tangent at P.
Show that this is $y = -\frac{1}{p^2}x + \frac{4}{p}$.

iii Find $\frac{dy}{dx}$ for the curve in terms of t.

iv Verify that when $t = p$, $\frac{dy}{dx}$ is the same as the gradient of the equation that you derived in part ii.

v Tangents from two points on the curve pass through the point (3, 1).
Find the coordinates of the points of contact with the curve.

Short answers on page 221

Full worked solutions online   CHECKED ANSWERS

Chapter 12 Differential equations

About this topic

An equation involving a derivative such as $\frac{dy}{dx}$, $\frac{d^2y}{dx^2}$ or $\frac{d\theta}{dt}$ is known as a differential equation. A first order differential equation only involves a first derivative, for example $\frac{dy}{dx}$. A differential equation describes the rates of change of one variable with another. Many real-life problems involve the rate of change of a quantity such as temperature, acceleration, velocity and displacement and so give rise to differential equations. The solutions of such equations are used to make predictions about the behaviour of the variables involved.

Before you start, remember

- how to integrate x^n, $\frac{1}{x}$, e^x and trigonometric functions
- curve sketching
- direct and indirect (inverse) proportion.

Solving differential equations

REVISED

> **Key facts**
>
> 1. **Solving by direct integration**
>
> $\frac{dy}{dx} = f(x)$ ← f is a function of x.
>
> $\Rightarrow y = \int f(x)\,dx + c$
>
> 2. **Using separation of variables**
>
> $\frac{dy}{dx} = f(x)g(y)$ ← f is a function of x. g is a function of y.
>
> $\Rightarrow \int \frac{1}{g(y)}\,dy = \int f(x)\,dx$ ← You have now got all the terms in y and dy on one side and the terms in x and dx on the other.
>
> 3. The **general solution** of a differential equation may be represented by a family of curves. For example, the general solution for $\frac{dy}{dx} = 2x$ is $y = x^2 + c$. So the solution is a family of parabolas. Each curve corresponds to a different value of c, the constant of integration.
>
>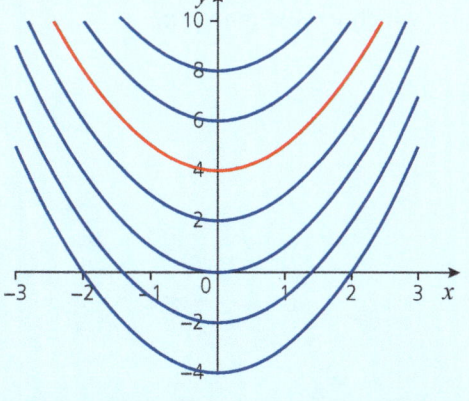
>
>   A **particular solution** is a single member of the family of curves corresponding to a particular value of c.
>
>   For example the red curve corresponds to $c = 4$, so $y = x^2 + 4$ is a particular solution.

Edexcel A Level Mathematics (Pure)

Worked examples

1 Using direct integration

i Solve the differential equation $\frac{dy}{dx} = x + 3$.

ii Sketch the family of solution curves, using $c = -2, -1, 0, 1$ and 2.

iii Find the particular solution for which $y = 9$ when $x = 2$. Indicate the curve on your graph corresponding to this particular solution.

Solution

i $\frac{dy}{dx} = x + 3$

$y = \int (x + 3) dx$ ← $\int \frac{dy}{dx} dx = \int dy = y$.

$y = \frac{1}{2}x^2 + 3x + c$

> **Common mistake**: When you carry out the integration only write '$+ c$' on the right-hand side of the equation.
> You do not need to add a constant to each side because these constants would simplify into a single constant.

ii

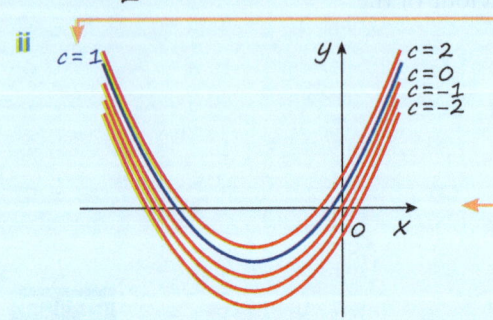

$c = 1$, $c = 2$, $c = 0$, $c = -1$, $c = -2$

> The blue curve is the curve of the particular solution in part **iii**.

> $y = \frac{1}{2}x^2 + 3x + c$ is the general solution and may be represented by a family of curves.

iii To find c substitute $x = 2$ and $y = 9$ into the equation for y

$y = \frac{1}{2}x^2 + 3x + c$

$9 = \frac{1}{2} \times 4 + 3 \times 2 + c$

$\Rightarrow c = 1$

The particular solution is $y = \frac{1}{2}x^2 + 3x + 1$.

> You are told that for this particular solution $y = 9$ when $x = 2$.

2 Using separation of variables to solve a differential equation

i Find the general solution of $\frac{dy}{dx} = \frac{\cos x}{3y^2}$.

ii Find the particular solution for which $y = 1$ when $x = \pi$.

Solution

i $\frac{dy}{dx} = \frac{\cos x}{3y^2}$

$\Rightarrow 3y^2 \frac{dy}{dx} = \cos x$

$\Rightarrow \int 3y^2 dy = \int \cos x \, dx$

$\Rightarrow y^3 = \sin x + c$

$\Rightarrow y = \sqrt[3]{(\sin x + c)}$

> First separate the variables.

> Remember that the integral of $\cos x$ is $\sin x$.

> You need to make y the subject, not y^3.

ii Substitute $x = \pi$ and $y = 1$ in the equation
$y^3 = \sin x + c$.

$\Rightarrow 1 = 0 + c$

so the particular solution is $y = \sqrt[3]{(\sin x + 1)}$.

Chapter 12 Differential equations

3 Use separation of variables to solve a differential equation

i Express $\dfrac{1}{y(y-1)}$ in partial fractions.

ii Find the general solution of $\dfrac{dy}{dx} = \dfrac{y(y-1)}{x}$, for cases where $x > 0$ and $y > 1$.

Solution

i $\dfrac{1}{y(y-1)} = \dfrac{A}{y} + \dfrac{B}{y-1}$ ← Choose the right form for the partial fractions – see page 162.

$\Rightarrow 1 = A(y-1) + By$ ← Multiply through by $y(y-1)$.

Let $y = 0$: $1 = A \times (-1)$
$\Rightarrow A = -1$

Let $y = 1$: $1 = B \times 1$
$\Rightarrow B = 1$

So $\dfrac{1}{y(y-1)} = -\dfrac{1}{y} + \dfrac{1}{y-1}$

ii Rearrange the equation by separating the variables.

$\dfrac{dy}{dx} = \dfrac{y(y-1)}{x}$

$\dfrac{1}{y(y-1)} \dfrac{dy}{dx} = \dfrac{1}{x}$

$\int \dfrac{1}{y(y-1)} dy = \int \dfrac{1}{x} dx$ ← You cannot integrate this as it stands; you need to use the result from part **i** $\dfrac{1}{y(y-1)} = -\dfrac{1}{y} + \dfrac{1}{y-1}$.

$\Rightarrow \int \left(-\dfrac{1}{y} + \dfrac{1}{y-1}\right) dy = \int \dfrac{1}{x} dx$

$\Rightarrow -\ln y + \ln(y-1) = \ln x + c$ ← **Hint:** Since $y > 1$ and $x > 0$, you do not need to include modulus signs.

$\Rightarrow \ln \dfrac{(y-1)}{y} = \ln x + c$ ← Remember the rules of logarithms: $\ln a - \ln b = \ln \dfrac{a}{b}$.

$\Rightarrow \dfrac{(y-1)}{y} = e^{\ln x + c}$

$\Rightarrow 1 - \dfrac{1}{y} = ke^{\ln x}$, where $k = e^c$

$\Rightarrow \dfrac{1}{y} = 1 - kx$ ← Remember that $e^{\ln x} = x$.

$\Rightarrow y = \dfrac{1}{1 - kx}$ ← You need to make y the subject.

Test yourself

TESTED

1 Which of the following is the general solution of the differential equation $\dfrac{dy}{dx} = \sqrt{x} + 3$?

A $y = \dfrac{3}{2}x^{\frac{3}{2}} + 3x + c$ B $y = x^{\frac{3}{2}} + 3x + c$ C $y = \dfrac{2}{3}x^{\frac{3}{2}} + 3x + c$ D $y = \dfrac{2}{3}x^{\frac{3}{2}} + c$

2 Find the general solution of the differential equation $\dfrac{dy}{dx} = x^2 y$, for $y > 0$.

A $y = Ae^{x^3}$ B $y = \dfrac{1}{3}e^{x^3 + c}$ C $y = Ae^{\frac{1}{3}x^3}$ D $y = e^{\frac{1}{3}x^3} + c$

3 Find the particular solution of the differential equation $(x^2 - 3)\dfrac{dy}{dx} = \dfrac{2x}{y}$ for which $y = 2$ when $x = 2$.

A $y = \pm\sqrt{(\ln(x^2 - 3)^2 + 4)}$ B $y = \ln(x^2 - 3) + 2$ C $y = \sqrt{\ln(x^2 - 3)^2}$ D $y = \sqrt{\ln(x^2 - 3)^2 + 2}$

Edexcel A Level Mathematics (Pure)

4 Three of the following statements about the differential equation $x\frac{dy}{dx} = y^2 - 1$ are false and one is true. Which one is true?

  A To integrate $\frac{1}{y^2 - 1}$ use partial fractions to write it as $\frac{1}{y-1} - \frac{1}{y+1}$.

  B The function $\ln\frac{y-1}{y+1}$ is defined for all values of y.

  C $e^{\ln x^2 + c}$ can be replaced by Ax^2.

  D The general solution of this differential equation is $y = \frac{1 - Ax^2}{1 + Ax^2}$, $A > 0$.

5 Which of the following graphs is the curve of the particular solution of the differential equation $\frac{dy}{dx} = \frac{3y}{x}$, $x > 0$ and $y > 0$ for which $y = 1$ when $x = 1$?

A $y = x^3$

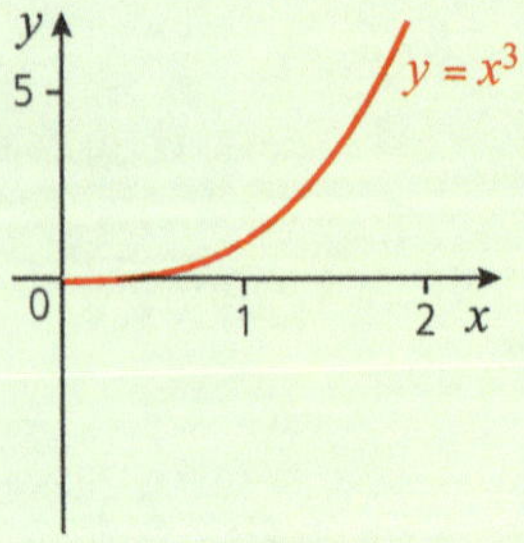

B $y = e^{3x}$

C $y = 3x$

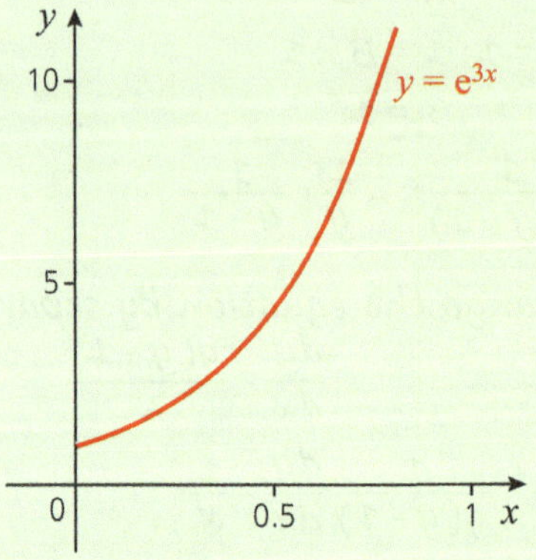

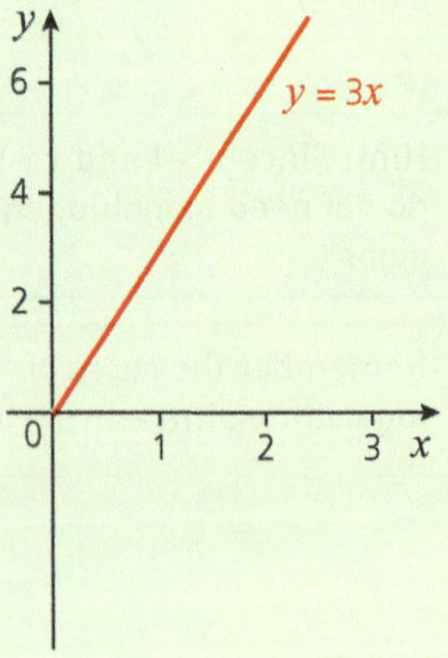

D $y = 3\ln x$

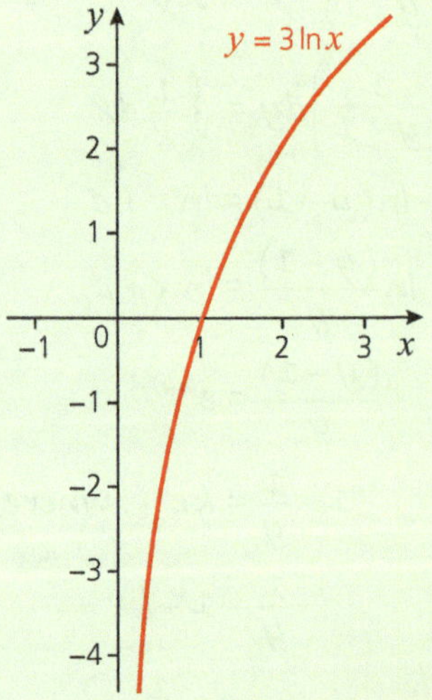

Full worked solutions online   CHECKED ANSWERS

Exam-style question

i Find the general solution of the differential equation $\frac{dy}{dx} = -xy$, given that $y > 0$.

ii Sketch the family of solution curves.

iii The curve of a particular solution passes through the point $(0, 5)$. Write down its equation.

Short answers on page 221

Full worked solutions online   CHECKED ANSWERS

Differential equations and problem solving

REVISED

Key facts

1. 'The rate of change of a quantity' usually means 'the rate of change of this quantity with respect to time'.

   For example, the rate of change of a volume, V, is $\dfrac{dV}{dt}$.

   > The words 'with respect to time' are often omitted.

   - Velocity, v, is the rate of change of position, s, of an object with respect to time.
     $$v = \dfrac{ds}{dt}$$
   - In motion along a straight line, the acceleration a is the rate of change of an object's velocity.
     $$a = \dfrac{dv}{dt}$$

2. It is possible to have a rate of change with respect to another variable.
   For example, the rate of change of temperature, T, with respect to distance x from a heat source is $\dfrac{dT}{dx}$.

Worked examples

1 Forming a differential equation

A shape has area A at time t. The variables A and t are related by the differential equation. $\dfrac{dA}{dt} = k\sqrt{A}$.

   i Explain the meaning of $\dfrac{dA}{dt}$.

   ii What does the differential equation tell you?

Solution

   i $\dfrac{dA}{dt}$ means the rate at which the area A, is changing with time t.

   ii The differential equation $\dfrac{dA}{dt} = k\sqrt{A}$ tells you that the rate of change of the area A is directly proportional to the square root of the area.

2 Using a differential equation to solve a problem

At time $t = 0$, a small ball is dropped into still water, forming a circular ripple.

The radius, r, of the ripple at time t seconds increases at a rate that is inversely proportional to its size.

When the radius is 20 cm the rate of increase of the radius is 2 cm per second.

   i Obtain the differential equation that represents this situation.

   ii The ball has radius 1 cm so that the initial radius of the ripple is 1 cm.

      a Solve the differential equation.

      b Sketch the graph of the radius against time and describe what it shows.

      c Find the radius 1 minute after the ball hits the water.

Edexcel A Level Mathematics (Pure)

Solution

i $\frac{dr}{dt}$ is the rate of increase of r

$\Rightarrow \frac{dr}{dt} \propto \frac{1}{r}$

$\Rightarrow \frac{dr}{dt} = \frac{k}{r}$

> You are told that the radius of the circle increases at a rate that is **inversely proportional** to its size.

> k is the constant of proportionality. It is positive because the radius is increasing. If the radius was decreasing then you should write $\frac{dr}{dt} = -\frac{k}{r}$

To find k use the fact that $\frac{dr}{dt} = 2$ when $r = 20$.

Substituting in the equation $\frac{dr}{dt} = \frac{k}{r}$ gives $2 = \frac{k}{20}$ and so $k = 40$.

So the differential equation is $\frac{dr}{dt} = \frac{40}{r}$.

> When the radius is 20 cm the rate of increase of the radius is 2 cm per second.

ii **a** Separating the variables gives: $\int r \, dr = \int 40 \, dt$

$\Rightarrow \frac{1}{2} r^2 = 40t + c$

> Integrate both sides. Remember to write '$+c$' on the right-hand side only.

To find t, use the additional information that when $t = 0$, $r = 1$.

Substituting these values gives

$\frac{1}{2} \times 1^2 = 40 \times 0 + c$

giving $c = \frac{1}{2}$

$\frac{1}{2} r^2 = 40t + \frac{1}{2}$

$r = \sqrt{80t + 1}$.

b The radius of the ripple increases, but at a decreasing rate.

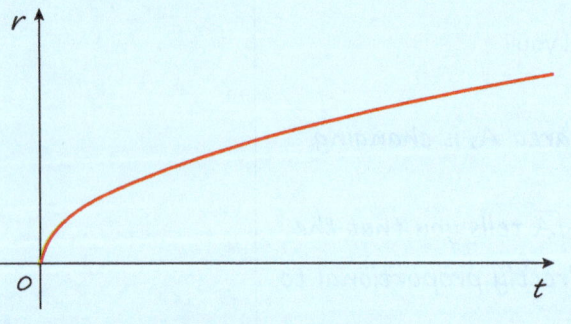

> Drawing a sketch graph is often very helpful when you are interpreting the solution of a differential equation.

c After half a minute, $t = 30$

$r = \sqrt{2401} = 49$

The radius is 49 cm.

> $t = 30$; t is the time in seconds.

Test yourself

TESTED

1 The population of a country increases at a rate that is proportional to the square root of the number of people present. Which of the following is the differential equation relating the population, x, to the time t?

A $\frac{dx}{dt} = kx^2$, $k > 0$ B $\frac{dx}{dt} = k + \sqrt{x}$, $k > 0$ C $\frac{dx}{dt} = k\sqrt{x}$, $k > 0$ D $\frac{dx}{dt} = -k\sqrt{x}$, $k > 0$

2 An object has velocity $v \, \text{m s}^{-1}$ at time t seconds. Its acceleration is proportional to the cube of its velocity and in the negative direction. Which one of these differential equations models the situation?

A $\frac{dv}{dt} = -k + v^3$, $k > 0$ B $\frac{dv}{dt} = -k\sqrt[3]{v}$, $k > 0$ C $\frac{dv}{dt} = -kv^3$, $k > 0$ D $\frac{dt}{dv} = -kv^3$, $k > 0$

3 Kevin starts on a diet. His weight, w kg after t days, decreases at a rate which is inversely proportional to the square root of his weight. Given that k is a positive constant, one of the following differential equations models this situation. Which one is it?

A $\dfrac{dw}{dt} = \dfrac{k}{\sqrt{w}}, k > 0$ B $\dfrac{dw}{dt} = -\dfrac{k}{\sqrt{w}}, k > 0$ C $\dfrac{dw}{dt} = -\dfrac{k}{w^2}, k > 0$ D $\dfrac{dt}{dw} = -k\sqrt{w}, k > 0$

4 A curve has equation $y = f(x)$. The gradient function of the curve is inversely proportional to the cube of x. The curve passes through $(1, 0)$ and $(2, 3)$. Find the equation of the curve.

A $y = 4 - \dfrac{4}{x^2}$ B $y = \dfrac{12}{x^2}$ C $5y = x^4 - 1$ D $y = 1 - \dfrac{1}{x^2}$.

5 This sketch graph shows the solution of a differential equation connecting P and t, for $t > 0$.

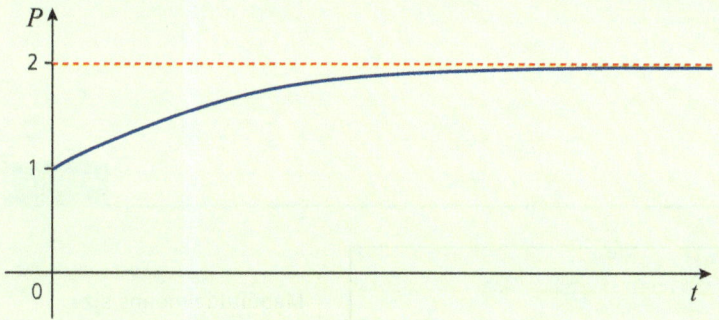

Three of the following statements are false and one is true. Which one is true?

A The equation of the solution could be $P = 1 + 0.1\sqrt{t}$.
B P decays exponentially.
C For larger values of t, P may start to decrease in value.
D The differential equation could be $\dfrac{dP}{dt} = 0.5P(2 - P)$.

Full worked solutions online CHECKED ANSWERS

Exam-style question

A quantity of oil is dropped into water. When the oil hits the water, it spreads out as a circle. The radius of the circle is r cm after t seconds.

When $t = 0$, $r = 0$ and when $t = 3$, the radius of the circle is increasing at the rate of $0.5\,\text{cm s}^{-1}$.

One observer believes that the radius increases at a rate which is proportional to $\dfrac{1}{t+1}$.

i Write down a differential equation to model this situation, using k as the constant of proportionality.
ii Show that $k = 2$.
iii Solve the differential equation and hence calculate the radius of the circle after 10 seconds according to this model.

Another observer suggests that the rate of increase of the radius of the circle is proportional to $\dfrac{1}{(t+1)(t+2)}$.

iv Write down a new differential equation for this model.
Using the same initial conditions as before, find the value of the new constant of proportionality.
v Hence solve the differential equation.
vi Calculate the radius of the circle after 10 seconds according to this model.

Short answers on page 221
Full worked solutions online CHECKED ANSWERS

Chapter 13 Vectors

About this topic

Vectors are important in many areas of mathematics and you will also use them in Mechanics when working with forces and velocities.

Before you start, remember

- Pythagoras' theorem
- coordinate geometry.

Vectors

REVISED

Key facts

1. A **scalar** quantity has **magnitude** only.
2. A **vector** quantity has both **magnitude** and **direction**.
3. A **unit vector** is a vector with magnitude 1.
4. Vectors are printed in bold, e.g. **a**.
   In handwriting vectors are underlined, e.g. $\underline{a}$.
   The vector from point A to point B is written as $\overrightarrow{AB}$.
5. You can describe vectors using **magnitude-direction form**.

   A vector in two dimensions is often represented by a straight line with an arrowhead, which shows the direction of the vector.

   The direction is usually taken to be the angle that the vector makes with the positive x-axis, measured in an anticlockwise direction.

   The length of the line represents the magnitude of the vector.

6. You can also describe vectors using **components**.

   Vectors can be described using the unit vectors **i**, **j** and **k** in the x, y and z directions.

   Vectors can also be written as column vectors.

7. Two vectors are **equal** if they have the same magnitude and the same direction.
8. Two vectors are **parallel** if they have the same direction.
9. The **position vector** of a point P is the vector from the origin, O, to P.
   This is written as $\overrightarrow{OP}$ or **p**.
   The point $P(a, b, c)$ has position vector $\overrightarrow{OP} = a\mathbf{i} + b\mathbf{j} + c\mathbf{k}$,
   or as a **column vector**, $\begin{pmatrix} a \\ b \\ c \end{pmatrix}$.

> Magnitude means size.
> Mass (kg) and speed (m s⁻¹) are scalar quantities.

> Weight (N) and velocity (m s⁻¹) are vector quantities.

> This vector has magnitude 8 units and direction $-20°$ to the positive x-axis.
> The direction can also be given as $340°$.

> The vector shown in red can be written as $-5\mathbf{i} + 2\mathbf{j}$, or as the column vector $\begin{pmatrix} -5 \\ 2 \end{pmatrix}$.

> Two vectors are parallel when they are scalar multiples of each other.

> The origin is not always shown in diagrams.

10 The **magnitude** of a vector is found using Pythagoras' theorem.
The vector $a\mathbf{i} + b\mathbf{j} + c\mathbf{k}$ has magnitude $\sqrt{a^2+b^2+c^2}$.
The magnitude of the vector $\overrightarrow{OP}$ is written $|\overrightarrow{OP}|$.

> The magnitude of the vector $-5\mathbf{i} + 2\mathbf{j}$ is $\sqrt{(-5)^2+2^2} = \sqrt{25+4} = \sqrt{29}$.

11 The unit vector in the direction of $a\mathbf{i} + b\mathbf{j} + c\mathbf{k}$ is
$$\frac{a}{\sqrt{a^2+b^2+c^2}}\mathbf{i} + \frac{b}{\sqrt{a^2+b^2+c^2}}\mathbf{j} + \frac{c}{\sqrt{a^2+b^2+c^2}}\mathbf{k}$$

> A unit vector in the same direction as $-5\mathbf{i} + 2\mathbf{j}$ is $-\frac{5}{\sqrt{29}}\mathbf{i} + \frac{2}{\sqrt{29}}\mathbf{j}$.

12 To find the **resultant** of two or more vectors you add the vectors. This is particularly useful in mechanics for finding the resultant of two or more forces.

> The resultant of the vectors **a**, **b** and **c** is **a** + **b** + **c**.

Worked examples

1 Geometry using vectors

The diagram shows a parallelogram ABCD.
$\overrightarrow{AB} = \mathbf{p}$ and $\overrightarrow{AD} = \mathbf{q}$.

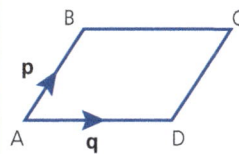

i Find, in terms of **p** and **q**, the vectors $\overrightarrow{AC}$ and $\overrightarrow{AD}$.

ii The point M is the mid-point of BD.
Find the vector $\overrightarrow{AM}$.

iii The point N is $\frac{1}{3}$ of the way along AC.
Find the vector $\overrightarrow{DN}$.

Solution

i $\overrightarrow{AC} = \overrightarrow{AB} + \overrightarrow{BC} = \mathbf{p} + \mathbf{q}$
$\overrightarrow{BD} = \overrightarrow{BA} + \overrightarrow{AD} = -\mathbf{p} + \mathbf{q} = \mathbf{q} - \mathbf{p}$

> As ABCD is a parallelogram, $\overrightarrow{BC} = \overrightarrow{AD} = \mathbf{q}$.
> $\overrightarrow{AC}$ is the **resultant** of **p** and **q**.

ii $\overrightarrow{BM} = \frac{1}{2}\overrightarrow{BD} = \frac{1}{2}(\mathbf{q} - \mathbf{p})$

$\overrightarrow{AM} = \overrightarrow{AB} + \overrightarrow{BM} = \mathbf{p} + \frac{1}{2}(\mathbf{q} - \mathbf{p}) = \mathbf{p} + \frac{1}{2}\mathbf{q} - \frac{1}{2}\mathbf{p} = \frac{1}{2}\mathbf{p} + \frac{1}{2}\mathbf{q}$

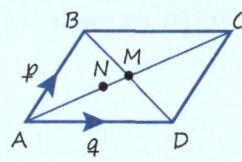

> **Hint:** When a vector is multiplied by a scalar (a number), then each component is multiplied by the scalar.
> Two vectors in component form can be added or subtracted by dealing with each component separately.

iii $\overrightarrow{AN} = \frac{1}{3}\overrightarrow{AC} = \frac{1}{3}(\mathbf{p} + \mathbf{q})$

$\overrightarrow{DN} = \overrightarrow{DA} + \overrightarrow{AN} = -\mathbf{q} + \frac{1}{3}(\mathbf{p} + \mathbf{q}) = -\mathbf{q} + \frac{1}{3}\mathbf{p} + \frac{1}{3}\mathbf{q} = \frac{1}{3}\mathbf{p} - \frac{2}{3}\mathbf{q}$

2 Working with 3D vectors

Three points A, B and C have coordinates (1, −2, 0), (1, 3, −2) and (3, 0, 2).

i Find the vectors $\overrightarrow{AB}$ and $\overrightarrow{BC}$.
ii Show that the triangle ABC is isosceles.
iii Find the angle BAC.
iv M is the midpoint of BC.
Find the position vector of M.

Solution

i $\vec{AB} = \vec{AO} + \vec{OB} = \vec{OB} - \vec{OA} = \begin{pmatrix} 1 \\ 3 \\ -2 \end{pmatrix} - \begin{pmatrix} 1 \\ -2 \\ 0 \end{pmatrix} = \begin{pmatrix} 0 \\ 5 \\ -2 \end{pmatrix}$

$\vec{OB}$ and $\vec{OA}$ are the position vectors of A and B.

$\vec{BC} = \vec{BO} + \vec{OC} = \vec{OC} - \vec{OB} = \begin{pmatrix} 3 \\ 0 \\ 2 \end{pmatrix} - \begin{pmatrix} 1 \\ 3 \\ -2 \end{pmatrix} = \begin{pmatrix} 2 \\ -3 \\ 4 \end{pmatrix}$

ii $|\vec{AB}| = \sqrt{5^2 + (-2)^2} = \sqrt{25 + 4} = \sqrt{29}$

$|\vec{BC}| = \sqrt{2^2 + (-3)^2 + 4^2} = \sqrt{4 + 9 + 16} = \sqrt{29}$

$\vec{AC} = \vec{AO} + \vec{OC} = \vec{OC} - \vec{OA} = \begin{pmatrix} 3 \\ 0 \\ 2 \end{pmatrix} - \begin{pmatrix} 1 \\ -2 \\ 0 \end{pmatrix} = \begin{pmatrix} 2 \\ 2 \\ 2 \end{pmatrix}$

$|\vec{AC}| = \sqrt{2^2 + 2^2 + 2^2} = \sqrt{4 + 4 + 4} = \sqrt{12} = 2\sqrt{3}$

Since ABC has two equal sides, the triangle is isosceles.

iii Use the cosine rule to find the angle BAC:

$\cos \angle BAC = \dfrac{|AB|^2 + |AC|^2 - |BC|^2}{2|AB||AC|}$

$= \dfrac{29 + 12 - 29}{2 \times \sqrt{29}\sqrt{12}}$

$= 0.321...$

$\Rightarrow \angle BAC = 71.23...°$

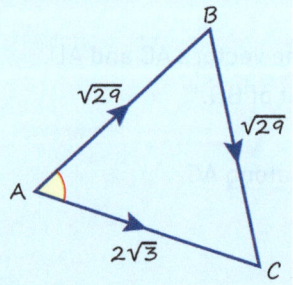

iv $\vec{BM} = \frac{1}{2}\vec{BC} = \frac{1}{2}\begin{pmatrix} 2 \\ -3 \\ 4 \end{pmatrix} = \begin{pmatrix} 1 \\ -\frac{3}{2} \\ 2 \end{pmatrix}$

Make sure you halve each component.

$\vec{OM} = \vec{OB} + \vec{BM} = \begin{pmatrix} 1 \\ 3 \\ -2 \end{pmatrix} + \begin{pmatrix} 1 \\ -\frac{3}{2} \\ 2 \end{pmatrix} = \begin{pmatrix} 2 \\ \frac{3}{2} \\ 0 \end{pmatrix}$

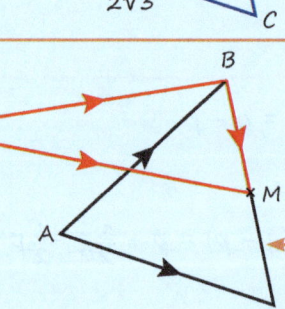

Draw a sketch to help you; it doesn't matter where you place the origin, O.

Imagine going on a vector walk: start at O, then walk to B and then to M.

Test yourself

TESTED

1. The vectors **p** and **q** are given by **p** = 2**i** − **j** + 3**k** and **q** = 3**i** + 2**j** − 4**k**.
   Find the vector 3**p** − 2**q**.
   - A   **j** − **k**
   - B   **j** + **k**
   - C   −7**j** + 17**k**
   - D   −7**j** + **k**

Questions 2, 3, 4 and 5 are about three points A, B and C which have coordinates (1, 0, 3), (3, 1, −4) and (−2, 6, 5) respectively.

2. Find the vector $\vec{BA}$.

   - A $\begin{pmatrix} 2 \\ 1 \\ -7 \end{pmatrix}$
   - B $\begin{pmatrix} 4 \\ 1 \\ -1 \end{pmatrix}$
   - C $\begin{pmatrix} 3 \\ 0 \\ -12 \end{pmatrix}$
   - D $\begin{pmatrix} -2 \\ -1 \\ 7 \end{pmatrix}$

3. Find the unit vector in the direction $\vec{AC}$.

   - A $\begin{pmatrix} -3 \\ 6 \\ 2 \end{pmatrix}$
   - B $\begin{pmatrix} -\frac{3}{7} \\ \frac{6}{7} \\ \frac{2}{7} \end{pmatrix}$
   - C $\begin{pmatrix} -\frac{3}{5} \\ \frac{6}{5} \\ \frac{2}{5} \end{pmatrix}$
   - D $\begin{pmatrix} -\frac{3}{\sqrt{31}} \\ \frac{6}{\sqrt{31}} \\ \frac{2}{\sqrt{31}} \end{pmatrix}$

4. Find the coordinates of the point D so that ABCD is a parallelogram.
   - A   (−5, 5, 9)
   - B   (6, −5, −6)
   - C   (0, 7, −2)
   - D   (−4, 5, 12)

5. Find the position vector of the midpoint M of BC.

   - A $\begin{pmatrix} 0.5 \\ 3.5 \\ 0.5 \end{pmatrix}$
   - B $\begin{pmatrix} -2.5 \\ 2.5 \\ 4.5 \end{pmatrix}$
   - C $\begin{pmatrix} 2 \\ 4 \\ -1.5 \end{pmatrix}$
   - D $\begin{pmatrix} -4.5 \\ 8.5 \\ 9.5 \end{pmatrix}$

Full worked solutions online   CHECKED ANSWERS

Exam-style question

Two points A and B have coordinates (4, −1, 3) and (0, 3, 1), respectively.
i   Find the vector $\vec{AB}$.

The point C is such that $\vec{AC}$ has the same magnitude as $\vec{AB}$ and is parallel to the vector **i** + 2**j** − 2**k**.
ii   Find the vector $\vec{AC}$.
iii   Hence find the coordinates of C.

D is the midpoint of AC.
iv   Find the position vector of D.
v   The median M of triangle ABC is the point $\frac{1}{3}$ of the way along DB.
   Find the coordinates of M.

Short answers on page 221

Full worked solutions online   CHECKED ANSWERS

Edexcel A Level Mathematics (Pure)

Chapter 14 Numerical methods

About this topic

Not all equations can be solved algebraically and not all functions can be integrated by the techniques you have met so far (or indeed at all!) so you may need to use a numerical method to solve an equation or find the approximate value of a definite integral.

Before you start, remember

- how to rearrange equations
- differentiation
- curve sketching.

Solving equations numerically

REVISED

Key facts

1. When f(x) is **continuous**, if f(a) and f(b) have opposite signs, there is at least one root in the interval [a, b].
2. The change of sign method fails if:
   - the function is discontinuous and has a vertical asymptote; for example, using this method with $f(x) = \dfrac{1}{x-2} + 3$ would lead to a false root at $x = 2$

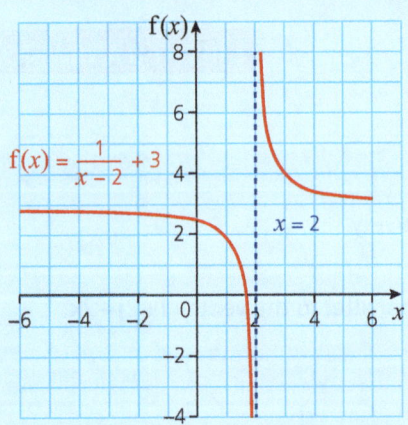

The graph of a **continuous function** is an unbroken line. You can draw the graph of a continuous function without lifting your pencil from the paper.

f(2.1) = 13 > 0
f(1.9) = −7 < 0

So it looks like there is a root in the interval [1.9, 2.1].
But $x = 2$ is not a root, it is an asymptote.

- there are roots close together, then it is easy to miss a pair; for example,

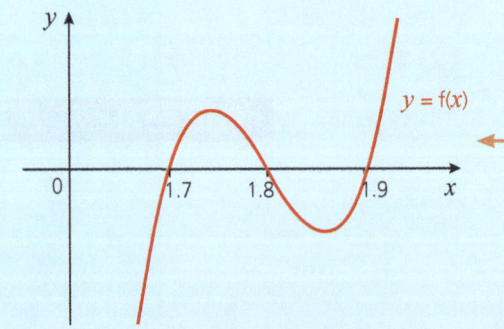

Using the change of sign method shows there is a root in the interval [1.5, 2] as f(1.5) < 0 and f(2) > 0. But there are actually 3 roots in this interval.

- there is a repeated root (so no change of sign); for example,

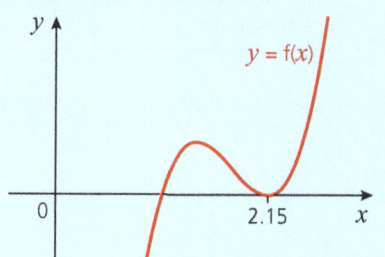

> There is a root at $x = 2.15$, but there is no change of sign on either side of the root.

3. You can use **fixed point iteration** to solve the equation $f(x) = 0$.
   - Rearrange $f(x) = 0$ into the form $x = g(x)$.
   - You now have an iterative formula: $x_{n+1} = g(x_n)$.
   - Use x_1 as your first estimate for the root and substitute this value into $x_{n+1} = g(x_n)$ to find x_2 and then substitute x_2 into $x_{n+1} = g(x_n)$ to find x_3 and so on.
   - Eventually the iterative procedure will either
     - ✓ converge to the root
     - ✗ diverge (the values increase).

> Iteration means repeating the same process again and again.

> The roots of $f(x) = 0$ are at the points where the line $y = x$ meets the curve $y = g(x)$.

> $x_2 = g(x_1), x_3 = g(x_2)$ and so on.

> The arrangement fails if the gradient of $y = g(x)$ is too steep near the root.

4. **Cobweb diagram**

   Successive values of x_n in fixed point interation oscillate about the root as they converge to the root. Graphically this looks like a **cobweb**.

   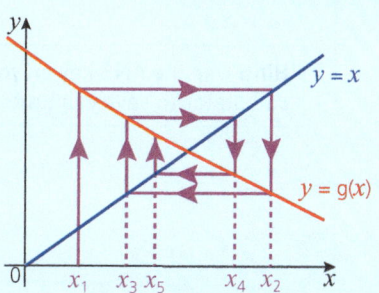

   Staircase diagram

   Successive values of x_n in fixed point interation get progressively closer to the root. Graphically this looks like a **staircase**.

   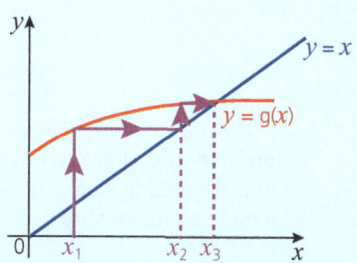

5. The **Newton–Raphson method**

   You can use the **iterative formula** $x_{n+1} = x_n - \dfrac{f(x_n)}{f'(x_n)}$ to find a root of $f(x) = 0$.

6. The Newton–Raphson method fails if:
   - x_1 is too far from the root you are looking for
   - x_1 is too close to a stationary point
   - $f(x)$ is discontinuous.

> In this case the process may converge to a different root.

> In this case $f'(x)$ may be too small.

Edexcel A Level Mathematics (Pure)

Worked examples

1 Use fixed-point iteration by rearranging the equation

i Show that the equation $4^x = x + 4$ has a root between $x = 1$ and $x = 2$.

ii Show that the equation $4^x = x + 4$ can be rearranged into the form $x = \dfrac{\ln(x+4)}{\ln 4}$.

iii Use the iteration $x_{n+1} = \dfrac{\ln(x_n + 4)}{\ln 4}$ with $x_1 = 1$ to find the values of x_2, x_3, x_4 and x_5, giving your answers to 5 decimal places.

Hint: Questions on numerical methods often start with asking you to use a change of sign method.

Solution

i $4^x = x + 4 \Rightarrow 4^x - x - 4 = 0$
When $x = 1$, $4^1 - 1 - 4 = -1$.
And when $x = 2$, $4^2 - 2 - 4 = 10$.

There is a change of sign in the interval [1, 2], so there must be a root between $x = 1$ and $x = 2$.

Rearrange the equation so it is equal to zero.

Then show there is a change of sign in the interval [a, b]

ii $4^x = x + 4$
$\Rightarrow \ln 4^x = \ln(x+4)$
$\Rightarrow x \ln 4 = \ln(x+4)$
$\Rightarrow x = \dfrac{\ln(x+4)}{\ln 4}$

Take natural logs of both sides.

Common mistake: The answer has been given to you so you must show all your working; otherwise you may lose marks.

iii $x_{n+1} = \dfrac{\ln(x_n + 4)}{\ln 4}$

$x_1 = 1 \Rightarrow x_2 = \dfrac{\ln(1+4)}{\ln 4} = 1.16096...$

$x_2 = 1.16096... \Rightarrow x_3 = \dfrac{\ln(1.16096... + 4)}{\ln 4} = 1.18382...$

$x_3 = 1.18382... \Rightarrow x_4 = \dfrac{\ln(1.18382... + 4)}{\ln 4} = 1.18701...$

$x_4 = 1.18701... \Rightarrow x_5 = \dfrac{\ln(1.18701... + 4)}{\ln 4} = 1.18745...$

Hint: Use the ANS key on your calculator to save you time.

2 Use the Newton–Raphson method

The equation $\sin x + x - 1 = 0$ has a single root, α, which lies between $x = 0$ and $x = 1$.

i Use the Newton–Raphson method together with $x_1 = 0$ to find the value of α correct to 5 decimal places.

ii Explain why using $x_1 = \pi$ leads to the Newton–Raphson method failing.

Solution

i The Newton–Raphson formula is $x_{n+1} = x_n - \dfrac{f(x_n)}{f'(x_n)}$.

Let $f(x) = \sin x + x - 1 \Rightarrow f'(x) = \cos x + 1$.

So $x_{n+1} = x_n - \dfrac{\sin x_n + x_n - 1}{\cos x_n + 1}$.

$x_1 = 0 \Rightarrow x_2 = 0 - \dfrac{\sin 0 + 0 - 1}{\cos 0 + 1} = 0.5$

$x_2 = 0.5 \Rightarrow x_3 = 0.5 - \dfrac{\sin 0.5 + 0.5 - 1}{\cos 0.5 + 1} = 0.510957..$

$x_4 = 0.510973...$

$x_5 = 0.510973...$

So $\alpha = 0.51097$ to 5 d.p.

Make sure you write the subscripts n.

Common mistake: You have differentiated to work out this formula so make sure you work in radians.

Common mistake: You have been asked to give your answer correct to 5 decimal places so make sure you show at least 6 decimal places.

The 5th decimal place has settled down so you can be confident of it.

Chapter 14 Numerical methods

ii When $x = \pi$, $f'(\pi) = \cos \pi + 1 = -1 + 1 = 0$.
So there is a stationary point at $x = \pi$.
Hence the Newton–Raphson method will fail as $f'(x_1) = 0$ and you can't divide by 0.

Test yourself

TESTED

1 The equation $\dfrac{3}{\cosec x} = 2x - 1$ where x is in radians, has a root in the interval:
   A [1.5, 2]
   B [1, 1.5]
   C [0, 0.5]
   D None of these intervals

2 Use the iterative formula $x_{n+1} = x_n - \dfrac{x_n^3 + 4x_n - 2}{3x_n^2 + 4}$ together with $x_1 = 0.2$ to find the value of x_4.
Give your answer correct to 5 decimal places.
   A $x_4 = 0.48607$
   B $x_4 = 0.47347$
   C $x_4 = 0.47354$
   D $x_4 = 0.48604$

3 Which of these is a correct rearrangement of the equation $3x^3 - x - 6 = 0$?
   i $x = \sqrt[3]{\dfrac{x+6}{3}}$
   ii $x = \sqrt{\dfrac{1}{3} + \dfrac{2}{x}}$
   iii $x = 3x^3 - 6$

   A i only
   B ii and iii
   C i and iii
   D i, ii and iii

4 Which of these iterative formulae can be used to find a root of $e^x - x = x^2 + 2$ near $x = 2$?
   A $x_{n+1} = e^{x_n} - x_n^2 - 2$
   B $x_{n+1} = \sqrt{e^{x_n} - x_n - 2}$
   C $x_{n+1} = \ln(x_n^2 + x_n + 2)$
   D All of these.

5 Give the Newton–Raphson iterative formula for the equation $2e^{2x} + 3x - 1 = 0$.
   A $x_{n+1} = x_n - \dfrac{2e^{2x_n} + 3x_n - 1}{4e^{2x_n} + 3}$
   B $x_{n+1} = x_n - \dfrac{4e^{2x_n} + 3}{2e^{2x_n} + 3x_n - 1}$
   C $x_{n+1} = x_n - \dfrac{2e^{2x_n} + 3x_n - 1}{e^{2x_n} + 3}$
   D $x_{n+1} = \dfrac{4e^{2x_n} + 3}{2e^{2x_n} + 3x_n - 1} - x_n$

Full worked solutions online

CHECKED ANSWERS

Exam-style question

You are given that $f(x) = 2x^3 - 3x - 5$.
The equation $f(x) = 0$ has a root at $x = \alpha$.
i Show that α lies between 1 and 2.
ii Show that the equation $2x^3 - 3x - 5 = 0$ can be rearranged into the form $x = \sqrt[3]{\dfrac{3x+5}{2}}$.
iii Use the iterative formula $x_{n+1} = \sqrt[3]{\dfrac{3x_n + 5}{2}}$ with $x_1 = 1$ to find the value of α correct to 2 decimal places.

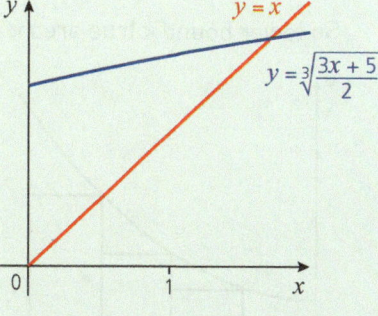

The diagram shows part of the graphs of $y = \sqrt[3]{\dfrac{3x+5}{2}}$ and $y = x$.

iv On a copy of the diagram, mark on the position of $x_1 = 1$ and draw a cobweb or staircase diagram to show how convergence takes place.

Short answers on page 222

Full worked solutions online

CHECKED ANSWERS

Numerical integration

REVISED

Key facts

1. You can use the trapezium rule to find an approximate value of a definite integral,

$$A = \int_a^b y\, dx \approx \frac{1}{2} \times h \times [y_0 + y_n + 2(y_1 + y_2 + \ldots + y_{n-1})],$$

where n is the number of strips and h is the strip width.

You can use $h = \dfrac{b-a}{n}$ to find h.

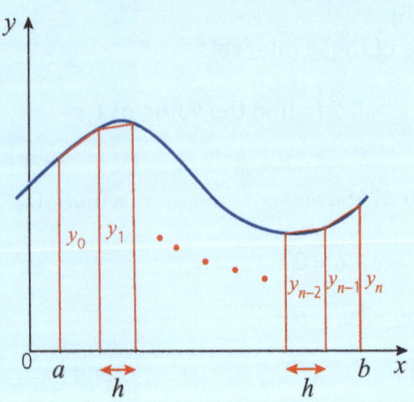

Hint:
- Not all functions can be integrated. When this is the case, you need to use numerical integration to evaluate a definite integral.
- You can think of the trapezium rule as '$\frac{1}{2}$ strip width × (ends + 2 × middles)'.
- Increasing the number of strips will, in general, increase the accuracy of your answer.

2. A sketch of the curve will show whether the trapezium rule gives an underestimate or an overestimate of the actual area.

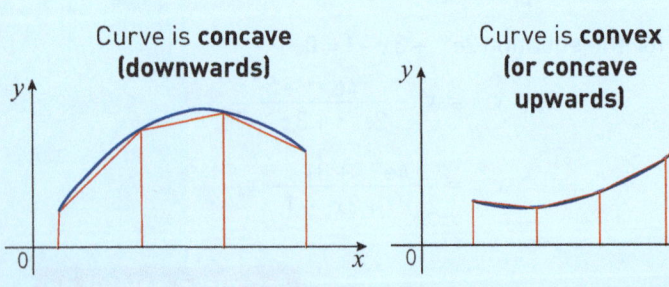

Trapezium rule **underestimates** area   Trapezium rule **overestimates** area

Hint: If the curve is convex in one place and concave in another, then you may not be able to tell whether the trapezium rule gives an under- or overestimate.

3. You can find the lower and upper bounds for the area under a curve by finding the area of rectangles that lie below and above the curve.

So lower bound < true area < upper bound.

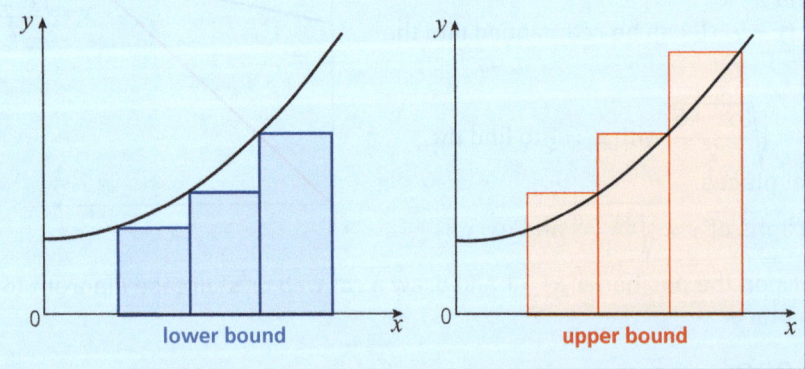

Hint: Increasing the number of rectangles tightens these area bounds and allows you to home in on the true area.

Worked examples

1 Using the trapezium rule

Use the trapezium rule with 4 strips to evaluate $\int_{-1}^{1} \sqrt{(x+1)}\,dx$.
Give your answer to 2 d.p.

Is your answer an underestimate or an overestimate?
Give a reason for your answer.

Hint: It is a good idea to use a table. You must show working to a greater accuracy than the final answer.

Solution

$n = 4$, $h = \dfrac{1-(-1)}{4} = \dfrac{1}{2}$, and there are 5 values.

x	$y = \sqrt{x+1}$		
-1	0	×1	0
-0.5	0.7071...	×2	1.4142...
0	1	×2	2
0.5	1.225	×2	2.45
1	1.4141...	×1	1.4142...
		Total	7.2784...

← y_0

← y_1, y_2 and y_3.

← y_4

Hence $A = \dfrac{1}{2} \times \dfrac{1}{2} \times 7.2784... = 1.8196...$
$= 1.82$ sq. units to 2 d.p.

Here is a sketch of the curve $y = \sqrt{(x+1)}$.

$y = \sqrt{(x+1)}$ is a translation of $y = \sqrt{x}$ by 1 unit to the left.

The curve is concave and so is above the lines forming the trapezia.

Hence this approximation is an underestimate.

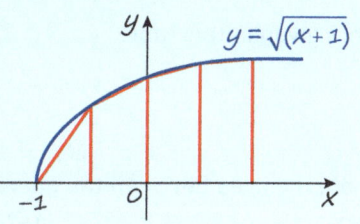

2 Using rectangles to find bounds for the area under a curve

This table gives the values of y for different values of x.

These could be results from an experiment or measurements.

x	1	3	5	7	9
y	45	34	25	18	11

Use the sum of a series of rectangles to find

i an upper bound ii a lower bound for $\int_{1}^{9} y\,dx$.

State the bounds for the integral as an inequality.

Solution

i **Upper bound**
 area of rectangles = 2×45
 $+ 2 \times 34 + 2 \times 25 + 2 \times 18$
 $= 244$

This is the upper bound for $\int_{1}^{9} y\,dx$.

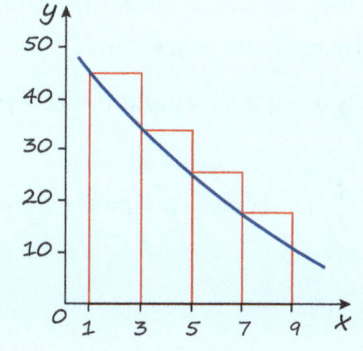

Edexcel A Level Mathematics (Pure)

ii Lower bound
area of rectangles = 2 × 34
+ 2 × 25 + 2 × 18 + 2 × 11
= 176

This is the lower bound for $\int_{1}^{9} y\,dx$.

As an inequality:
$176 < \int_{1}^{9} y\,dx < 244$.

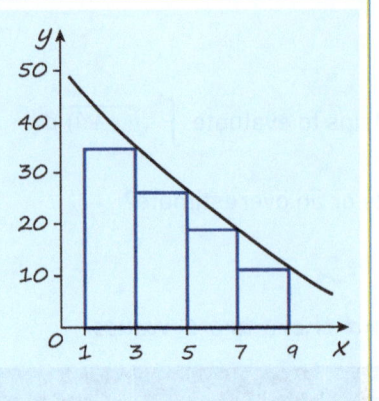

Test yourself

TESTED

1. It is necessary to estimate the area between the x-axis and the curve $y = f(x)$ between $x = 0$ and $x = 12$. This is done by using the trapezium rule on a given number of strips.
Which of the following is a correct application of the trapezium rule?

   A 2 strips: $A = \frac{6}{2}[y_0 + y_1 + y_2]$

   B 3 strips: $A = \frac{4}{2}[y_0 + 2(y_1 + y_2)]$

   C 4 strips: $A = 3[y_0 + y_4 + 2(y_1 + y_2 + y_3)]$

   D 5 strips: $A = \frac{12}{10}[y_0 + 2(y_1 + y_2 + y_3 + y_4) + y_5]$

   E 6 strips: $A = \frac{12}{12}[y_0 + y_6 + 2(y_1 + y_2 + y_3 + y_4)]$

2. A curve passes through the points given in this table.

x	1	4	7
y	0	12	42

   Use the trapezium rule to estimate the area between the curve and the x-axis.

   A 198 B 81 C 162 D 99

3. Use repeated applications of the trapezium rule to estimate the value of $\int_{0}^{\pi} \sqrt{2 - \cos^2 x}\,dx$ correct to 2 decimal places.

   A 15.28 B 2.30 C 3.82 D 4.11

4. The diagram below shows the graph of $y = f(x)$.

   Three of the following statements are false and one is true.
   Which statement is true?

   A $\int_{a}^{b} f(x)\,dx$ is estimated using the trapezium rule with 4 strips and again with 8 strips. The 8-strip estimate is greater.

   B $\int_{c}^{d} f(x)\,dx$ is estimated using the trapezium rule with 4 strips and again with 8 strips. The 8-strip estimate is greater.

   C $\int_{a}^{d} f(x)\,dx$ can be found exactly using the trapezium rule if you take enough strips.

   D Using the trapezium rule for $\int_{a}^{c} f(x)\,dx$ gives an underestimate.

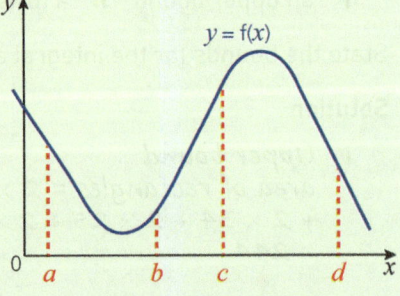

5 The graph shows the curve $y = \sqrt{x^2 + 3}$.

By using the method of finding the sum of a series of rectangles, find an upper and lower bound for $\int_0^5 \sqrt{x^2 + 3}\,dx$.

Use rectangles of width 1 in your calculations.

A $14.20 < \int_0^5 \sqrt{x^2+3}\,dx < 19.49$

B $17.76 < \int_0^5 \sqrt{x^2+3}\,dx < 19.49$

C $14.20 < \int_0^5 \sqrt{x^2+3}\,dx < 17.76$

D $11.06 < \int_0^5 \sqrt{x^2+3}\,dx < 12.16$

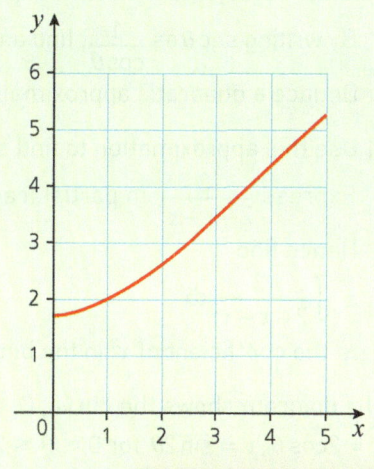

Full worked solutions online

CHECKED ANSWERS

Exam-style question

i Use the trapezium rule with
   a 2 strips and
   b 4 strips
   to obtain an approximation for $\int_1^2 \frac{12}{1+x^2}\,dx$.

   In each case, write down the first 6 significant figures from your calculator display.

ii By comparing your two answers to part **i** state with a reason whether the trapezium rule is likely to be an underestimate or an overestimate for the value of the integral.

The trapezium rule with 8 strips gives 3.86632... .
The trapezium rule with 16 strips gives 3.86233... .

iii Show that the error decreases with each application of the trapezium rule.

iv Without applying the trapezium rule again, give the value of $\int_1^2 \frac{12}{1+x^2}\,dx$ to as many decimal places as you can justify.

Short answers on page 222

Full worked solutions online

CHECKED ANSWERS

Edexcel A Level Mathematics (Pure)

Review questions (Chapters 9–14)

1. The fourth term of a geometric sequence is 4 and the sixth term is 100.
   Find the value(s) of the common ratio and hence the two possible values of the first term.

2. **i** By writing $\sec\theta$ as $\dfrac{1}{\cos\theta}$, find an approximate expression for $\sec\theta$ in the form $a+b\theta^2$ when θ is small.
   ii Deduce a quadratic approximation for $\sqrt{\sec\theta}$ for small values of θ.
   iii Use this approximation to find an estimate for $\int_0^{0.2}\sqrt{\sec\theta}\,d\theta$.

3. **i** Express $\dfrac{3}{2-x-x^2}$ in partial fractions.
   ii Hence find
     a $\displaystyle\int \dfrac{3}{2-x-x^2}\,dx$
     b the coefficient of x^3 in the binomial expansion of $\dfrac{3}{2-x-x^2}$ when $|x|<1$.

4. The diagram shows the curve, C, defined by the parametric equations
   $x=2\cos\theta, y=\sin 2\theta$ for $0\leqslant\theta\leqslant 2\pi$.
   i Find $\dfrac{dy}{dx}$.
   ii Find the equation of the normal to the curve at the point with parameter $\theta=\dfrac{\pi}{3}$.
   iii Find the coordinates of the stationary points.
   iv Find the total area of the shaded region.
   v Prove that the Cartesian equation of C is $4y^2=x^2(4-x^2)$.

5. A pan of water is heated to 100°C. It is then placed in a room with a constant temperature of 18°C. The rate at which the temperature of a body placed in room falls is proportional to the difference between the body temperature and the room temperature. (This is called Newton's law of cooling.)
   i Write this information as a differential equation.
   ii After 5 minutes the water temperature falls by 20°C.
   Solve the differential equation and sketch the graph of the solution.
   iii Calculate the temperature after a further 5 minutes.
   iv Find the total area of the shaded region.
   v Is the model likely to be accurate in the long term? Give a reason for your answer.

6. The diagram shows a sketch of the triangle ABC.
   Given that $\overrightarrow{AB}=2\mathbf{i}-3\mathbf{j}+\mathbf{k}$ and $\overrightarrow{AC}=\mathbf{i}-\mathbf{j}+2\mathbf{k}$, find
   i $\overrightarrow{BC}$
   ii the value of $\cos A$, giving your answer as a simplified surd.

   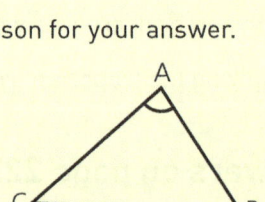

7. The diagram is a map of a length of coastline.
   It can be modelled by the curve $y=5+2x-x^2$.
   Units are kilometres. It is planned to build a wall from A (−1, 2) to B (2, 5). The shaded region, between the wall and the coast will become reclaimed land.
   i Verify that the equation of the retaining wall is $y=x+3$.
   ii Calculate the area of land to be reclaimed.
   iii A new map of the area is issued, showing that part of the coastline has been eroded. Use the trapezium rule and the following points taken from the map to estimate the new area that will now be reclaimed.

   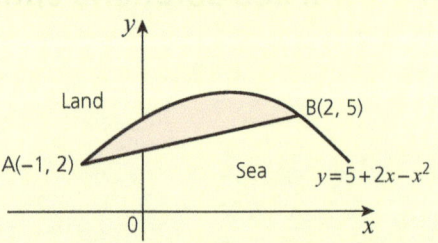

x	−1	−0.5	0	0.5	1	1.5	2
y	2	4.3	5.2	5.9	6.2	5.8	5

Short answers on page 222
Full worked solutions online

CHECKED ANSWERS

Exam preparation

Before your exam

REVISED

- *Start revising early*…half an hour a day for 6 months is better than cramming in all-nighters in the week before the exam. Little and often is the key.
- *Don't procrastinate* – you won't feel more like revising tomorrow than you do today!
- Put your phone on *silent* while you revise – don't get distracted by a constant stream of messages from your friends.
- Make sure your *notes are in order* and nothing is missing.
- *Be productive* – don't waste time colouring endless revision timetables. Make sure your study time is actually spent revising!
- Use the '**Target your revision**' sections to *focus your revision* on the topics you find tricky – remember you won't improve if you only answer the questions you could do anyway!
- Don't just read about a topic. *Maths is an active subject* – you improve by answering questions and actually *doing* maths, not just reading about it.
- Cover up the solution to an example and then try and answer it yourself.
- Answer as many past exam questions as you can. Work through the '**Review questions**' first and then move on to past papers.
- Try *teaching a friend* a topic. Teaching something is the best way to learn it yourself – that's why teachers know so much!

The exam papers

REVISED

You must take **all** of Papers 1, 2 and 3 to be awarded the Edexcel A Level in Mathematics.

Paper	Title	No. of marks		
1	Pure Mathematics 1	100	2 hours Written paper	33.33% of total A Level
2	Pure Mathematics 2	100	2 hours Written paper	33.33% of total A Level
3	Statistics and Mechanics	100	2 hours Written paper	33.33% of total A Level

The content of this book covers all the Pure Mathematics content for all papers

Make sure you know these formulae for your exam

From GCSE Maths you need to know...

REVISED

Topic	Formula
Circle	Area = πr^2 Circumference = $2\pi r$, where r is the radius
Parallelogram	Area = base × vertical height
Trapezium	Area = $\frac{1}{2}h(a+b)$
Triangle	Area = $\frac{1}{2}$ base × vertical height
Prism	Volume = area of cross section × length
Cylinder	Volume = $\pi r^2 h$ Area of curved surface = $2\pi rh$ Total surface area = $2\pi rh + 2\pi r^2$, where r is the radius and h is the height
Pythagoras' theorem	$a^2 + b^2 = c^2$
Trigonometry	$\cos\theta = \dfrac{\text{adjacent}}{\text{hypotenuse}}$ $\sin\theta = \dfrac{\text{opposite}}{\text{hypotenuse}}$ $\tan\theta = \dfrac{\text{opposite}}{\text{adjacent}}$

Topic	Formula
Circle theorems	The angle in a semi-circle is a right-angle.
	The perpendicular from the centre of a circle to a chord bisects the chord.
	The tangent to a circle at a point is perpendicular to the radius through that point.

From A Level Pure Maths you should know...

REVISED

Topic	Formula
Laws of indices	$a^m \times a^n = a^{m+n}$ $\dfrac{a^m}{a^n} = a^{m-n}$ $(a^m)^n = a^{mn}$ $a^{-n} = \dfrac{1}{a^n}$ $\sqrt[n]{a} = a^{\frac{1}{n}}$ $\sqrt[n]{a^m} = a^{\frac{m}{n}}$ $a^0 = 1$
Quadratic equations	The quadratic equation $ax^2 + bx + c = 0$ has roots $x = \dfrac{-b \pm \sqrt{b^2 - 4ac}}{2a}$.

Coordinate geometry	• For two points (x_1, y_1) and (x_2, y_2):
	    Gradient $= \dfrac{y_2 - y_1}{x_2 - x_1}$
	    Length $= \sqrt{(x_2 - x_1)^2 + (y_2 - y_1)^2}$
	    Midpoint $= \left(\dfrac{x_1 + x_2}{2}, \dfrac{y_1 + y_2}{2} \right)$.
	 • The equation of a straight line with gradient m and y-intercept $(0, c)$ is $y = mx + c$.

Topic	Formula
	• The equation of a straight line with gradient m and passing through (x_1, y_1) is $y - y_1 = m(x - x_1)$. • Parallel lines have the same gradient. • For two perpendicular lines $m_1 m_2 = -1$. • The equation of a circle, centre (a, b) and radius r is $(x - a)^2 + (y - b)^2 = r^2$.
Trigonometry	For any triangle ABC **Area:** Area $= \frac{1}{2} ab \sin C$ **Sine rule:** $\dfrac{a}{\sin A} = \dfrac{b}{\sin B} = \dfrac{c}{\sin C}$ or $\dfrac{\sin A}{a} = \dfrac{\sin B}{b} = \dfrac{\sin C}{c}$ **Cosine rule:** $a^2 = b^2 + c^2 - 2bc \cos A$ or $\cos A = \dfrac{b^2 + c^2 - a^2}{2bc}$ **Identities:** $\sin^2 \theta + \cos^2 \theta \equiv 1$ $\tan \theta \equiv \dfrac{\sin \theta}{\cos \theta}, \quad \cos \theta \neq 0$ $\sec \theta \equiv \dfrac{1}{\cos \theta}$ $\operatorname{cosec} \theta \equiv \dfrac{1}{\sin \theta}$ $\cot \theta \equiv \dfrac{\cos \theta}{\sin \theta}, \quad \sin \theta \neq 0$ $\sec^2 \theta \equiv 1 + \tan^2 \theta$ $\operatorname{cosec}^2 \theta \equiv 1 + \cot^2 \theta$ $\sin 2A \equiv 2 \sin A \cos A$ $\cos 2A \equiv \cos^2 A - \sin^2 A \equiv 2\cos^2 A - 1 \equiv 1 - 2\sin^2 \theta$ $\tan 2A \equiv \dfrac{2 \tan A}{1 - \tan^2 A}$
Mensuration	For θ in radians: Arc length, $s = r\theta$ Area of sector, $A = \frac{1}{2} r^2 \theta$
Transformations	$y = f(x + a)$ is a translation of $y = f(x)$ by $\begin{pmatrix} -a \\ 0 \end{pmatrix}$. $y = f(x) + b$ is a translation of $y = f(x)$ by $\begin{pmatrix} 0 \\ b \end{pmatrix}$. $y = f(ax)$ is a one-way stretch of $y = f(x)$, parallel to x-axis, scale factor $\dfrac{1}{a}$. $y = af(x)$ is a one-way stretch of $y = f(x)$, parallel to y-axis, scale factor a. $y = f(-x)$ is a reflection of $y = f(x)$ in the y-axis. $y = -f(x)$ is a reflection of $y = f(x)$ in the x-axis.

Topic	Formula	
Polynomials and binomial expansions	**The Factor theorem:** If $(x-a)$ is a factor of $f(x)$ then $f(a)=0$ and $x=a$ is a root of the equation $f(x)=0$.	
	Conversely, if $f(a)=0$ then $(x-a)$ is a factor of $f(x)$.	
	Pascal's triangle:	
	$$\begin{array}{c} 1 \\ 1 \quad 1 \\ 1 \quad 2 \quad 1 \\ 1 \quad 3 \quad 3 \quad 1 \\ 1 \quad 4 \quad 6 \quad 4 \quad 1 \end{array}$$	
	Factorials: $n! = n \times (n-1) \times (n-2) \times \ldots \times 1$	
Differentiation	**Function:**	**Derivative:**
	$y = kx^n$	$\dfrac{dy}{dx} = knx^{n-1}$
	$y = e^{kx}$	$\dfrac{dy}{dx} = ke^{kx}$
	$y = \ln x$	$\dfrac{dy}{dx} = \dfrac{1}{x}$
	$y = f(x) + g(x)$	$\dfrac{dy}{dx} = f'(x) + g'(x)$
	The chain rule for differentiating a function of a function is $\dfrac{dy}{dx} = \dfrac{dy}{du} \times \dfrac{du}{dx}$.	
	The product rule: $y = uv$, $\dfrac{dy}{dx} = u\dfrac{dv}{dx} + v\dfrac{du}{dx}$.	
Integration	**Function:**	**Integral:**
	$\int kx^n \, dx$	$\dfrac{kx^{n+1}}{n+1} + c$
	$\cos kx$	$\dfrac{1}{k}\sin kx + c$
	$\sin kx$	$-\dfrac{1}{k}\cos kx + c$
	e^{kx}	$\dfrac{1}{k}e^{kx} + c$
	$\dfrac{1}{x}$	$\ln|x| + c \quad x \neq 0$
Sequences and series	**Arithmetic sequence:**	**Geometric sequence:**
	The kth term is $a_k = a + (k-1)d$	The kth term is $a_k = ar^{k-1}$
	The last term, $l = a_n = a + (n-1)d$	The last term, $a_n = ar^{n-1}$
Vectors	$\overrightarrow{AB} = \overrightarrow{OB} - \overrightarrow{OA}$	
	If $\mathbf{a} = x\mathbf{i} + y\mathbf{j} + z\mathbf{k}$ then $|\mathbf{a}| = \sqrt{x^2 + y^2 + z^2}$	
Exponentials and logarithms	$y = \log_a x \Leftrightarrow a^y = x$ for $a > 0$ and $x > 0$	
	$\log xy = \log x + \log y \qquad \log \sqrt[n]{x} = \dfrac{1}{n}\log x \quad \log_a a = 1$	
	$\log \dfrac{x}{y} = \log x - \log y \qquad \log \dfrac{1}{x} = -\log x \quad e = 2.718\ldots$	
	$\log x^n = n\log x \qquad \log 1 = 0 \qquad \log_e x = \ln x$	

Make sure you know these formulae for your exam

Integration at a glance

The flowchart below provides an overview of all the different integration methods. It will help you decide which method to use.

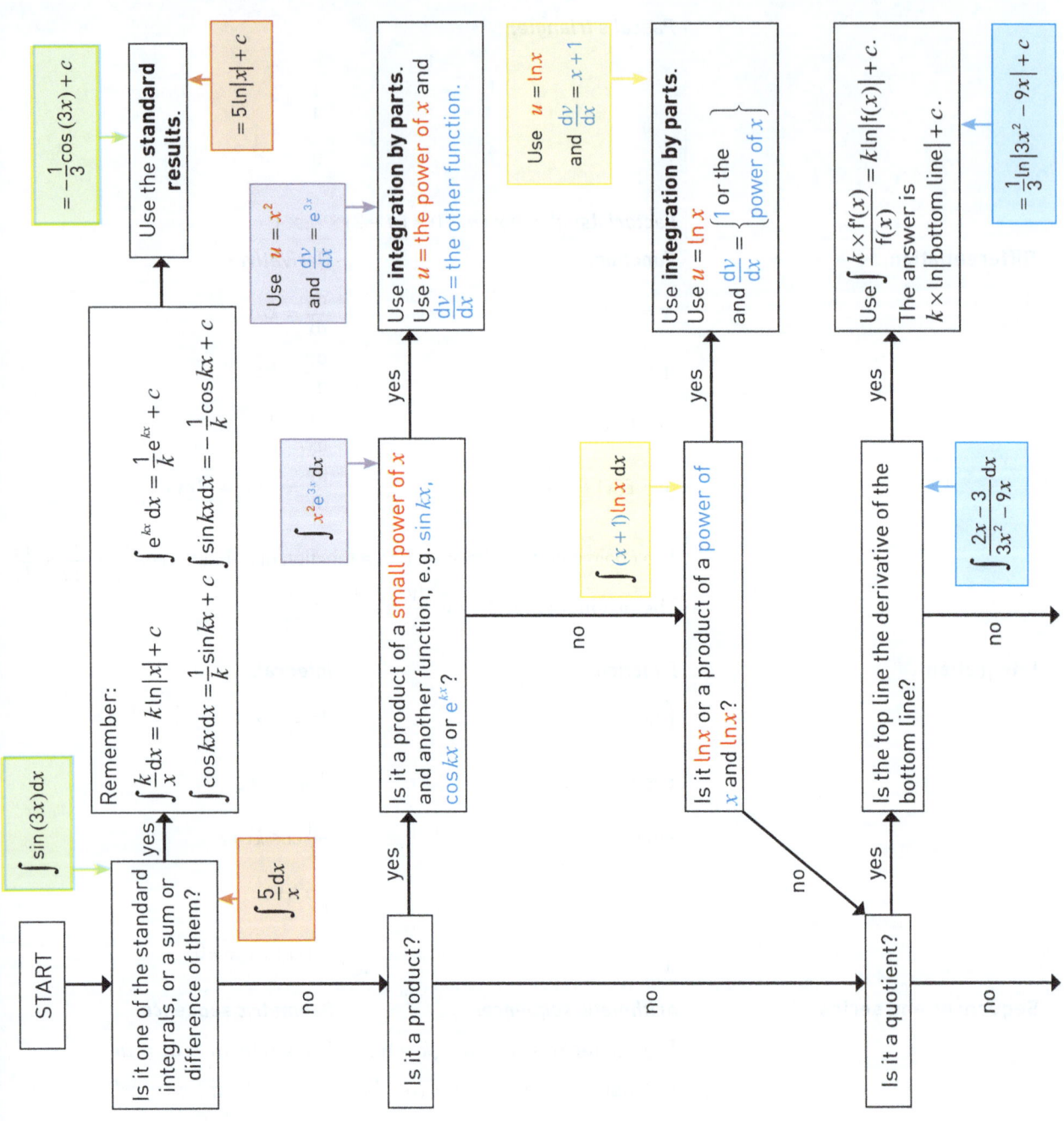

Integration at a glance

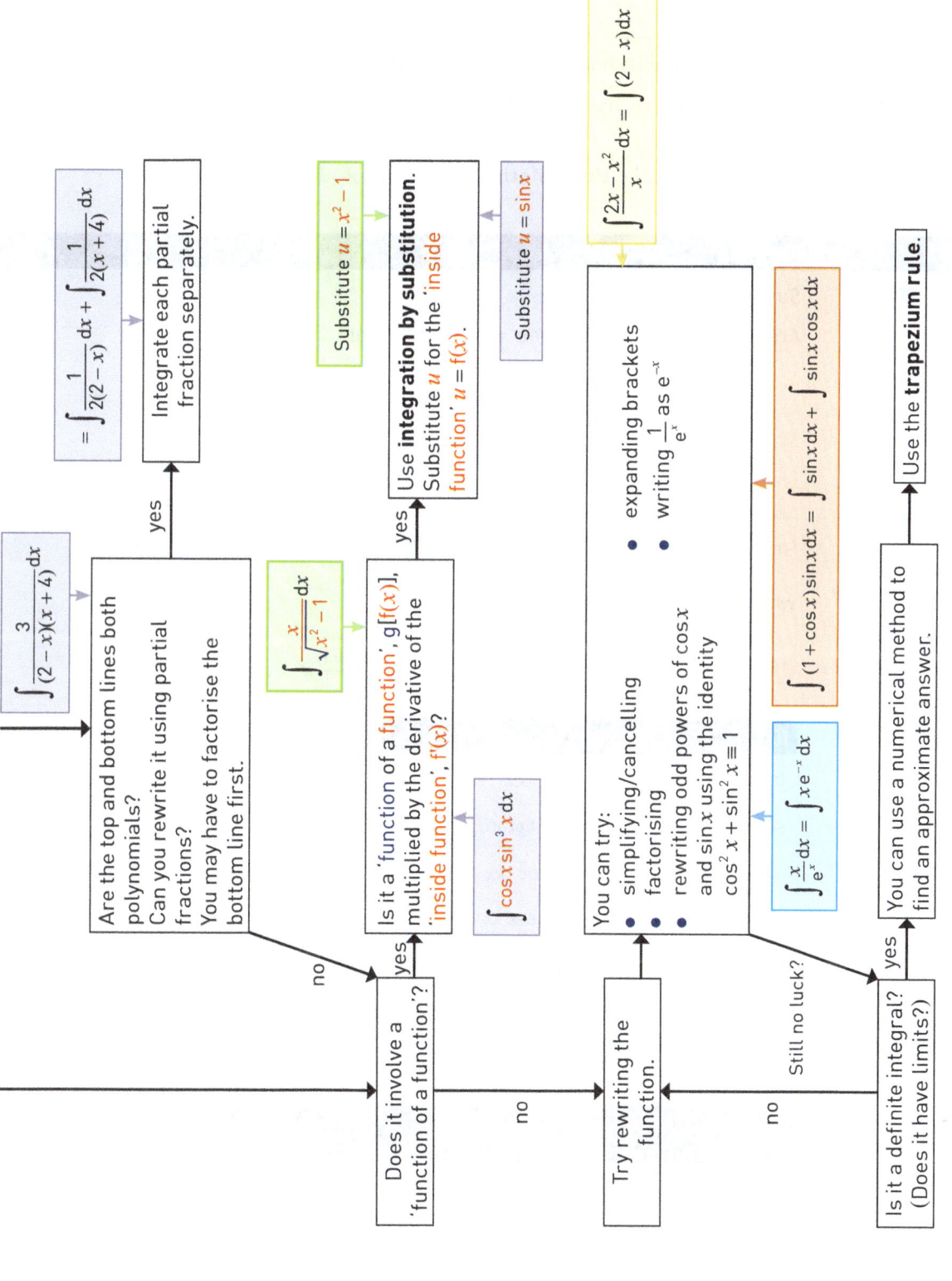

Formulae that will be given

Make sure you are familiar with the formula book you will use in the exam.

The formulae sheet is subject to change so always check the latest version on the exam board's website.

Here are the formulae you are given for the **Pure Mathematics** part of the exam.

Topic	Formula
Mensuration	**Surface area of sphere** $= 4\pi r^2$ **Area of curved surface of cone** $= \pi r \times$ slant height
Arithmetic series	$S_n = \tfrac{1}{2}n(a + l) = \tfrac{1}{2}n\{2a + (n-1)d\}$
Geometric series	$S_n = \dfrac{a(1-r^n)}{1-r}$ $S_\infty = \dfrac{a}{1-r}$ for $\lvert r \rvert < 1$
Binomial series	$(a+b)^n = a^n + {}^nC_1 a^{n-1}b + {}^nC_2 a^{n-2}b^2 + \ldots + {}^nC_r a^{n-r}b^r + \ldots + b^n \quad (n \in \mathbb{N})$, where ${}^nC_r = \binom{n}{r} = \dfrac{n!}{r!(n-r)!}$ $(1+x)^n = 1 + nx + \dfrac{n(n-1)}{2!}x^2 + \ldots + \dfrac{n(n-1)\ldots(n-r+1)}{r!}x^r + \ldots \quad (\lvert x \rvert < 1, n \in \mathbb{R})$
Differentiation	<table><tr><th>$f(x)$</th><th>$f'(x)$</th></tr><tr><td>$\tan kx$</td><td>$k \sec^2 kx$</td></tr><tr><td>$\sec kx$</td><td>$k \sec kx \tan kx$</td></tr><tr><td>$\cot kx$</td><td>$-\csc^2 kx$</td></tr><tr><td>$\csc kx$</td><td>$-k \csc kx \cot kx$</td></tr></table> $\dfrac{f(x)}{g(x)}, \quad \dfrac{dy}{dx} = \dfrac{f'(x)g(x) - f(x)g'(x)}{(g(x))^2}$
First principles	$f'(x) = \displaystyle\lim_{h \to 0} \dfrac{f(x+h) - f(x)}{h}$
Integration (+ constant)	<table><tr><th>$f(x)$</th><th>$\int f(x)\,dx$</th></tr><tr><td>$\sec^2 kx$</td><td>$\tfrac{1}{k}\tan kx$</td></tr><tr><td>$\tan kx$</td><td>$\tfrac{1}{k}\ln\lvert\sec kx\rvert$</td></tr><tr><td>$\cot kx$</td><td>$\tfrac{1}{k}\ln\lvert\sin kx\rvert$</td></tr><tr><td>$\csc kx$</td><td>$-\tfrac{1}{k}\ln\lvert\csc kx + \cot kx\rvert,\ \tfrac{1}{k}\ln\lvert\tan(\tfrac{1}{2}kx)\rvert$</td></tr><tr><td>$\sec kx$</td><td>$\tfrac{1}{k}\ln\lvert\sec kx + \tan kx\rvert,\ \tfrac{1}{k}\ln\lvert\tan(\tfrac{1}{2}kx + \tfrac{1}{4}\pi)\rvert$</td></tr></table> $\displaystyle\int u\dfrac{dv}{dx}\,dx = uv - \int v\dfrac{du}{dx}\,dx$

Topic	Formula
Logarithms and exponentials	$\log_a x = \dfrac{\log_b x}{\log_b a}$ $e^{x \ln a} = a^x$
Small angle approximations	$\sin \theta \approx \theta$, $\cos \theta \approx 1 - \dfrac{1}{2}\theta^2$, $\tan \theta \approx \theta$ where θ is measured in radians.
Trigonometric identities	$\sin(A \pm B) = \sin A \cos B \pm \cos A \sin B$ $\cos(A \pm B) = \cos A \cos B \mp \sin A \sin B$ $\tan(A \pm B) = \dfrac{\tan A \pm \tan B}{1 \mp \tan A \tan B} \quad (A \pm B \neq (k + \tfrac{1}{2})\pi)$ $\sin A + \sin B = 2\sin\dfrac{A+B}{2}\cos\dfrac{A-B}{2}$ $\sin A - \sin B = 2\cos\dfrac{A+B}{2}\sin\dfrac{A-B}{2}$ $\cos A + \cos B = 2\cos\dfrac{A+B}{2}\cos\dfrac{A-B}{2}$ $\cos A - \cos B = 2\sin\dfrac{A+B}{2}\sin\dfrac{A-B}{2}$
Numerical methods	**Trapezium rule:** $\displaystyle\int_a^b y\,dx \approx \tfrac{1}{2}h\{(y_0 + y_n) + 2(y_1 + y_2 + \ldots + y_{n-1})\}$, where $h = \dfrac{b-a}{n}$ **The Newton-Raphson iteration for solving** $f(x) = 0$: $x_{n+1} = x_n - \dfrac{f(x_n)}{f'(x_n)}$

Formulae that will be given

Edexcel A Level Mathematics (Pure)

During your exam

Watch out for these key words

REVISED

- **Exact** … leave your answer as a simplified surd, fraction or power.

  Examples: $\ln 5$ ✓ 1.61 ✗ e^2 ✓ 7.39 ✗

  $2\sqrt{3}$ ✓ 3.26 ✗ $1\frac{5}{6}$ ✓ 1.83 ✗

- **Give/State/Write down** … no working is expected – unless it helps you.

  The marks are for the answer rather than the method.

  Example: The equation of a circle is $(x + 2)^2 + (y - 3)^2 = 13$

  Write down the radius of the circle and the coordinates of its centre.

  > Make sure you give both answers!

- **Prove/Show that** … the answer has been given to you. You must show full working otherwise you will lose marks. Often you will need the answer to this part to answer the next part of the question. Most of the marks will be for the method.

  Example: i Prove that $\sin x - \cos^2 x \equiv \sin^2 x + \sin x - 1$

- **Hence** … you **must** follow on from the given statement or previous part. Alternative methods may not earn marks.

  Example: ii Hence solve $\sin x - \cos^2 x = -2$ for $0° \leq x \leq 180°$

  > Remember that if you couldn't answer part **i** you can still go on and answer part **ii**.

- **Hence or otherwise** … there may be several ways you can answer this question but it is likely that following on from the previous result will be the most efficient and straightforward method.

  Example: Factorise $p(x) = 6x^2 + x - 2$

  Hence, or otherwise, solve $p(x) = 0$

Watch out for these common mistakes

REVISED

- ✗ Miscopying your own work or misreading/miscopying the question.
- ✗ Not giving your answer as coordinates when it should be e.g. when finding where two curves meet.
- ✗ Giving your answer as coordinates inappropriately e.g. writing a vector as coordinates.
- ✗ Not finding y coordinates when asked to find coordinates.
- ✗ Not finding where the curve cuts the x **and** y axes when sketching a curve.
- ✗ Using a ruler to draw curves.
- ✗ Not using a ruler to draw straight lines.
- ✗ Spending too long drawing graphs when a sketch will do.
- ✗ Not stating the equations of asymptotes when sketching an exponential or reciprocal curve.
- ✗ Not simplifying your answer sufficiently.

- ✗ Rounding answers that should be **exact**.
- ✗ Rounding errors – don't round until you reach your final answer.
- ✗ Giving answers to the wrong degree of accuracy – use 3 s.f. unless the questions says otherwise.
- ✗ Not showing any or not showing enough working – especially in 'show that' or 'proof' questions.

Once you have answered a question **re-read the question** making sure you've answered **all** of it. It is easy to miss the last little bit of a question.

Check your answer is ...

REVISED

- ✓ To the correct accuracy
- ✓ in the right form
- ✓ complete ... have you answered the whole question?

If you do get stuck ...

REVISED

Keep calm and don't panic.
- ✓ **Reread the question** ... have you skipped over a key piece of information that would help? Highlight any numbers or key words.
- ✓ **Draw** ... a diagram. This often helps. Especially in questions on graphs, coordinate geometry and vectors, a sketch can help you see the way forward.
- ✓ **Look** ... for how you can re-enter the question. Not being able to answer part **i** doesn't mean you won't be able to do part **ii**. Remember the last part of a question is not necessarily harder.
- ✓ **Move on** ... move onto the next question or part question. Don't waste time being stuck for ages on one question, especially if it is only worth one or two marks.
- ✓ **Return later** ... in the exam to the question you are stuck on – you'll be surprised how often inspiration will strike!
- ✓ **Think positive!** You are well prepared, believe in yourself!

Good luck!

Answers

Here you will find the answers to the 'Target your revision' exercises, 'Exam-style questions' and 'Review questions' in the book. Full worked solutions for all of these (including 'show' questions that do not have short answers in the book) are available online at www.hoddereducation.co.uk/MRNEdexcelALPure

Note that answers for the 'Test yourself' multiple choice question are available online only.

SECTION 1

Target your revision (pages 1–3)

1. i $\Leftrightarrow$
   ii $\Leftarrow$
   iii $\Rightarrow$
2. Go online for the full worked solution.
3. Go online for the full worked solution.
4. Go online for the full worked solution.
5. Go online for the full worked solution.
6. $2\sqrt{5}$
7. $a = -9$ and $b = 5$
8. $\dfrac{2a}{c^3}$
9. $\dfrac{11}{3}\sqrt{3}$
10. i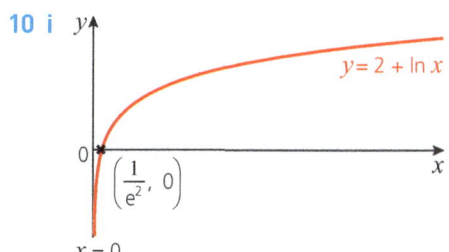
    ii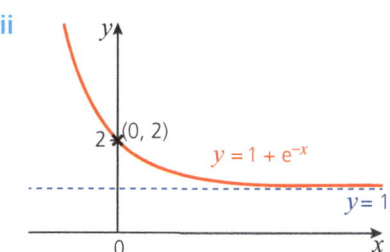
11. $\log 3x^{\frac{3}{2}}$
12. i 9
    ii $x = 10\sqrt{2}$
13. i Go online for the full worked solution.
    ii $A = 200$ and $k = \dfrac{1}{2}$
    iii 2×10^7
    iv -0.6
14. $x = \dfrac{1}{3}$ or $x = \dfrac{3}{2}$

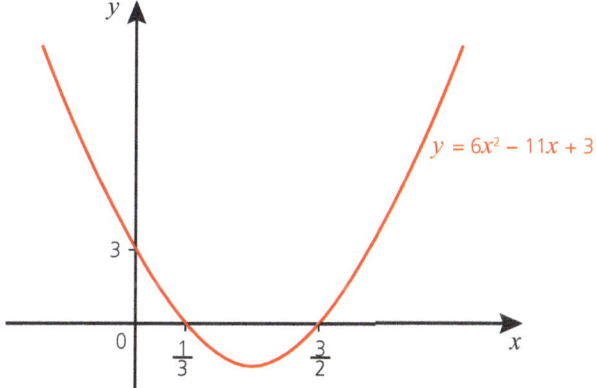

15. $y = (x-3)^2 - 4$
16. $k = \pm 6$
17. $X\left(\dfrac{13}{9}, -\dfrac{1}{3}\right)$
18. $\left(-\dfrac{2}{9}, -\dfrac{8}{3}\right)$ and $(2, 4)$
19. $(1, 2)$ and $\left(-\dfrac{3}{2}, \dfrac{23}{4}\right)$
20. $-1 \leq x < \dfrac{7}{2}$
21. $x \leq -2$ or $x \geq 5$
22.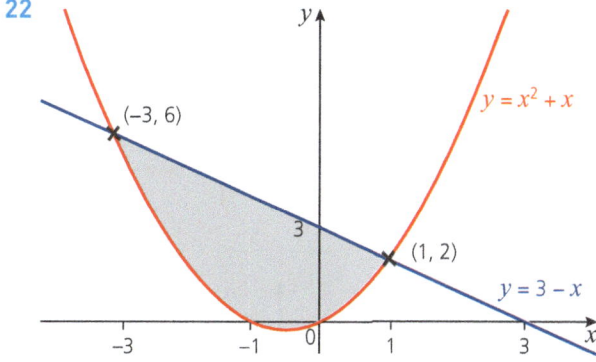
23. i $5x^3 - 7x^2 + 5x + 3$
    ii $x^3 + 7x^2 - 9x + 7$
    iii $3x^4 - 6x^3 - 2x^2 + 9x - 10$
    iv $2x^2 - 3x + 1$
24. i $\dfrac{x-1}{2x-1}$
    ii $\dfrac{2-x}{x-1}$

Edexcel A Level Mathematics (Pure)   213

25 i Go online for the full worked solution.
   ii Go online for the full worked solution.
   iii $(3x-1)(x+1)(2-x)$; $x=-1$, $x=\frac{1}{3}$, $x=2$
   iv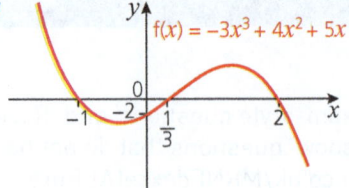

26 i $(1, -1)$ ii $-\frac{4}{3}$ iii 10

27 i $y=-\frac{1}{2}x-4$
   ii $(0, -4)$ and $(-8, 0)$

28 $-\frac{3}{2}$

29 i $(x+2)^2+(y-1)^2=5$
   ii $(x+1)^2+(y-5)^2=13$

30 $4y+3x=14$

31 i A 10 B $gf(x)=(3x-2)^2$
   ii $x=\frac{1}{2}$ and $x=1$

32 i $P''(0, -2)$ and $Q''(2, -3)$
   ii $P''(0, 1)$ and $Q''(4, 4)$
   iii $P''\left(-\frac{1}{2}, 2\right)$ and $Q''\left(\frac{3}{2}, 3\right)$

33 i Domain: $x \in \mathbb{R}$; range: $f(x) > 1$, $f(x) \in \mathbb{R}$
   ii $f^{-1}(x)=\frac{1}{2}\ln(x-1)$, Domain: $x > 1$, $x \in \mathbb{R}$;
      range: $f^{-1}(x) \in \mathbb{R}$
   iii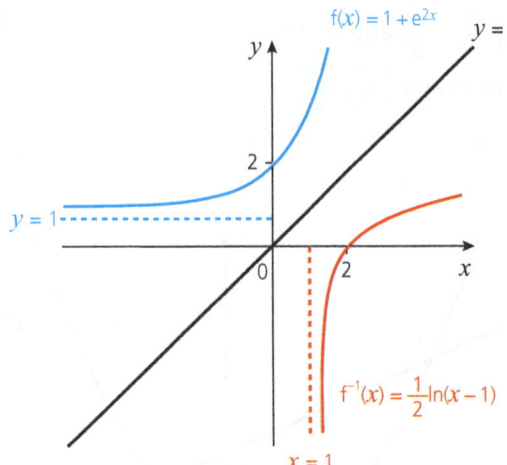

34 i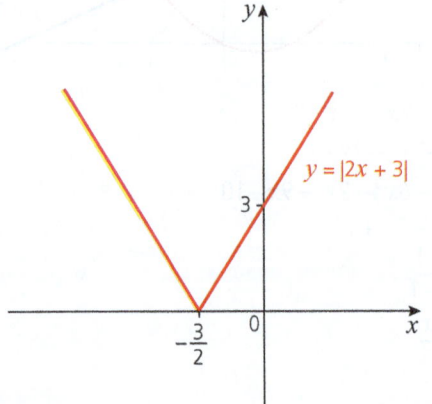

ii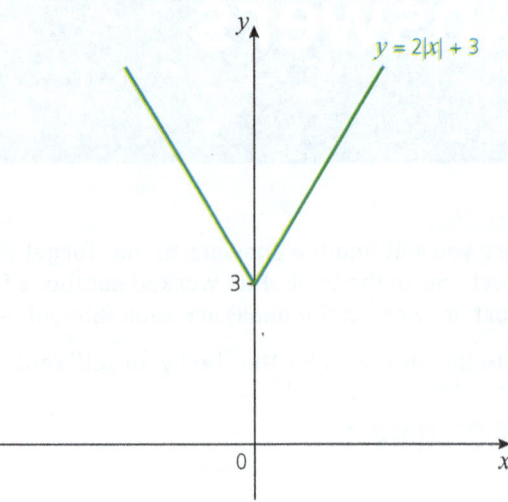

35 i $x=-1$, $x=5$
   ii $x=\frac{4}{5}$
   iii $x=-10$, $x=-\frac{2}{5}$

36 i $|x-5| < 2$
   ii $x \leq -1$ and $x \geq 2$

Chapter 1 Proof

Proof (page 8)

Go online for the full worked solution.

Chapter 2 Indices, surds and logs

Surds and indices (page 12)

i $a=-1$ and $b=\frac{1}{4}$

ii $a=9$ and $b=15$

Exponential functions and logarithms (page 18)

i a 8221
  b 4000
ii 2127
iii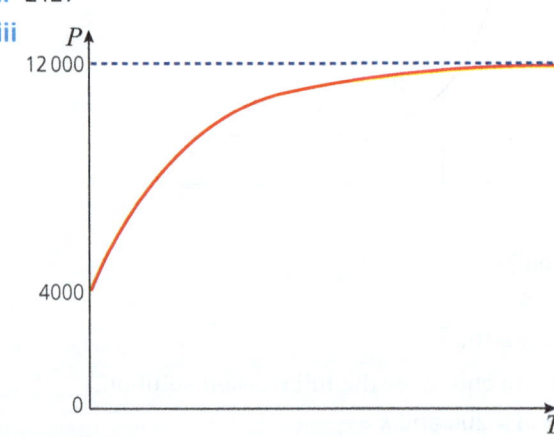

iv 1768–1769

Modelling curves (page 23)

i

Year (x)	1	2	3	4	5	6
Profit (P)	7800	9400	11200	13500	16200	19400
$\log_{10} P$	3.89	3.97	4.05	4.13	4.21	4.29

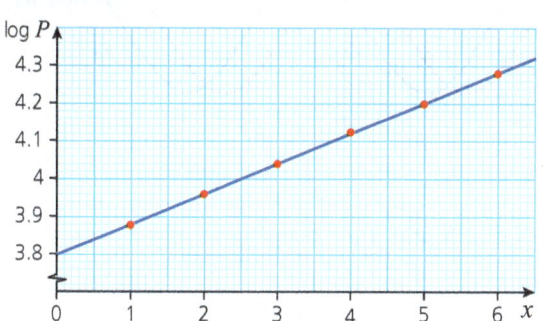

ii $P = 6500 \times 1.20^x$
iii £40 700
iv 28 years

Chapter 3 Algebra

Quadratic equations (page 30)

i $f(x) = 2(x+3)^2 - 8$
ii a P(0, 10) and Q(−3, −8)
b

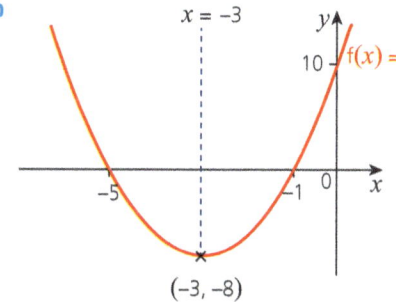

Simultaneous equations (page 35)

i A (−1, 0) and B $\left(-\frac{1}{5}, \frac{12}{5}\right)$
ii $k = 7$; C $\left(\frac{1}{2}, \frac{17}{2}\right)$

Inequalities (page 38)

i $x \geq \frac{3}{7}$
ii $x < -\frac{1}{2}$ or $x > 4$

Working with polynomials and algebraic fractions (page 42)

i a $\dfrac{x-2}{x+3}$
  b $\dfrac{x^2+4}{x(x^2-4)}$
ii $a = 2, b = -1$ and $c = 4$

The factor theorem and curve sketching (page 46)

$a = -32, b = 16$

Chapter 4 Coordinate geometry

Straight lines (page 52)

i (1, 3)
ii 9.15 square units

Circles (page 56)

i C (4, −3) and radius 5
ii A $(4 - 2\sqrt{6}, -2)$ and B $(4 + 2\sqrt{6}, -2)$
iii $3x + 4y + 25 = 0$

Chapter 5 Functions

Functions (page 61)

i $-1.6 \leq y \leq 1.6$
ii Go online for the full worked solution. The curve has rotational symmetry of order 2 about the origin.
iii $fg(x) = \dfrac{(x-1)^3}{(x+2)(x-4)}$, domain is $-1 \leq x \leq 1$ and range is $0 \leq y \leq 1.6$

Graphs and transformations (page 66)

i $gf(x) = 2\ln x$
ii A one way stretch parallel to the y-axis, scale factor 2.

Inverse functions (page 69)

i and iii

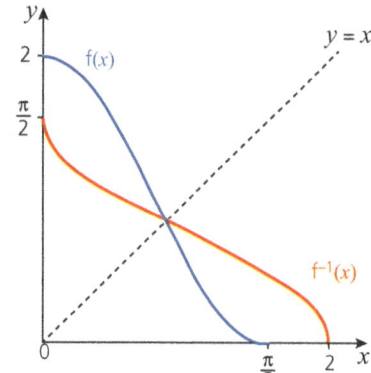

ii $f^{-1}(x) = \frac{1}{2} \arccos(x - 1)$
Domain of f^{-1} is $0 \leq x \leq 2$
Range of f^{-1} is $0 \leq f^{-1}(x) \leq \frac{1}{2}\pi$

The modulus function (page 73)

i

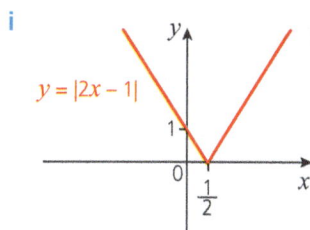

ii a $x \geqslant -\frac{1}{5}$
    b $x = -3, x = -\frac{1}{5}$

Review questions (page 74)

1 a Go online for the full worked solution.
  b i   2, 3, 5 and 7.
     ii Go online for the full worked solution.
     iii Go online for the full worked solution.
2 $x = 3$ and $y = 2$
3 a i $x = 1.71$
     ii $x = 4$
  b 5
4 i $-2\left(x - \frac{5}{4}\right)^2 + \frac{49}{8}$

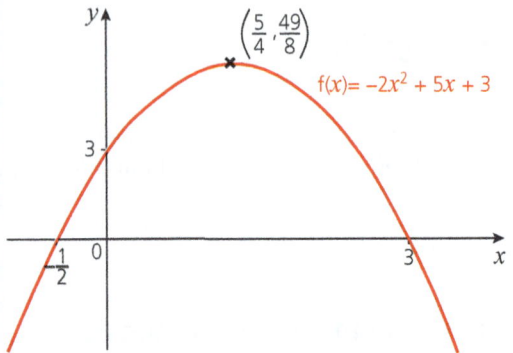

  ii a $x = \pm\sqrt{3}$
     b $x = 9$
     c $x = 1$
5 $\frac{14x - 9}{4x(2x - 1)}$;
   $x = \frac{3}{4}$ or $x = \frac{3}{2}$
6 i $f(x) = (x - 5)(2x - 1)(x + 2)$
  ii $x = 1, x = \frac{7}{2}$ and $x = 8$
7 i $AB = 2\sqrt{5}, BC = \sqrt{65}$
  ii 15 square units
8 i $x = -9$ and $x = 3$
  ii Go online for the full worked solution.
9 i $fg(x) = \cos(1 - x)$
      $gf(x) = 1 - \cos x$
  ii Reflection in the x-axis, followed by a translation by the vector $\begin{pmatrix} 0 \\ 1 \end{pmatrix}$

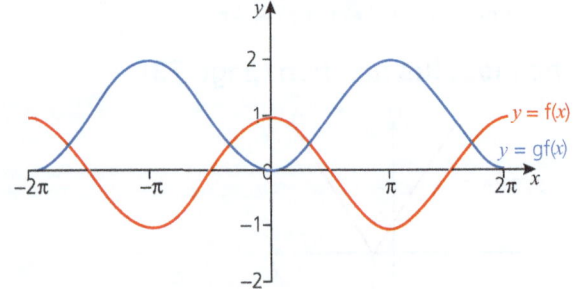

iii Translation by the vector $\begin{pmatrix} -1 \\ 0 \end{pmatrix}$, followed by a reflection in the y-axis.

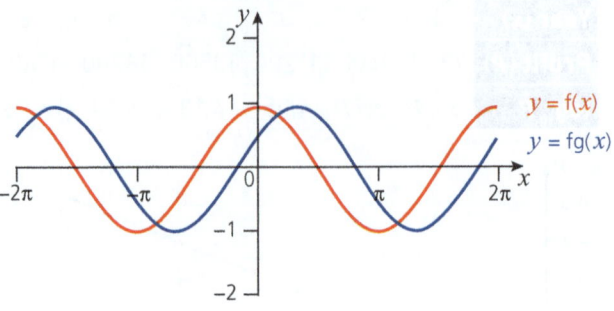

SECTION 2

Target your revision (pages 75–77)

1 i $\frac{1}{2}$
  ii $\frac{\sqrt{2}}{2}$
  iii -1
2 i $x = 60°, 120°, 240°$ or $300°$
  ii $x = 15°, 105°, 195°$ or $285°$
  iii $x = 105°$ or $345°$
3 i $x = 0.927$ rads or 2.21 rads
  ii $x = \frac{\pi}{12}, \frac{11\pi}{12}, \frac{13\pi}{12}, \frac{23\pi}{12}$
  iii $x = \frac{5\pi}{6}$ or $\frac{11\pi}{6}$
4 Go online for the full worked solution.
5 i $49.3°$ or $131°$
  ii $20.6\,\text{cm}^2$
6 $3(2\pi - 3\sqrt{3})\,\text{cm}^2$; $(6 + 2\pi)\,\text{cm}$
7 i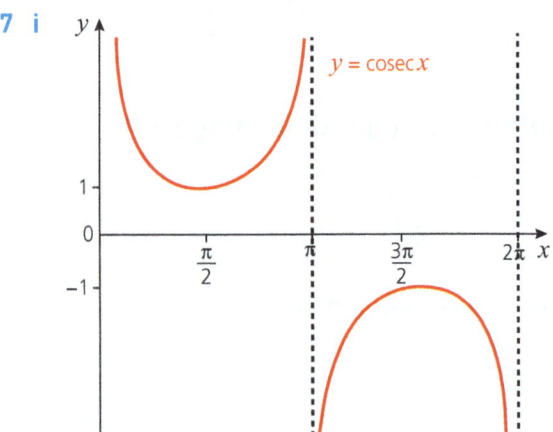
  ii $x = \frac{7\pi}{6}$ or $\frac{11\pi}{6}$
8 Go online for the full worked solution.
9 i $\frac{\theta}{2}$
  ii $\frac{2 - \theta^2}{\theta}$
10 $\frac{\sqrt{6} + \sqrt{2}}{4}$

11 i $\cos(x + \frac{\pi}{6}) = \frac{\sqrt{3}}{2}\cos x - \frac{1}{2}\sin x$;

$\sin(\frac{\pi}{6} - x) = \frac{1}{2}\cos x - \frac{\sqrt{3}}{2}\sin x$

ii $\frac{3\pi}{4}, \frac{7\pi}{4}$

12 Go online for the full worked solution.

13 Go online for the full worked solution: 34

14 i $\frac{dy}{dx} = 10x - \frac{4}{x^2} - 3$

ii $\frac{dy}{dx} = \frac{1}{2\sqrt{x}} - \frac{3}{x^4} - 1$

15 i 5.5

ii (36, 216)

16 i $y = 11x - 18$

ii $y = 2x - 3$

17 $f''(x) = 2 + \frac{2}{x^3} + \frac{1}{4\sqrt{x^3}}$

18 $-3 < x < 2$

19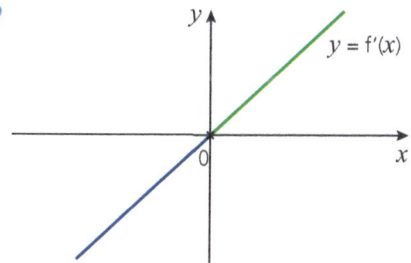

20 i $x^3 + 3x^2h + 3xh^2 + h^3$

ii $6x^2h + 6xh^2 + 2h^3$

iii $f'(x) = 6x^2$

21 $\frac{dy}{dx} = 20(4x - 5)^4$

22 $\frac{dy}{dx} = 3(x\cos x + \sin x)$

23 i $\frac{dy}{dx} = \frac{3(2x^3 - x^2 - 1)}{(3x-1)^2}$

ii 1.32

24 i a $\frac{dy}{dx} = 2e^{2x}$

b $\frac{dy}{dx} = (2x+1)e^{2x}$

c $\frac{dy}{dx} = \frac{e^{2x}(2x-1)}{x^2}$

ii a $\left(-\frac{1}{2}, -\frac{1}{2e}\right)$; minimum

b $\left(\frac{1}{2}, 2e\right)$; minimum

25 i $\frac{dy}{dx} = \frac{1}{x}$

ii $\frac{dy}{dx} = 1 + \ln 2x$

iii $\frac{dy}{dx} = \frac{1 - \ln 2x}{x^2}$

26 i $\frac{dy}{dx} = 3\cos(3x + \pi)$

ii $\frac{dy}{dx} = -8x\sin(4x^2)$

iii $\frac{dy}{dx} = 2\sec^2 2x$

27 i P(0, 1)

ii $\frac{dy}{dx} = 1 - 2\sin 2x$

$\frac{d^2y}{dx^2} = -4\cos 2x$

iii $Q\left(\frac{\pi}{12}, 1.13\right)$ and $R\left(\frac{5\pi}{12}, 0.443\right)$

iv $S\left(\frac{\pi}{4}, \frac{\pi}{4}\right)$; go online for the full worked solution

28 i and ii $\frac{dy}{dx} = -2\cos x \sin x$

Go online for the full worked solution.

29 i $\frac{dy}{dx} = \ln 3 \times 3^x$

ii $\frac{dy}{dx} = 2\ln 3 \times 3^{2x}$ or $\frac{dy}{dx} = \ln 9 \times 9^x$

30 $\frac{dy}{dx} = \frac{2x - y}{x - 2\sin 2y}$

31 i $\frac{2x^3}{3} + \frac{7x^2}{2} - 15x + c$

ii $\frac{2\sqrt{x^3}}{3} - \frac{1}{x} + c$

32 i $2\ln|x| - \frac{1}{x} + c$

ii $2e^{5x} - \frac{1}{e^x} + c$

33 $y = 2\ln|x| + \frac{e^{2x}}{2} + \frac{e^2}{2}$

34 i $-10\frac{1}{3}$

ii $\frac{1}{2}e^2 - \frac{1}{2}e - \ln 2$

35 i 15.75 square units

ii $5\frac{1}{3}$ square units

iii $21\frac{1}{12}$ square units

iv $10\frac{5}{12}$

36 i Go online for the full worked solution.

ii a $30\frac{2}{3}$ square units   b $9\frac{1}{3}$ square units

iii $21\frac{1}{3}$ square units

37 $17\frac{1}{3}$ square units

38 $\frac{1}{2}e^{x^2} + c$

39 1

40 i Go online for the full worked solution.

ii $\frac{1}{3}\cos^3\theta - \cos\theta + c$

41 $\frac{1}{2}xe^{2x} - \frac{1}{4}e^{2x} + c$

Chapter 6 Trigonometry

Working with trigonometric functions (page 83)

i $-2\cos^2 x - \cos x + 2$
ii 30°, 90°, 150°

Triangles without right angles (page 86)

i 10.6 cm
ii 54.9°
iii 6.05 cm
iv 77.9 cm²

Radians and circular measure (page 90)

i Go online for the full worked solution.
ii 59.6 cm²; 55.3 cm

Reciprocal trig functions and small angle approximations (page 94)

i a $\theta = 0.253$ rads or 2.89 rads
   b Go online for the full worked solution.
ii Go online for the full worked solution.
iii $P\left(\frac{\pi}{2}, 0.4\right)$ and $Q\left(\frac{3\pi}{2}, 6\right)$
    Range: $0.4 \leq f(\theta) \leq 6$

Compound angle formulae (page 98)

i a A $\frac{1}{2}ab\sin(A+B)$
      B $\frac{1}{2}ab\sin A\cos B$
      C $\frac{1}{2}ab\cos A\sin B$
   b Go online for full worked solution.
ii 15°, 195°

The forms $r\cos(\theta \pm \alpha)$ and $r\sin(\theta \pm \alpha)$ (page 102)

i Go online for the full worked solution.
ii $y = 13\sin(\theta + 22.6°)$
iii a and b

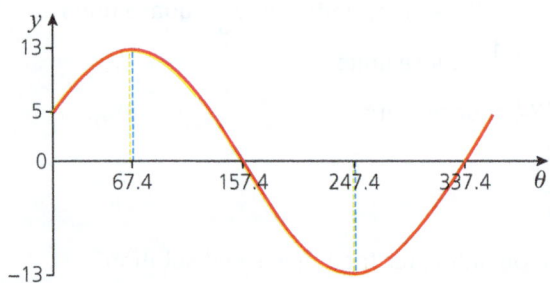

iv a 10.3 cm
    b 27.7°

Chapter 7 Differentiation

Tangents and normals (page 107)

i Go online for the full worked solution.
ii $Q(-2, -7)$
iii $12y + x + 86 = 0$

Curve sketching and stationary points (page 114)

i P(0, 5), P is a stationary point of inflection.
   $Q\left(\frac{4}{3}, -\frac{121}{27}\right)$, Q is a non-stationary point of inflection.
ii R(2, –11); a local minimum
iii

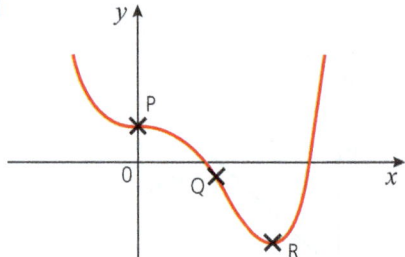

First principles and differentiating sin x, cos x and tan x (page 117)

$f'\left(\frac{\pi}{3}\right) = \frac{1}{2}$

Differentiating ln x and eˣ (page 118)

$f\left(\frac{1}{2}\right) = -\frac{3}{e}$, $f'\left(\frac{1}{2}\right) = 2 + \frac{6}{e}$, $f''\left(\frac{1}{2}\right) = -4 - \frac{12}{e}$

The chain rule (page 123)

$2\sqrt{3} - 4$

The product and quotient rules (page 126)

i (–2, 0) and (0, 2)
ii S(–1, e)
iii maximum
iv

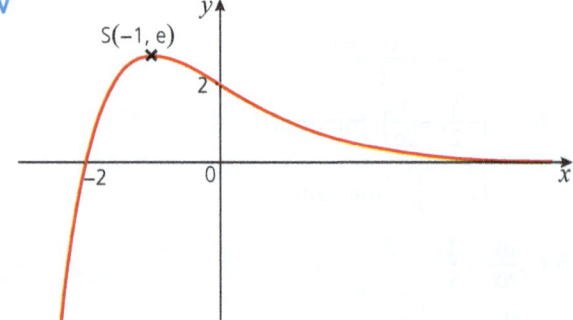

Implicit differentiation (page 130)

i Go online for the full worked solution.

ii $\frac{dx}{dy} = \frac{3y^2 - 2x}{2(x+y)}$; $\frac{7}{4}$

iii $y = x$ is not a tangent to the curve at (3, 3) as the gradient of $y = x$ is 1 and the gradient of the curve at A is $\frac{4}{7}$

Chapter 8 Integration

Integration as the reverse of differentiation (page 134)

$y = e^x + \frac{1}{e^{2x}} + 4$

Finding areas (page 140)

i $P(\ln p, p)$

ii $(p-1)$ square units; $(p \ln p - p + 1)$ square units

iii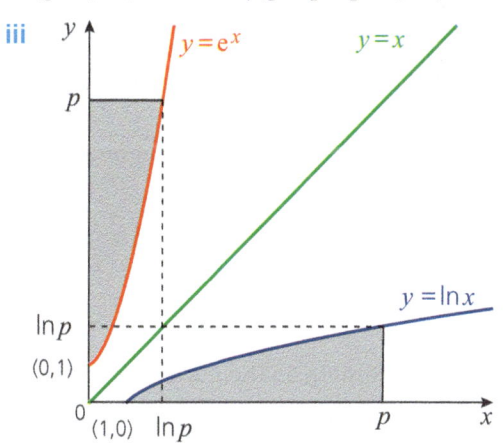

iv $(p \ln p - p + 1)$ square units

Integration by substitution (page 144)

i Go online for the full worked solution.

ii Go online for the full worked solution.

Integrating trigonometric functions (page 147)

i $-\frac{1}{e^{\sin x}} + c$

ii Go online for the full worked solution.

Integration by parts (page 151)

i Go online for the full worked solution.

ii $(-1, 0)$ and $(1, 0)$

iii $\frac{4}{e}$

Review questions (page 152)

1 $\frac{\sqrt{8}}{8}$

2 $\theta = 60°$ or $300°$

3 The transformations are a one-way stretch scale factor 5, parallel to the y-axis and a translation by the vector $\begin{pmatrix} -53.1 \\ 0 \end{pmatrix}$, carried out in either order.

4 $4240 \, cm^2$, go online for the full worked solution

5 i $\frac{dy}{d\theta} = 5\cos\theta + 4\sin\theta$

ii $(-0.896, -6.40)$ and $(2.25, 6.40)$

iii $(-0.896, -6.40)$ is a minimum point.
$(2.25, 6.40)$ is a maximum point.

iv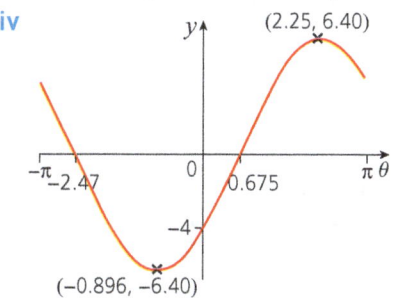

6 i a $\frac{dy}{dx} = -\frac{x}{\sqrt{100-x^2}}$

b $\frac{dy}{dx} = \frac{100 - 2x^2}{\sqrt{100-x^2}}$

ii Go online for the full worked solution.

iii Go online for the full worked solution.

iv $0.0175 \, m^2 \, s^{-1}$

7 $e^{1-\cos x} + c$

8 i $\frac{1}{k} x \sin(kx) + \frac{1}{k^2} \cos(kx) + c$

ii Go online for full worked solution.

iii $\frac{x}{4}\sin 2x + \frac{1}{8}\cos 2x + \frac{x^2}{4} + c$

iv Go online for the full worked solution.

v Go online for the full worked solution

SECTION 3

Target your revision (pages 153–155)

1 i $-x, x^2, -x^3, x^4$

ii $S_6 = -x + x^2 - x^3 + x^4 - x^5 + x^6$

2 i $\frac{1}{3}$

ii 3 terms

3 i -15

ii $3n - 18$

iii 2925

4 $\frac{1}{x} - \frac{1}{(2x-1)} - \frac{1}{(3x+2)}$

5 i $\frac{1}{3}\ln 6$

ii $\frac{1}{3}\ln\left(\frac{10}{9}\right)$

Edexcel A Level Mathematics (Pure)

6 a $1024 - 640x + 160x^2 - 20x^3 + \frac{5x^4}{4} - \frac{x^5}{32}$

  b $-1008x^5$

7 a $1 + \frac{3x}{2} - \frac{9x^2}{8}; |x| < \frac{1}{3}$

  b $\frac{1}{25} + \frac{2x}{125} + \frac{3x^2}{625}$, when $|x| < 5$

8 i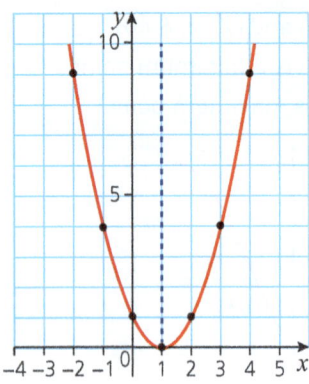

  ii $t = 0$

9 i a $y = \frac{2(1-x)}{2x-1}$ b $x = \frac{9}{3 + \ln x}$

  ii a $x = 4\cos\theta, y = 2\sin\theta$

     b $x = 4\sin 2\theta, y = 2\cos 2\theta$

10 i centre $(-3, 2)$ ii radius

11 $\frac{1}{3}$

12 i Go online for the full worked solution.

   ii a $2y = 3\sqrt{2}x - 6$

      b $3y + \sqrt{2}x - 13 = 0$

   iii Go online for the full worked solution.

13 i $-\frac{3\cos 3\theta}{2\sin 2\theta}$

   ii $\left(\frac{1}{2}, 1\right)$

   iii Go online for the full worked solution.

14 i $x = 3\sin 2t + c$

   ii $x = 3\sin 2t + 6$

   iii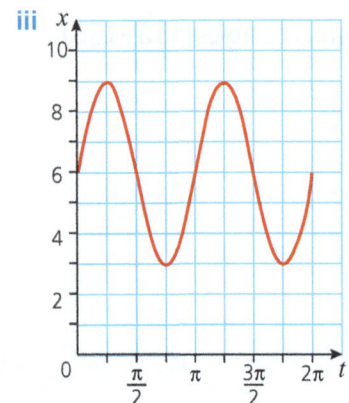

15 $y = Ae^{x^2}$

16 i $x = -1, y = 2$ and $z = -4$

   ii $\mathbf{c} = -\mathbf{j} + 2\mathbf{k}; 116.6°$

17 $\begin{pmatrix} 8 \\ -6 \\ -10 \end{pmatrix}$

18 i $\sqrt{6}$

   ii $3\mathbf{i} - \mathbf{k}$

   iii $7\mathbf{i} + 3\mathbf{j} + 8\mathbf{k}$

19 Go online for the full worked solution.

20 i Go online for the full worked solution.

   ii $x_2 = 1.76517, x_3 = 1.88029, x_4 = 1.89643$ and $x_5 = 1.89867$

21 $\alpha = 0.33134$

22 i 1 The function has a vertical asymptote.

     2 The roots are very close together.

     3 There is a repeated root (so no change of sign).

   ii 1 The choice of x_1 is too far from the required root.

      2 The choice of x_1 is close to a stationary point.

      3 f(x) is discontinuous.

   iii The gradient of g(x) is too steep near the root.

23 i a An underestimate as the curve is concave.

     b An overestimate as the curve is convex.

     c Not possible to tell as curve both convex and concave.

   ii $a = -\frac{\sqrt{2}}{2}$ and $b = \frac{\sqrt{2}}{2}$

24 i 0.931001; underestimate as curve is concave.

   ii a 0.931839

      b 0.932049

   iii 0.932 to 3 d.p.

25 i 0.93857

   ii 0.88558

   $0.88558 < \int_0^1 \sqrt{\cos x}\, dx < 0.93857$

Chapter 9 Sequences and series

Definitions and notation (page 159)

i $a_1 = a_6 = a_{11} = a_{16} = 120$

  $a_2 = a_7 = a_{12} = a_{17} = 110$

  $a_3 = a_8 = a_{13} = a_{18} = 100$

  $a_4 = a_9 = a_{14} = a_{19} = 90$

  $a_5 = a_{10} = a_{15} = a_{20} = 80$

ii Periodic with period 5.

   Divergent.

iii 70
iv 95
v 270 units
vi 04 00 the following day.

Sequences and series (page 162)

i a £76
   b £909
ii £2051.96

Chapter 10 Further algebra

Partial fractions (page 166)

i $\dfrac{2}{(x-1)} - \dfrac{1}{(x-1)^2} - \dfrac{6}{(2x-3)}$

ii $\ln\left(\dfrac{4}{27}\right) - \dfrac{1}{2}$

The binomial theorem (page 171)

i $1 + \dfrac{1}{2}y - \dfrac{1}{8}y^2$

ii The expansion is only valid when $|y| < 1$
   $4 > 1$ so the expansion is not valid when $y = 4$.

iii $2 + \dfrac{1}{4}x - \dfrac{1}{64}x^2$, Valid when $|x| < 4$

iv 2.234375; 0.076% to 3 d.p.

Chapter 11 Parametric equations

Parametric equations (page 176)

i $3y = 4x - 8$

ii $t = -\dfrac{1}{2}$ or $t = 2$; $\left(\dfrac{1}{2}, -2\right)$ and $(8, 8)$

iii Go online for the full worked solution.

iv The lines p and q are parallel.

v $y^2 = 8x$

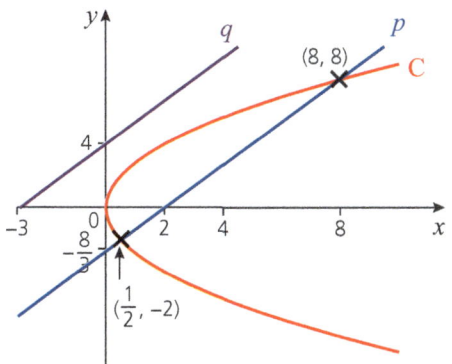

Calculus with parametric equations (page 180)

i $-\dfrac{1}{pq}$; $y = -\dfrac{1}{pq}x + 2\left(\dfrac{1}{p} + \dfrac{1}{q}\right)$

ii $y = -\dfrac{1}{p^2}x + \dfrac{4}{p}$

iii $\dfrac{dy}{dx} = -\dfrac{1}{t^2}$

iv Go online for the full worked solution.

v $(2, 2)$ and $\left(6, \dfrac{2}{3}\right)$

Chapter 12 Differential equations

Solving differential equations (page 184)

i $y = A\mathrm{e}^{-\frac{1}{2}x^2}$

ii

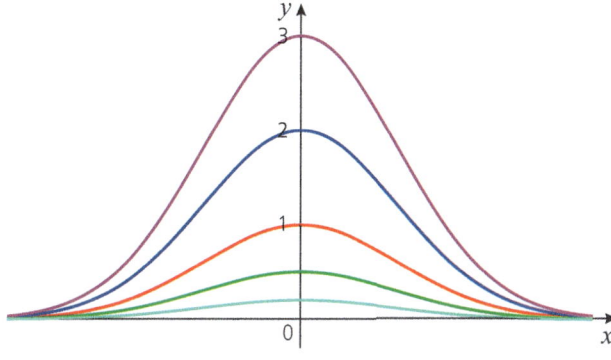

iii $y = 5\mathrm{e}^{-\frac{1}{2}x^2}$

Differential equations and problem solving (page 187)

i $\dfrac{dr}{dt} = \dfrac{k}{t+1}$

ii Go online for the full worked solution.

iii $r = 2\ln(t+1)$; 4.80 cm

iv $\dfrac{dr}{dt} = \dfrac{K}{(t+1)(t+2)}$
   $K = 10$

v $r = 10\ln\left[\dfrac{2(t+1)}{(t+2)}\right]$

vi 6.06 cm

Chapter 13 Vectors

Vectors (page 191)

i $\begin{pmatrix} -4 \\ 4 \\ -2 \end{pmatrix}$

ii $\begin{pmatrix} 2 \\ 4 \\ -4 \end{pmatrix}$

iii $(6, 3, -1)$

iv $\begin{pmatrix} 5 \\ 1 \\ 1 \end{pmatrix}$

v $\left(\dfrac{10}{3}, \dfrac{5}{3}, 1\right)$

Edexcel A Level Mathematics (Pure)

Chapter 14 Numerical methods

Solving equations numerically (page 195)

i Go online for the full worked solution.
ii Go online for the full worked solution.
iii 1.72
iv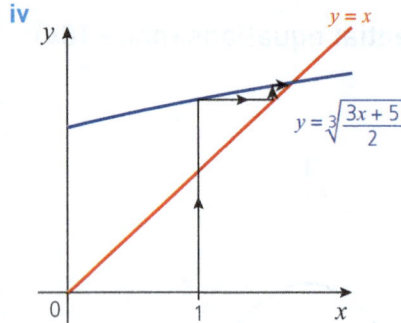

Numerical integration (page 199)

i a 3.94615
  b 3.88227
ii Overestimate.
iii

	Approximation	Improvement
2 strips	3.94615...	
4 strips	3.88227...	0.06388...
8 strips	3.86632...	0.01595...
16 strips	3.86233...	0.00399...

iv 3.86

Review questions (page 200)

1 $r = 5$ and $a = \frac{4}{125}$ or $r = -5$ and $a = -\frac{4}{125}$.

2 i $1 + \frac{\theta^2}{2}$
   ii $1 + \frac{\theta^2}{4}$
   iii 0.201 (3 s.f.)

3 i $\frac{1}{(1-x)} + \frac{1}{(2+x)}$

   ii a $\ln\left|\frac{2+x}{1-x}\right| + c$
      b $\frac{15}{16}$

4 i $-\frac{\cos 2\theta}{\sin \theta}$
   ii $2y + 2\sqrt{3}x = 3\sqrt{3}$
   iii $(\sqrt{2}, 1), (-\sqrt{2}, -1), (-\sqrt{2}, 1), (\sqrt{2}, -1)$
   iv $\frac{16}{3}$
   v Go online for the full worked solution.

5 i $\frac{dT}{dt} = -k(T - 18)$
   ii $T = 18 + 82e^{-0.0559t}$

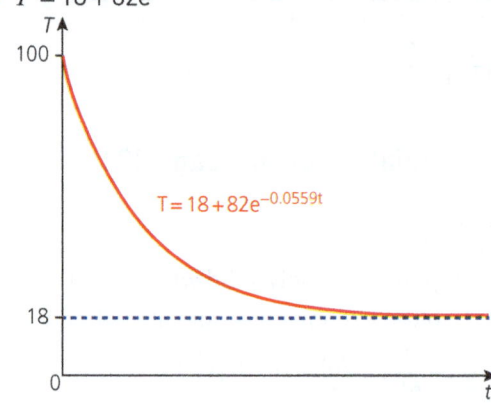

   iii 64.9°
   iv Go online for the full worked solution.
   v The model will not be valid in the long term. The temperature of the room is unlikely to remain at 18°C indefinitely which will affect the temperature of the water. Eventually all the water will evaporate.

6 i $-\mathbf{i} + 2\mathbf{j} + \mathbf{k}$
   ii $\frac{\sqrt{21}}{6}$

7 i Go online for the full worked solution.
   ii 4.5 km²
   iii 4.95 km²

Printed and bound by CPI Group (UK) Ltd, Croydon, CR0 4YY
22/03/2026
02076152-0001